Information and Life

Gérard Battail

# Information and Life

Gérard Battail
E.N.S.T., Paris, France (retired)
France

ISBN 978-94-007-9335-4 ISBN 978-94-007-7040-9 (eBook)
DOI 10.1007/978-94-007-7040-9
Springer Dordrecht Heidelberg London New York


Softcover re-print of the Hardcover 1st edition 2014


Printed on acid-free paper

Springer is part of Springer Science+Business Media (www.springer.com)

# Foreword

So much has happened since Watson and Crick gave us the structure of DNA in 1953 and a revolutionary new comet—the discipline "molecular biology"—burst into the bioscience skies. Six decades later, we feel confident that we now understand the principles of how the string of DNA base letters conveys, from generation to generation, the information that determines so much of our structure and function. Yet, as I write this Foreword to Professor Battail's timely book, the biosciences are in crisis over the simple four-letter word "junk." Since only 1–2 % of the string of DNA base letters seems to be necessary for conveying the classical genes needed for many of our day-to-day functions, it was for long convenient to dismiss the remainder as mere "junk"—DNA that had come along for the ride, but was not critical for our long-term survival. However, over the last decade growing evidence that the extra DNA does not remain silent, but is transcribed into RNA copies, has shaken the junksters from their complacency and strengthened the arguments of those who regard "junk" as a miasmic comfort-word that has served both to mislead the credulous and to justify an undue focus of resources on genes.

When a discipline is in crisis, there are two broad strategies to follow. (1) Go "back to square one" and trace step-by-step how the crisis came about – that is, examine the history. (2) Seek cross-boundary insights from other disciplines. Of course, over the decades numerous disciplines have jostled to bridge the trans-disciplinary gap for molecular biology. Biochemists complained that the molecular biologists were merely practicing their biochemical arts without licence. Mathematicians, foreseeing easy pickings, migrated into many areas, especially genetics. But instead of unification, the boundaries seemed to grow greater. Unhappy with the complex calculations of the mathematical geneticists, the biochemists retreated to their laboratories, where single bands on gels could provide clear, all-or-none, answers. Unhappy with the biological and chemical details, the geneticists invoked fancy fudge-factors and cosy coefficients, to cope with the discrepancies that so often appeared.?

Meanwhile, slowly and relentlessly, what is now recognized as the information revolution overtook us all. We were reminded of its relevance to the biological sciences by texts such as Werner Loewenstein's *The Touchstone of Life. Molecular Information, Cell Communication, and the Foundations of Life* (1999) and Hans Christian von Baeyer's *Information. The New Language of Science* (2003). These have recently been joined by James Gleick's popular *The Information. A History, A Theory, A Flood* (2012). To these works, directed to wide audiences, there is now need for the addition of more advanced texts, ideally authored by people deeply immersed in both the information sciences and the biosciences. But given the human difficulty of achieving such broad mastery, at this time we must look to workers at the interface between these two areas, who invariably have strength mainly in one. My text, *Evolutionary Bioinformatics* (2006, 2011) provided a bioscientist's perspective. Now we have Gérard Battail's *Information and Life* which covers the same area from the perspective of an information scientist. These two complementary books, backed by many others, herald the emergence of a new interface science for which I have tentatively suggested the name "Evolutionary Bioinformatics."

Of course, from speaking of information it is but a short step to what many would regard as the ultimate frontier in the biosciences, the understanding of cognition and the fundamental brain biochemistry and physiology that underlies it. Yet, as neuroscientist Stuart Firestein makes clear in *Ignorance, How It Drives Science* (1912), despite the hype by "the spike analysis industry" that "the language of the brain" is within our grasp, the discipline of neuroscience is also in crisis. While Battail does not directly address this, there can be little doubt that the information theory he so carefully outlines will be no less important for neuroscience than for the other biosciences.

Finally, it should be noted that, while his earlier book—*Théorie de l'information* (1997)—was in French, the present is in English. Battail has first-hand knowledge of scientific literatures other than the rather inward-looking English literature. Thus, his transdisciplinary perspective carries an international flavor. His title *Information and Life* may seem ambitious to some, but I can think of no better way of expressing the broad swath of ideas that so richly adorn his pages.

Queen's University, Kingston, Canada Donald R. Forsdyke
March 2013

# Preface

An engineer's approach to fundamental biology could well define the content of this book. I taught during 23 years theoretical disciplines of communication engineering at the Ecole nationale supérieure des Télécommunications (ENST) in Paris, especially *information theory* and *error-correction coding*; the latter was my main field of research. As a faculty member of a college of engineering, I limited my research activities to the engineering field. When I retired in 1997, I felt free to escape it and I undertook applying information theory and error-correcting codes to topics of broader significance, especially to biology. The reader will hopefully soon realize that, contrary to what may seem, there is a true continuity between communication engineering and some very basic biological problems.

The present book is thus a fruit of my retirement. My motivation for writing it was twofold. On the negative side, a sharp dissatisfaction about the trend of science towards increasing specialization, which results in its fragmentation into narrow subdisciplines which eventually become autistic to each others. Nothing prevents such subdisciplines to reach mutually incompatible conclusions. This situation makes extremely difficult to have a general view on science; even most philosophers seem to have given up. I think however, with Murray Gell-Mann, that 'it is vitally important that we supplement our specialized studies with serious attempts to take a crude look at the whole'. On the positive side, I think that information theory is as general and transdisciplinary as to help acquiring such a look and, moreover, that its use is mandatory in natural sciences, especially for understanding life.

My background is thus communication engineering. I nevertheless dare question biology and to some extent physics since I think that these sciences need the lessons of communication engineering and information theory, all the more they can be a strong antidote to their ever increasing specialization. Since I am not a biologist nor a physicist, my questions about biology and physics are naive. For lack of working experience, my knowledge of these sciences mainly relies on popularizing books. Such documents should be cautiously used and cannot reflect the latest researches. However, I compensate my lack of practical knowledge of these sciences, as far as I can, with my own experience. My position with respect to biology and physics is that of an observer from outside, similarly to that of a philosopher, except that my background in communication engineering hopefully provides usable concepts, perspective and methods. This is a rather uncomfortable position. It turns out that letting biologists and physicists accept lessons from engineering is especially difficult. It may be a legacy of the Greco-Roman antiquity when engineering was the lot of slaves, while philosophy (including science in its modern meaning) was a noble activity of the free citizens. Prejudices are long-lived, so 25 centuries later science is most often perceived as creating ideas to be later exploited by engineers, while the upholders of 'pure' scientific disciplines ignore or refuse that concepts possibly useful to them could originate in engineering.

A major difficulty I met when I began writing this book is that information theory has never been adequately popularized, to a large extent because this is extremely difficult. Information theory is a mathematical discipline, hence basically abstract. Moreover its main object—information—is rather elusive. My first task has thus been to popularize the main topics of information theory which can be useful to biologists, and it is what I do in the first part of the book. The word 'information' has become very common and polysemic. Restrictions with respect to the usual meanings of a word are necessary for defining a scientific entity, and the definition I propose is not straightforward and rather abstract. Such lexical difficulties are unavoidable since a precise vocabulary is mandatory.

Besides attempting to popularize information theory, the remainder of the book intends to improve the understanding of the *living world*. At the light of information theory, mainstream biology is shown to inadequately account for the most important and specific phenomenon of life: *heredity*. Reconstructing biology in an information-theoretic perspective thus appeared to me as necessary, leading to a formidable research agenda as regards both the broadness of its scope and the amount of difficult work it demanded. Only a small part of it has actually been performed. This book can thus by no means be claimed to be exhaustive, but is hopefully forerunner. It could at best skim over the matter: introduce information theory as simply and intuitively as possible and merely deal with a few easy examples of its application to biology. I hope that it will prompt broader and deeper researches, at the risk of its own obsolescence. My first project was to also discuss a few applications to physics, but there was too much work still to be done, so I stopped at the border between the living and the inanimate. As I progressed in my writing, I discovered that many other topics were worth being dealt with so the completion of the book was for me a kind of horizon seeming to move back as I moved forward. No wonder if the finished book looks incomplete and calls for further researches.

In its part intended to popularize information theory, this book contains few formal proofs, which can be found in more technical works. However, everywhere it is possible, remarks are intended to help the reader to intuitively understand why the written statements are true. Such remarks cannot be substituted for formal proofs but, as based on them, hopefully show the logical necessity of the statements and their coherence.

No great efforts have been made to avoid redundancy. Besides being very useful in communication engineering, it has also the didactic virtue of helping the reader's understanding. I tried of course to avoid mere repetition and to present the same concepts from changing points of view in order to hopefully provide more insight into them.

As a Frenchman, I have an easy access to documents written in my first language, and it is why the bibliography contains a rather unusual number of papers and books in French. No bibliography can be claimed to be exhaustive so, inevitably, it is somehow biased. In any way, besides prompting biologists to get interested in information theory and making their access to it as easy as possible, my project was to expound my own researches and not to objectively compile references. Most works referred to in this bibliography are those that I read when I wrote the book, but some of them are referred to because of their historical importance. Compilations like (Favareau 2010, Sloane and Wyner 1993, and Slepian 1974) were especially useful in this respect.

I would like to sincerely thank all those who manifested interest in my work or gave me opportunities to expound and exchange ideas, especially Claude Berrou, John D. Enderle, Paddy Farrell, Donald R. Forsdyke, David Haccoun, Joachim Hagenauer, Jean-Michel Labouygues, Vittorio Luzzati, Elebeoba E. May, Olgica Milenkovic, Mark E. Samuels, Tom Schneider, Karoline Wiesner and Hubert P. Yockey. I am indebted to Antoine Danchin who prompted me to contact Daniel Mange at the Ecole Polytechnique Fédérale in Lausanne, Switzerland, who let me meet Marcello Barbieri of the University of Ferrara, Italy. Marcello Barbieri friendly invited me to expound my research at the 2005 Biosemiotics Gathering in Urbino and prompted me later to write the present book. Still more important, his ideas had a determining influence on my work. His bold statement that 'Nature is artefact-making' closely agrees with my own intuition that biology and engineering should go hand in hand. Besides, at least two very important concepts in his work were especially appealing to me, as being in resonance with some of my own ideas: that of *nominable entity* and that of *organic codes*. I found no better way than the former to qualify an information. As regards the latter, I independently suggested the existence in the living world of 'nested soft codes' which turn out to be very similar to Barbieri's 'organic codes'. This convergence is all the more interesting since he and I arrived at these concepts from entirely different backgrounds, by entirely different ways.

I am extremely grateful to Professor Donald R. Forsdyke who accepted to write a foreword to this book.

# Contents

# Chapter 1
# Introduction

**Abstract** Chapter 1 expounds the aim of the book: defining information as a scientific entity and using it for explaining life phenomena. The science of communication engineering, Shannon's information theory, is available to this end but must be somewhat extended, especially as regards the identification of the specific properties which can be used for defining information as a scientific entity. As a mathematical discipline, information theory can hopefully provide biology with a yet lacking theoretical basis. Communication engineering is only concerned with literal communication so information theory excludes semantics. The relation of semantics with information must thus be explicated.

## 1.1 Aim of the Book

It is commonplace to refer to the current civilization as the information era. News from everywhere are broadcast everywhere in the world. Electronic mail connects each of us with people regardless of distance. If we think of information as 'what is communicated', then information is one of the most familiar and omnipresent entities of our daily lifes.

If we think of information as the expected result of any investigation, and especially of scientific research, we may wonder whether information itself can be an object of research. The question may look incongruous, or even logically circular. Does a science of information exist? If it does, is it connected with the sciences of nature, physics and biology and, if it is, what is and what can be its impact on them? This book intends to take up such questions.

A science of information actually exists, initiated in 1948 when Claude Shannon (1916–2001) published his seminal paper *A mathematical theory of communication* (Shannon 1948). It is referred to as *information theory*. As a theory of communication *means*, it is at the heart of communication engineering. Together with the semi-conductor technology which enabled its implementation, it is the main event from which the information era originated. Despite its conceptual and practical importance, however, information theory remains ill-known outside the communication engineering community, even by educated people. Although information theory is embodied in electronic objects of daily use, it remains invisible. We think that information is a scientific entity of capital importance. Its lack of visibility is

G. Battail, *Information and Life*, DOI 10.1007/978-94-007-7040-9_1,

unfortunate because it could be extremely useful in many branches of science besides communication engineering. This book will hopefully help bridging this gap.

Information theory has indeed never been properly popularized, to a large extent because it is a very difficult task. Besides the difficulties intrinsic to the abstractness of a mathematical discipline, the polysemy of the word 'information' as well as its often loose use are among the main obstacles in popularizing information theory. The main difficulty comes from the need of restricting the meaning of the word 'information' in order to make it as precise and unambiguous as to define a scientific entity. We tried to do so in the first part of this book. With the tremendous development of communications, the word 'information' has become trivial and almost nobody cares of what it really means. Everyone believes he/she understands it, just like Augustine believed he understood the meaning of 'time' untill he realized he was unable to explicate it. It is also quite difficult to communicate the *experience* gained by communication engineers during the last decades, during which the formal concepts of the theory were found to perfectly fit the needs of communication practice, and its quantitative predictions were accurately checked.

Probably the main obstacle met when attempting to popularize information theory is that it demands a methodological precaution often perceived as a sacrifice: *semantics should be discarded from information.* Is not meaning for all and sundry the very essence of information? What is information without semantics? Communication engineers know by experience that semantics is foreign to their activity. Discarding semantics even appears for them as having cut the Gordian knot, making their discipline possible and successful. But can this reasonable faith be communicated to the layman? Philosophers and even scientists of other disciplines often reject this basic tenet and therefore ignore information theory. For lack of adequate popularization of communication engineering and information theory, they do not realize that doing so makes them blind to a major scientific and engineering achievement. One of the main goals of this book is to make the principles of communication engineering and information theory understandable without too many technical details, and especially with a limited use of mathematical formalism.

Of course, the scientific concept of information notably differs from the usual meanings of the word. The radical dissociation of information from semantics has been accepted by engineers only inasmuch as it is fully successful. In this respect, the status of information is not so different from that of other scientific entities. Think for instance of the fundamental physical concept of *energy*. Energy can be mechanical (potential or kinetic), electrical, electromagnetic, chemical, thermal, ..., but these various forms are considered as diverse embodiments of a single entity on the basis of an equivalence principle: they can be converted into each others and thus can be quantitatively measured using a same unit. Unlike the concept of matter which can be thought of as directly deriving from the sensible intuition, it actually took centuries before the concept of energy emerges and becomes a fundamental physical entity. If we now accept it, it is because it has demonstrated in innumerable instances its necessity for describing and predicting physical facts. It has become familiar with time, but its scientific meaning is definitely different from the looser and broader usual meaning of the word 'energy'.

Besides the goal of popularizing the principles of communication engineering and information theory, this book intends to use them in order to improve the understanding of the *living world*. At first sight, life may seem quite foreign to engineering. But is the evolution process which created the living world so different from engineering? Evolution actually met and necessarily solved engineering problems, admittedly using methods foreign to those of humans. Both human engineers and Nature created objects which *perform functions*, fulfilling an *a priori* given purpose in the former case, or in the latter carrying out specific tasks which together result in the perpetuation of living processes. How each of these tasks is performed can be analyzed in engineering terms. Moreover, these tasks must be closely coordinated, which implies communications between the individual objects which perform them (organelles, cells, tissues, organs, . . . ). Nature thus necessarily evolved extremely varied and complex communication means and it turns out that describing the living world must increasingly often involve communication at all scales from the molecules up to the ecosystems and advanced animal societies, including the human ones. The results of information theory, which often state the limits of what is possible, are as general as to apply to any communication system regardless of its origin and its implementation means. Information theory perfectly fits communication engineering, and its generality makes it relevant to natural communication means as well. As a consequence, they are subjected to the limits stated by information theory. Then, as a fundamental scientific entity, *information* appears as necessarily having a central role in biology. Mainstream biology ignores the concepts and results of information theory although the word 'information' has become omnipresent in the biological literature, but it is almost always used in a loose meaning. Reconstructing biology in an information-theoretic perspective appears as a necessary (and immense) task.

My experience in communication engineering and information theory led me soon to realize that genetics and evolution could not be understood unless a genomic error-correction system is assumed to exist. When I started writing this book, I mainly intended to develop this idea. It became however increasingly clear to me that, much more generally, information is a fundamental scientific entity that biology needs but as yet ignores, and moreover ignores that it needs it. My initial project was a three-part book, the third one being devoted to applications of information theory to physics. Having almost completed the first two parts, respectively devoted to the science of information and to some of its biological applications, I realized that dealing with physics would need much more further research, so I gave up, at least provisionally. It is why the present book is limited to two parts. However, it will be argued that whether information is used or not actually delineates the border between the living and the inanimate, so physics will be present at its interface with biology. It will be shown that any living thing actually acts as Maxwell's demon (I resumed here an old project of continuing the work of Léon Brillouin (Brillouin 1956), and it turns out that I reached conclusions very different from his). A physical measurement, moreover, can be interpreted as a means for an observer to acquire information about an object of the inanimate world. As the observer belongs to the living world, the border between these two worlds must be crossed. It is why an informational theory of physical measurement, hopefully containing the seeds of future researches, is outlined in Chap. 10.

Further exploiting the idea that information is a fundamental scientific entity needed for understanding life leads to conclude that it actually delineates the border between the living and the inanimate: living things bear, process and use information, inanimate objects do not (except of course those which have their origin in the living world, mostly as products of the human industry fabricated for this purpose). Since the living world is but a part of the physical world, this implies that physics itself must accept information as a fundamental scientific entity. Physicists realized since the beginning of the XX-th century that the human observer must be considered as an actor of any physical measurement. An observer belongs of course to the living world and a measurement is a means for acquiring information from the physical world. The revolution which we claim to be necessary in biology thus cannot leave physics itself unchanged. Any theory of measurement in physical sciences *must* integrate information theory. It is why this book was initially intended to the integration of information theory within physics as well as biology. However I gave up due to the amount of work required by physics, so the book is mainly restricted to biology, which has a more straightforward relation to information.

Information is not only a scientific entity yet almost completely overlooked, but it possesses properties radically different from those of matter and energy. The main one is that it does not obey any conservation law since an information can be copied, hence shared, as well as annihilated. We show below that these properties directly result in the main specific features of the living world. Ignoring information, mainstream biology thus ignores the main entity which governs life! It must be realized that, as a branch of mathematics, information theory is a normative science: its statements are as compelling as those of arithmetic. As an expected result, the mathematical character of information theory will hopefully endow biology with a level of rigour and structure comparable to that attained by physics in its own field. Most of the results of information theory concern the limits of what is possible, which are sharply defined and impassable. Today's physics does not ignore information and many physicists recognize it as fundamental, but deal with it as a physical quantity. Strangely, they often deny life as if the laws of physics would suffice to explain everything in the universe. To say the least, the relationship of physics and life lacks clarity.

Information will appear to us as providing the unique bridge between the abstract and the concrete. Writing about information thus demands extreme care for avoiding any confusion between both, a concern which does not always inspire writing scientific treatises. Reading this book needs to keep in mind that information is dealt with as an abstract entity.

Reflections about information contained in this book result in a critical look at the bases of contemporary science. Far from purely negative, such criticisms fortunately open to research the immense field of integrating information within science, as a fundamental entity. A new world is to be discovered, with unforeseeable consequences.

## 1.2 About the Method

Bringing to light the fundamental importance in biology of a mathematical entity, information, entails that methods using *simplified models* dealt with by mathematics and computer simulation, which proved extremely successful in physics and engineering, should be useful in biology, too. Some current beliefs in biology which are only justified by an argument expressed in textual form may well not resist the test of a computation based on appropriate models. This book contains two examples of this approach, but it could be useful in many other instances.

As a first example, an important statement in this book is that the existence of DNA mutations in somatic cells within time intervals as short as a human life is absolutely incompatible with the conservation of genomes at the timescale of geology. We perceive this contradiction as obvious but many biologists do not, and deny it without any rational reason. Since this incompatibility can be solved only by assuming that genomes are endowed with error-correcting codes, they also reject this conclusion although it is inescapable once the incompatibility is established. In Sect. 8.1.2, we intend to confirm this incompatibility by means of information-theoretic arguments. It turns out that precisely describing the hereditary communication channel is impossible for lack of knowing by what kind of errors the DNA nucleotides are affected. However, we want only to show that the quantity which measures the ability of a channel to convey information, its *capacity* as defined by information theory, decreases with time and eventually vanishes. In order to show that it does, it suffices to prove that an *upper bound* to the actual capacity behaves so. An upper bound results from assuming in the capacity computation that the errors are of the mildest possible type, namely erasures, although we do not know the kind of the errors which actually occur.

As a second example, we develop in Sect. 8.3 some computations concerning an extremely simplified model of heredity we name 'toy living world'. These calculations show that only the assumption that genomic error-correcting codes exist can explain the existence of discrete species. Although extremely simplified, this model hints at actual properties of the true living world. Would such codes not exist, the living world would be an incomprehensible chaos populated with chimeras.

In both cases we make use of very simplified and easily computable models so as to acquire knowledge about a far more complex reality. Such methods are commonplace in physics and engineering, but are not familiar to biologists although they could be extremely useful to them. Moreover, designing a model demands a reflection about what is important or not, and considering a few possible models can help choosing the one which most closely approaches reality.

Some comments about orders of magnitude will perhaps be useful to the reader. A book devoted to astronomy or cosmology refers to distances, durations and numbers of objects as large as completely escaping our sensible intuition. Still much larger numbers of objects, however, will be met in this book when considering the many different *sequences* that the genetic material can assume. Inconceivably numerous combinations can indeed arise from comparatively small genomes. Only very few of them are actually realized, which hints at the pervasive presence of *redundancy*, an extremely important but ill-known property of living objects. We do not perceive this situation as strange because the same is true in language where, among all the possible combinations of phonemes, letters, words, only very few are actually compatible with linguistic constraints. High redundancy is so present in our daily life that we do not even notice its existence.

Readers may deem that some parts of the book, especially in Chaps. 3, 4 and 5 are too technical when describing transmission and reception processes. Other readers, or possibly the same ones, may deem that the last chapters of the book, especially Chaps. 9 and 10, are too philosophical as questioning the epistemological status of information. The technical details may be justified by the aim of showing that the principles of communication engineering and information theory are actually implemented in operational devices now become of daily use, hence are not mere intellectual speculations. As regards the status of information, recognizing it as an abstract entity is mandatory for understanding that it provides a bridge between the abstract and the concrete. It can thus act on physical objects, which suggests an 'explanation' to life. Readers, of course, may skip what they want.

## 1.3 On the Book Content

Most of the topics dealt with in this book have already been published in a number of papers (Battail 1997, 2001, 2004, 2005, 2006a, b, 2007a, b, 2008a, b, 2009a, b, 2010, 2011, 2012) and a short book (Battail 2008c). An easy way to carry out my present book project could have been to merely collect the papers. For the sake of unity, I preferred however to gather their content into an entirely rewritten synthesis hopefully avoiding too technical details. This book thus contains many parts of previous publications, more or less transformed to fulfill its goals. However, it also contains new material, and the need of synthesis led me on some important points to more general and more radical conclusions than those previously published.

The first part starts from a common use of the word 'information' to examine features which can be exploited for endowing information with the status of a scientific entity (Chap. 2). The next chapter does not directly deal with information, but states the basic principles of communication engineering and examines how literal communication can be ensured (Chap. 3). Emphasis is laid on the fact that reception, far from being a passive function, cannot be understood outside a probabilisic framework. It thus results in decisions *which may be wrong*. Then, Chaps. 4 and 5 introduce information theory as the science of literal communication, according to Shannon's original probabilistic version. The last chapter of the first part, Chap. 6, deals with information as a fundamental entity. It begins with an algorithmic version of information theory which complements the probabilistic one, each giving some insight about the other. Since any application of information theory needs that the relationship of information with other conceptual and physical entities is properly understood, this chapter questions the relationship of information with semantics, in a very broad meaning including its interaction with the physical world. Information then appears as a bridge between the abstract and the concrete.

Chapter 7 introduces the second part. Then, as a typical example of how results of information theory can be applied to a fundamental problem of life, Chap. 8 is devoted to heredity. It shows that the faithful conservation of genomes at the geological timescale cannot result from the mere replication of DNA molecules, as currently believed. A blatant contradiction between the assumed conservation of genomes and the presence of mutations is pointed out, and it is shown that information theory demands that genomes, as digital messages, are actually endowed with error-correcting codes. Assuming their existence then suffices for explaining many actual basic properties of the living world. Chapter 9 states that information is specific to life and actually delineates the border between the living and the inanimate. Chapter 10 considers the place of life within the physical world, and interprets a physical measurement as a means for a living observer to acquire information beyond the living-inanimate border. Chapter 11 concludes the whole book.

There are three appendices besides the index. Appendix A, 'Tribute to Shannon', is a slightly corrected version of a text I wrote just after Shannon's death in February 2001, which has not been published in print. It may look somewhat redundant with respect to the account of information theory presented in the first part of this book, but I preferred keeping it unabridged as expressing my thoughts at that time, some of which were the seeds of the present book. Some footnotes have been appended to the original text. The second one, Appendix B, 'Some comments about mathematics', questions the relevance of mathematics for describing natural features; it is also intended to recall some definitions which are necessary for understanding the few mathematical concepts used in the book, and to explicate the notations. The third Appendix C, is a short glossary intended to sketchily recall the main topics of molecular genetics.

## References


Battail, G. (1997). Does information theory explain biological evolution? *Europhysics Letters, 40*(3), 343–348.

Battail, G. (2001). Is biological evolution relevant to information theory and coding? *Proceeding ISCTA '01* (pp. 343–351). Ambleside, UK.

Battail, G. (2004). An Engineer's view on genetic information and biological evolution. *BioSystems, 76,* 279–290.

Battail, G. (2005). Genetics as a communication process involving error-correcting codes. *Journal of Biosemiotics, 1*(1), 103–144.

Battail, G. (2006a). Should genetics get an information-theoretic education? *IEEE Engineering in Medicine and Biology Magazine, 25*(1), 34–45.

Battail, G. (2006b). Error-correcting codes and genetics. *tripleC* (http://triplec.uti.at/), *4*(2), 217–229.

Battail, G. (2007a). Information theory and error-correcting codes in genetics and biological evolution. In M. Barbieri (Ed.), *Introduction to Biosemiotics* (pp. 299–345). Dordrecht: Springer.

Battail, G. (2007b). Impact of information theory on the fundamentals of genetics. In G. Witzany (Ed.), *Biosemiotics in Transdisciplinary Contexts*. Helsinki: Umweb.

Battail, G. (2008a). Genomic error-correcting codes in the living world. *Biosemiotics, 1,* 221–138. doi:10.1007/s12304-008-9019-z.

Battail, G. (2008b). Can we explain the faithful communication of genetic information? *Advances in information recording*. In P. H. Siegel, E. Soljanin, van Wijngaarden A. J. & B. Vasic (Eds.), DIMACS Series, No. 73, pp. 79–103.

Battail, G. (2008c). *An outline of informational genetics*. San Rafael: Morgan & Claypool. doi:10.2200/S00151ED1V01Y200809BME023

Battail, G. (2009a). Applying semiotics and information theory to biology: A critical comparison. *Biosemiotics, 2*(3), 303–320. doi:10.1007/s12304-009-9062-4.

Battail, G. (2009b). Living versus inanimate: The information border. *Biosemiotics, 2*(3), 321–341. doi:10.1007/s12304-009-9059-z.

Battail, G. (2010). Heredity as an encoded communication process. *The IEEE Transactions on Information Theory, 56*(2) 678–687. doi:10.1109/TIT.2009.2037044.

Battail, G. (2011). An answer to Schrödinger's *What is life? Biosemiotics, 4*(1), 55–67. doi:10.1007/s12304-010-9102-0.

Battail, G. (2012). Biology needs information theory. *Biosemiotics, 6*(1), 77–103. doi:10.1007/s12304-012-9152-6.

Brillouin, L. (1956), *Science and Information Theory*, New York: Academic Press.

Favareau, D. (2010). *Essential readings in biosemiotics*. Dordrecht: Springer.

Shannon, C. E. (1948). A mathematical theory of communication. *The Bell System Technical Journal, 27,* 379–457, 623–656. (Reprinted in Shannon and Weaver 1949, Sloane and Wyner 1993, pp. 5–83 and in Slepain 1947, pp. 5–29).

Shannon, C. E., & Weaver, W. (1949). *The mathematical theory of communication*. Urbana: University of Illinois Press.

Slepian, D. (Ed.). (1974). *Key papers in the development of Information Theory*, IEEE Press.

Sloane, N. J. A., & Wyner, A. D. (Eds.). (1993), *Claude Elwood Shannon, Collected papers*, IEEE Press.

# Part I
# Information as a Scientific Entity

# Chapter 2
# What is Information?

**Abstract** Chapter 2 examines how the most current use of the word 'information' can lead to outline an axiomatic definition of information. Its most specific features are that it has no existence unless it is physically inscribed as a sequence of symbols on some medium, which however has no influence on it besides ensuring its existence, and that it can be defined only as an equivalence class, with respect to transformations like alphabet change and coding. It is thus an abstract entity which resides in the physical world. An information meets Barbieri's concept of 'nominable entity', which refers to a singular object. This concept is explicated and illustrated. A natural number can be used, besides its usual meanings of representing a quantity (cardinal number) or a rank in a sequence (ordinal number), as a label uniquely representing a nominable entity. The uniqueness of nominable entities entails that their representatives do not suffer any change and thus must be protected against any perturbation. A short history of communication engineering, which developed the means of such a protection referred to as 'error-correcting codes', is briefly presented. It is also stated that the theoretical tools needed in order to deal with communication at a distance can be used as well for communication over time such as biological heredity.

## 2.1 Information in a Usual Meaning

We think it is helpful to begin a discussion of the information concept with examining a usual meaning of the word 'information' so as to determine what of its features endow it with the status of a scientific entity. Maybe the most familiar modern use of the word concerns mass media where an information consists of telling that some event has occurred and/or of reporting its circumstances. Some source then transmits some spoken or written text, or a succession of sounds and/or images, i.e., some *message*, in order to let know something to some recipient. A characteristic feature of an information is that it is *new* for its recipient or, more precisely, that a message is perceived as an information only if it is new. (Indeed, the word 'news' is used as a synonymous of 'information' in this meaning.) We may think of an information as increasing the recipient's knowledge inasmuch as he/she is able and willing to memorize it. The main feature of information, the *meaning* conveyed by the message, escapes any measurement and can be thought of as intrinsically qualitative. What can be measured, however, is how infrequent or unexpected it is: for instance, 'a man bites a dog' is much more informative than 'a dog bites a man' which refers

G. Battail, *Information and Life,* DOI 10.1007/978-94-007-7040-9_2,

to a more frequent event. If the probability of the reported event can be assessed, its unexpectedness can be measured by its improbability. This is how information is quantitatively measured according to Shannon (1948) (see Sect. 4.2.1 below).

A characteristic feature of an information in this meaning is that the reported event and its circumstances are only perceived by the recipient through the agency of a message conveyed by a channel. In the important case where this message consists of a sequence of symbols[1], like a spoken or written text, we will refer to the information as 'symbolic'. Then, the message consists of a sequence of symbols which can evoke the reported event in the recipient's mind, although there is no causal relation between the message and the event in the physical world. This is possible only insofar as the recipient can understand this message, i.e., provided the source and the recipient share a common linguistic system which consists of a set of conventional rules. The message has then a meaning within, and only within, this system. At the recipient's end, a communication thus involves two successive steps: the message has first to be received; then, using linguistic rules enables perceiving its intended meaning. The first step is performed by processing the channel output and results in making the message available to the recipient. The second step involves using the linguistic rules obeyed by the source for recovering the intended meaning, given the received message. Clearly, the first step is a mandatory prerequisite to the second one. It is easily overlooked as seemingly trivial, but communication engineers know by experience that it is far from being so. It should be emphasized that these two steps concern entirely different functions. Dealing separately with such unrelated problems is not only possible, but it is a methodological necessity. The competence of information theory is restricted to the first step of delivering the message to its intended recipient. We refer to this function as 'literal communication'. As not involved at this step, the recipient is no longer necessarily a living being (especially a human), but may be a machine as well. In any way, the literal communication between the source and the destination can entirely ignore linguistic and semantic aspects, hence the possible meaning of the message does not matter for it.

Still another fundamental property of an information is that it is non-autonomous. It is necessarily embodied within some physical medium which can be made of several different substances, devices or waves, and assume several forms. For instance a text can be written or spoken. The written text is made of a succession of visible marks of conventional shapes on some sheet of paper or computer screen, while the spoken text is represented by a succession in time of acoustic waveforms by the agency of which a listener can perceive the text. Admittedly, the spoken text has specific features like pitch, timbre, rhythm, intonation, accent, ... that the written one lacks, but as an information the text itself is common to both. In the technical field, a symbolic sequence can be recorded in the form of a binary sequence in the memory of a computer, then read and broadcast, for instance, as a 4-phase modulated electromagnetic wave. Then, both the alphabet size and the very nature of the physical medium which bears an information can be modified: the computer memory and the electromagnetic wave then bear the same information. In the absence of

[1] A symbol is an element of some given finite set of distinct objects, referred to as an alphabet.

any physical medium, however, an information cannot have any interaction with any material device or observer. Even our most abstract thoughts manifest themselves by the activity of neurons in our brain. We may say that no information exists unless it is borne by a physical medium, this remark being used for explicating what is meant here by the *existence* of an information. In what follows we refer to the physical bearer of an information as its *support*.

This way of defining existence meets a fundamental concept of Buddhist philosophy, according to Matthieu Ricard and Xuan Thuan Trinh (Trinh 2011, Ricard and Trinh 2002). It also complies with Carlo Rovelli's relational point of view on physics (Rovelli 2004) which has roots in ancient Eastern philosophy, too. Ref. (Ricard and Trinh 2002, p. 46) contains indeed the following quotation from the Indian philosopher Nagarjuna, who lived in the second century: 'Phenomena draw their nature from a mutual dependence and are nothing by themselves.'

## 2.2 Features of Information as a Scientific Entity

The few simple remarks above suffice for endowing the concept of information with the status of a scientific entity, and especially for founding a quantitative theory of information. As stated above, only finite alphabets are contemplated. This restriction entails that only discrete information is considered in this book although the theory has been fruitfully extended to continuous information. We prefer not to deal with this extension, which is not needed in the examples considered in this book, because it involves mathematical difficulties and some of its results are weaker than the homologous ones of the discrete case. Other applications to biology or physics would nevertheless need the extension of information to continuous random variables.

**Dissociating information and meaning** The first remark above enabled us to distinguish literal and semantic communication, and to state they are unrelated. Then, literal communication constitutes in itself a function which can be dealt with independently of any semantic consideration. In what follows, the word 'information' will be restricted to designate what can be *literally* communicated, and the mathematical theory of literal communication will be referred to as *information theory*.

This founding divide has been clearly stated by Shannon. He did not deny that information has something to do with meaning, of course, but he realized that communication engineering is entirely foreign to semantics. He wrote in the very first page of his seminal paper (Shannon 1948):

> The fundamental problem of communication is that of reproducing at one point either exactly or approximately a message selected at another point. Frequently the messages have *meaning*; that is they refer to or are correlated according to some system with certain physical or conceptual entities. These semantic aspects of communication are irrelevant to the engineering problem. The significant aspect is that the actual message is one *selected from a set* of possible messages. The system must be designed to operate for each possible selection, not just the one which will actually be chosen since this is unknown at the time of design. (Shannon's italics.)

'Communication' is intended here in its *engineering* meaning of literal communication, foreign to the *philosophical* problems of semantics. Information processing devices and quantitative information measures are relevant only to the former. Indeed, a messenger has not to know about the content of the message he or she carries, and the same is true for communication machines, all the more semantics is intrinsically foreign to them. Ignoring semantics simply made information theory and its innumerable engineering applications possible, enabling the use of *mathematical* means for dealing with literal communication.

**An information needs a physical medium but does not otherwise depend on it** Indeed, an information has no existence (in the above meaning) unless it is borne by a physical medium, but a given information can be borne by any medium. In other words, an information is invariant with respect to the medium which bears it. It is thus an entity in itself. A medium is needed for embodying an information, but has no influence on it beyond securing its existence.

That an information does not exist unless it is borne by some physical medium, one of our basic postulates, sharply contradicts the common perception of informations (or of ideas) as purely abstract entities. For instance, no physical medium is as large as to memorize the immense knowledge that Laplace's omniscient demon is assumed to possess (see Sect. 6.3 below), so this demon cannot be but a pure spirit, hence foreign to the physical world. Stating that 'Information is physical', Landauer has to be credited for having challenged this opinion[2] (Landauer 1996). We are far from endorsing Landauer's statement, however, deeming instead that information needs to be physically inscribed. This statement is quite different from Landauer's but departs from idealism and similarly intends to anchor information within the physical world. We criticize Landauer's statement in Sect. 2.3 below.

**Information is not conserved and can be shared** Since its very existence depends on the medium which bears it, an information is *annihilated* if this medium is destroyed or incurs any change which alters the information it bears (altering an information, according to our viewpoint, replaces it by another one). Thus, contrary to many entities met in physics, information is not conserved. On the other hand, an information written on some medium can be copied on another one without being lost. An information can be copied several times so it can *proliferate*, meaning that *the same* information is borne by *several* different supports in increasing number. Proliferation does not mean any increase of information quantity, but that an information can be simultaneously borne by several supports. It is only when *differences* among the set of copies of a same information are created that the quantity of information it bears increases.

The ability of information to proliferate entails in the living world the ability of individuals to proliferate, which is a characteristic attribute of life. It is only because we deal with information as an *abstract* entity that we can reach this conclusion.

[2] In the information-theoretic literature. Physicists do not even ask the question whether information is a physical entity but deal with it as such, following Schrödinger and Brillouin.

Dealing with information as physical as did Landauer has not this consequence, hence is not adequate for biological applications.

**An information as an equivalence class** We notice that an information is invariant with respect to:

a) the physical nature of the medium which bears it (e.g., computer memory, acoustic or electromagnetic wave, sheet of paper, ...);
b) the alphabet size, which can be arbitrarily chosen for its practical convenience. For instance the binary alphabet is very convenient in operations performed by computers, but quite inconvenient for humans who prefer alphabets of larger size like the Latin one which are much better fitted to their perceptive performance. It is why programming languages contain instructions written in the Latin alphabet which are later converted into sequences of binary digits which control the machine;
c) for a given alphabet size, the possible transformation of a sequence into another equivalent one, related to the original sequence by a one-to-one correspondence according to some *encoding rule*.

Properties of invariance are, by essence, cumulative. The invariance stated by (a) and (b) is just a rewording of the above remarks. That stated by (c) is the most important but it is also the less obvious. It is why we will lay emphasis on it, especially in its form referred to as channel coding, in Sect. 3.4 and in Chap. 5 below. An entity defined as the set of elements which are invariant with respect to a transformation is referred to as *an equivalence class*. (Defining an equivalence class is a standard means for creating a mathematical object.)

Then *an* information is an equivalence class of sequences with respect to transformations a), b) and c). Dealing with an equivalence class implies designating a *representative* of it, which may be any of its elements. The most convenient representative of an information is the *shortest binary sequence* which belongs to its equivalence class, to be referred to as its *information message* (the physical medium needs not be specified as irrelevant to information theory, which by essence is mathematical). The symbols of the information message are necessarily independent[3] because, if they were not, source coding could transform it into a shorter one (see Sect. 4.3 below).

It should be kept in mind that an information thus defined as an equivalence class is quite an abstract entity, all the more it obviously contains infinitely many elements. As a collective object, a symbolic information may by no means be reduced, or likened, to a single sequence. Such a sequence may only act as its representative. Defining as we did an information as an equivalence class is a necessary consequence of the basic fact that an information must be *physically inscribed*. The multiplicity of possible physical supports of a same information suffices to show that it must be dealt with as an equivalence class. However, the necessity of encodings of different

[3] In any possible meaning of the word: there should not be a causal relation between them and, if they are random, they should be mutually independent in the probabilistic meaning of the word, i.e., their joint probability should be the product of their individual probabilities.

kinds which generally modify the length of a message is a still stronger incentive to do so. It is thus practical necessities of handling information which demand such an abstract definition, which is by no means gratuitous.

An information cannot be reduced to a number of any kind. It is an entity in itself, which meets Barbieri's concept of *nominable entity* (Barbieri 2007) (see Sect. 2.4.3 below). However, numbers can be associated with it. First of all and more important, the information message, its representative, can be interpreted as a natural number expressed in the binary numeration system and this may be useful, say, for classification purpose. Moreover, information theory enables associating a quantitative measure with an information, and it turns out that this measure equals the length of its information message. An information should by no means be confused with its measure but a difficulty arises as regards the vocabulary because the word 'information' is often used in papers dealing with information theory as an abridgement for 'information quantity'. Due to the dissociation of information and semantics, no ambiguity results in the engineering literature. When trying as we do here to make explicit the relationship of information with objects foreign to communication engineering, however, the reader must be warned against this confusion. We try to avoid it, sometime at the expense of rather lengthy periphrases.

Being defined as equivalence classes and not being numbers, informations cannot be ordered although their quantitative measures can be. Each information is an abstract object which has no other property than its *uniqueness*. No topology thus exists among informations, and the information quantity which measures an information is a mere attribute of it. As regards sequences, the Hamming metric to be defined in Sect. 2.4.4 and used later defines a topology among the sequences which represent informations, not among informations themselves.

## 2.3 Comments on the Definitions of Information

Shannon's approach in (Shannon 1948) was purely empirical. He defined an information *quantity* (see Sect. 4.2.1) but did not attempt to define information as an entity. It is such a definition which is proposed in the previous section, and the reader should be warned that this definition is not the only possible one. Indeed, defining information has not been necessary for developing information theory and was not needed for its engineering applications. We attempt here to define information in order to help applying it to objects foreign to communication engineering, i.e., so as to explicate its relationship with semantics (in a very broad sense).

We stated above that an information is *physically inscribed*. Let us emphasize how this is different from Landauer's statement, who wrote that 'information is physical' (Landauer 1996). He arrived at this conclusion by studying the physics of objects which can bear a binary digit, i.e., of two-state machines. But why should the physical properties of the support of an information be likened to that of information itself? Our statement that an information is physically inscribed leads to very different conclusions. For us, physical objects bear an information to which we attribute properties which have no physical counterpart, the most specific one being the possibility of its copy on a new support while keeping it on the initial one. Another very important property of information is that coding processes can transform a given sequence into another strictly equivalent one but having a different length. These two sequences obviously bear the same information, so an information cannot be likened to a single sequence, and still less to the physical support of such a sequence. The discrepancy between the status of information according to Landauer and to us is actually of capital importance, and it is only our information concept which will enable the conclusions of Chap. 10 regarding the relationship of the living and inanimate worlds.

We define an information, abstractly, as an *equivalence class* with respect to the possible supports which can bear symbols. Moreover, we consider as equivalent symbolic sequences deriving from each other by coding of any kind, i.e., by abstract transformations. We think that an information, instead of being itself a physical entity, has merely a mandatory relationship with physics due to the necessary physical inscription of a representative of it as an equivalence class. This will be examined in Sect. 3.1 and 3.2. The main parameter expressing the dependency of information on the physical world is the signal-to-noise ratio which determines according to Eq. (3.10) a symbol error probability. Symbol errors do not directly affect an information when it is represented by a word of a redundant code as efficient as to ensure its conservation. Instead of dealing with information as a physical entity, we propose in Sect. 6.3 to consider information as a fundamental entity from which we can derive the physical entropy, and not the other way round. Doing so is the exact contrary of what Schrödinger and Brillouin did (Sect. 6.3.4). Endowing information with the status of a fundamental entity will be especially useful in biology, as we shall see in the second part. It is the abstract definition of information that we propose, and only it, which can account for the specific properties of life. Using this definition entails that information appears as bridging the abstract and the concrete, as shown in Sect. 6.4. Only this definition can account for the fact that the abstract content of a symbolic sequence instructs the assembly and the maintenance of concrete objects as does a genome. Far from expressing a purely philosophical disagreement, opposing the statement 'information is abstract' to Landauer's 'information is physical' is of fundamental importance for our project of refounding biology on the science of information.

The confusion of an information with its necessary support, which reifies information and hence denies its abstract facet, is not limited to Landauer's work. Most physicists implicitly accept it without a serious examination of the issue. I am moreover afraid that the comparatively new discipline of quantum information theory which intends to integrate quantum physics into information theory relies on the same confusion (for instance, the acronym 'qubit' was coined for designating a physical system, while 'bit' designates the quantitative unit of information in conventional information theory). I am even reluctant as regards this very approach, and it is why this book ignores quantum information theory. On the contrary, I wonder if quantum physics could *derive from* information theory, although no attempt to answer this question is made in the present book.

## 2.4 An Information as a Nominable Entity

### 2.4.1 Naming and Counting

Any human society is made of individuals, each of them is *unique*. Similarly, we are surrounded with objects which can be uniquely identified as possessing some distinctive properties. We refer to objects or beings which can be unambiguously identified as *singular*. *Naming* a singular object means associating with it a vocal or written label which unambiguously designates, evokes or represents it. This label or tag is a sequence of a finite number of signs which belong to some finite repertoire given once and for all. Naming is an act of language, hence specific to the human culture. The first naming systems were probably vocal, or some combination of vocal signs and gestures. The signs of the repertoire should be mutually distinct but they are otherwise arbitrary. We mostly restrict ourselves in the sequel to *written* texts, i.e., to sequences of readable signs (often intended to represent vocal signs, or phonemes). Then, the repertoire of signs is referred to as the *alphabet* and its elements as *letters*.

A wide variety of objects can be named. They may be living beings, or singular objects of the physical world, or a set of physical objects which share some common specific property within an equivalence class. Relations between objects or sets of objects can be named, as well as these objects or sets. Sets of objects and relations of any kind belong to the abstract world. As perceived by the human consciousness, mental objects too can be named.

Among abstract objects, certain equivalence classes referred to as *numbers* soon needed to be specifically represented by symbol sequences in order to incur a kind of processing referred to as *computation*. *Cardinal* numbers just tell *how many* objects of some kind are present in some given set and *ordinal* numbers tell the place *where* given objects of some given ordered sequence are located. Letters, i.e., the same signs as used for transcribing phoneme sequences, have sometimes been used for denoting numbers when combined according to rules specific to this purpose, as for instance in Roman numeration. Such representations of numbers were rather cumbersome and their use for computing was quite complicated. Using symbols specifically intended to represent numbers, the Arabic *digits*, much better fits the needs of computation. Together with the modern numeration system, it enabled performing computation by simple machines as that invented in 1652 by Blaise Pascal (then 19-year old) as well as by nowadays computers. The representation of numbers by numeration systems will be considered in more detail later (Sect. 2.4.2). The choice of the base of a numeration system is just a matter of convention: it should be such that its digits are conveniently distinguished. We inherited the base ten from the Greco-Roman antiquity and, in accordance with this chosen base, the Arabic digits of our numeration system are ten. The bases twenty and sixty have been used in certain cultures, however, and we still use the base sixty for measuring certain time intervals (hours, minutes, seconds), a legacy of the Babylonians. Humans easily distinguish signs which belong to alphabets of this order of magnitude. Computers generally use much smaller bases, especially the simplest possible one, 2, although calculation circuits using the base 3 were implemented in early Soviet computers.

We have thus now in Western culture two main sets of written symbols: the letters of, say, the Latin alphabet, which properly combined in sequences represent the words of a language; and the set of signs, referred to as digits, which represent numbers. There is however no intrinsic difference between letters as used for writing texts in some language and digits used for denoting numbers and computing, since both are elements of an arbitrary finite set of symbols. What differentiates written representations of words and of numbers by sequences of signs is merely that they are interpreted according to different rules. A written word and a written number may thus be considered both as sequences of symbols of some alphabet; moreover, a same alphabet can be used for both. This alphabet can be assumed to be the simplest possible, i.e., binary, without loss of generality. Sequences of binary digits are used in computers for representing both texts and numbers. We use from now on the acronym 'bit' for *binary digit*.

As an example, the letters of the Latin alphabet are currently represented in computers according to the American Standard Code for Information Interchange (ASCII). Each letter is denoted by a 7-bit word according to a one-to-one correspondence: a lower case letter is represented by '11' followed by the 5-bit sequence which represents its rank in the alphabet according to the binary natural numeration. For instance, 1100001 denotes 'a', 1100010 denotes 'b', etc. Capital letters use the prefix '10' instead of '11', so 1000001 denotes 'A', 1000010 denotes 'B', etc. Another representation of the Latin letters uses the 8-bit words which result from appending to a 7-bit word as previously defined a single bit such that the total number of '1's in the word is even: then 11000011 denotes 'a', 10000010 denotes 'A', 11000101 denotes 'b', 11000110 denotes 'c', etc. Appending this eighth bit provides a rudimentary means of error control: if an error affects a single bit in the 8-bit word, the number of '1's becomes odd so counting the '1's in each word enables detecting that an error has affected a single symbol, but not correcting it. More sophisticated means, using longer words, can result in locating the error in the word and correcting it.

The previous remark was concerned with the problem of representing the letters of an alphabet (hence words of a language combining several letters) by means of digits. At variance with such words, which are semantically related with outer objects, numbers have intrinsic properties and are endowed with structures of their own. Their study constitutes an important part of the mathematical science. The representation of numbers of any kind relies on the structure of the most basic ones, referred to as the 'natural integers'. Let us now have a look at it.

### *2.4.2 Defining and Representing Natural Integers*

The mathematical definition of a natural integer is given in many textbooks. It is most often defined as an element of a set $\mathbb{N}$ which satisfies Peano's axioms, namely:

1. $\mathbb{N}$ contains a particular element named 'one' and denoted by 1.
2. For any element $a$ of $\mathbb{N}$, there exists in $\mathbb{N}$ an element $b$, referred to as the successor of $a$, denoted by $b = a'$. Then $a$ is said the predecessor of $b$.

3. Any element of $\mathbb{N}$ has a predecessor, except 1 which has none.
4. Two natural integers are equal if, and only if, their successors are equal.
5. $\mathbb{N}$ obeys the axiom of recursivity, namely:
   (a) If a property stated in terms of some integer $n$ is true for the number 1;
   (b) If it can be proved that, if this property is true for any integer $m$ larger than 1, then it is also true for the successor $m' = m + 1$ of $m$ (an integer $b$ is said to be larger than an integer $a$, $b > a$, if $b$ belongs to the successors of $a$; the 'successors' of $a$ should be understood here as $a'$, the successor of $a'$, the successor of this successor, etc.);
   (c) Then this property is true for any integer larger than or equal to 1.

Two operations, addition and multiplication, are defined on natural integers: addition (denoted by +) defined as $a+1 = a'$ and $(a+b)' = a+b'$, and multiplication (denoted here by $*$) defined as $a * 1 = a$ and $(a * b) + a = a * b'$. These operations are associative and commutative. Multiplication is distributive with respect to addition, i.e., $(a + b) * c = a * c + b * c$.

Less formally, Henri Poincaré assumes that the operation $x + 1$, which consists of adding the number 1 to a given number $x$, is firstly defined and he notices that this definition, *whatever it is*, does not play any role in the reasonings to follow (Poincaré 1902). He defines the operation $x+a$, which consists of adding the number $a$ to a given number $x$, assuming that the operation $x + (a - 1)$ has been defined. Then, the operation $x + a$ is recursively defined by the equality

$$x + a = [x + (a - 1)] + 1. \tag{2.1}$$

In other words, we know what $x + a$ means when we know what $x + (a - 1)$ means, and since we already know the meaning of $x + 1$, it is possible to successively define $x + 2$, $x + 3$, etc.

Being *recursive*, the definition using Eq. (2.1) actually contains infinitely many distinct definitions, each of them becoming meaningful only when the meaning of the preceding one is known. Having thus defined the addition of integers, Poincaré recursively shows that it is associative, i.e., $a + (b + c) = (a + b) + c$ for any integers $a$, $b$ and $c$, and commutative, i.e., $a + b = b + a$. He then defines the multiplication of integers by the equalities

$$a * 1 = a \tag{2.2}$$

and

$$a * b = [a * (b - 1)] + a. \tag{2.3}$$

Once $a * 1$ has been defined by Eq. (2.2) and (2.3) enables successively defining $a * 2$, $a * 3$, etc. Poincaré also recursively establishes the properties of multiplication, showing it is distributive with respect to addition, i.e., $(a + b) * c = (a * c) + (b * c)$, and commutative, i.e., $a * b = b * a$, for any integers $a$, $b$ and $c$.

If we ignore some subtleties of Peano's derivation which are motivated by the exigence of mathematical rigour, we may simply summarize how the natural integers are introduced. First, take an element 'one', denoted by 1, which is left undefined although we have a sensible intuition of its meaning. Then, define the operation 'add one', denoted by '+ 1'. Once an integer $n$ has been defined, $n + 1$ defines another integer. Thus, starting from $1 + 1 = 2$, the infinite series of integers results.

From a practical point of view, integers are useful for reckoning and labelling objects. Assume that I have some collection of arbitrary material objects which I can individually handle. I may take off one of them and say 'one', take off another object and say 'two', and so on, uttering the name of the next natural integer every time I take off an object. The last uttered number, say $n$, tells how many objects are present in the collection. This number is referred to as 'cardinal' and expresses a *quantity*: my collection contains $n$ objects. I just used numbers for *counting* the objects. If I am not interested in the peculiarities which possibly make these objects different from each other, I am satisfied with this result. However, if each object is unique and if I want to easily distinguish it from the other ones, I can use the number I utter when I take it so as to indicate its place in the series of objects I took off, and it is why it is then referred to as 'ordinal'. If I took off the objects in an arbitrary order, this process enables *identifying* each of them by tagging each object with its ordinal number according to the chosen order. Numbers are now used for *naming* the objects of my collection. Doing so provides a naming system which is both *open* since the set of integers is unlimited and as *universal* as the use of numbers.

Using an integer as a label is possible thanks to a capital feature left somewhat implicit as yet: its *uniqueness*. Any integer newly introduced in $\mathbb{N}$ by the recursive process described by Peano or Poincaré is different from those which were previously introduced. Indeed, any integer can be shown to be equal to the product (i.e., the result of multiplication) of a unique set of prime numbers, each of which is unique itself.

How then is it possible to denote each of these potentially infinitely many elements? Creating a new symbol for each newly introduced number would be highly impractical, so it is mandatory to use only a finite number of symbols. This problem has cleverly been solved by numeration systems, which use a set (the alphabet) of a finite number of signs (the digits), which can be combined so as to represent arbitrarily large integers. It is first necessary to introduce a number foreign to $\mathbb{N}$, namely 0, such that $0 + a = a$ and $0 * a = 0$ for any $a$ in $\mathbb{N}$. Then, an integer $n$ is represented in a numeration system of base $b \geq 2$ by the sequence $d_{\ell-1} \dots d_1 d_0$, where the digits $d_0, d_1, \dots, d_{\ell-1}$ denote the $\ell$ numbers $n_0, n_1, \dots, n_{\ell-1}$ which all belong to the alphabet $\{0, 1, \dots, b-1\}$ and are such that $n = n_0 + b * n_1 + \dots + b^{\ell-1} * n_{\ell-1}$. The number of digits $\ell$ needed for representing the integer $n$ is such that $b^{\ell-1} \leq n < b^{\ell}$. The number $n$, hence the number $\ell$, may increase without limit. Then an *arbitrarily large* integer can be represented using a *finite* set of symbols, combined into an orderly sequence of *unlimited* length. Infinitely many integers can thus be represented by this means.

A more convenient way for introducing the set of integers consists of starting from the set of non-negative integers consisting of the previously defined one by appending to it the element 'zero'. (The usual notation of this new set is $\mathbb{N}$, and the firstly introduced set is then denoted by $\mathbb{N}^*$.) This new set contains both the neutral element of addition, 0, ($a + 0 = a$ for any $a$ in $\mathbb{N}$) and that of multiplication, 1, ($a * 1 = a$). In $\mathbb{N}$, 0 is the predecessor of 1 but 0 itself has none. Then any element of $\mathbb{N}$ can be represented by a numeration system as described above.

Notice that such a numeration system can be extended to a *fractional* number $q = 1/a$ such that $q * a = 1$, $a$ being a natural integer as originally defined, i.e., different from 0. The number $q$ is also denoted by $a^{-1}$. Then, the sequence $0 \cdot d_{-1}d_{-2}\ldots$ is interpreted as meaning $q = n_{-1} * b^{-1} + n_{-2} * b^{-2} + \ldots$, where the numbers $n_{-1}, n_{-2}, \ldots$, which all belong to the set $\{0, 1, \ldots, b - 1\}$, are denoted by the digits $d_{-1}, d_{-2}, \ldots$, respectively. The representation of certain factional numbers can then involve an infinite number of digits. For instance, in decimal numeration, $0 \cdot 111 \ldots = \sum_{i=1}^{\infty} 10^{-i}$ represents 1/9. In such a case, a finite number is represented by the sum of infinitely many finite terms[4]: each of them is as small relatively to the previous one as to keep the sum finite.

More about extensions of the number concept can be found in Appendix B, Sect. B.2.

### 2.4.3 Concept of Nominable Entity

We have seen in Sects. 2.4.1 and 2.4.2 two possible ways of using sequences, for naming objects hence for a semantic purpose, or for counting objects as needed for computing. Sequences which represent integers in some numeration system can even be used for labelling objects, which is another way for naming them. But sequences exist by themselves, regardless of how they are interpreted. Indeed, every sequence is unique. Substituting one of its symbols for another one suffices to transform it into another sequence which is unique, too. Barbieri referred to objects or classes of objects which can be unambiguously named as *nominable entities* (Barbieri 2007). Any entity which cannot suffer any change without losing its very identity (or ceasing to be itself) is a nominable entity. The sequences intended to name them must be unique, too, so as to match the singularity of the named objects. Sequences are then pure, prototypic nominable entities, independently of any interpretation. They have no other property than their *uniqueness*. We already noticed that any integer can be uniquely expressed as a product of primes and is thus unique. A numeration system associates with it a unique sequence of digits.

Being the science of sequences, information theory may be thought of as the science of nominable entities. More precisely, we defined above an information as an equivalence class of sequences, and it is such a class which possesses the property of uniqueness. We may thus interpret *an* information, defined as an equivalence class of sequences, as *a nominable entity in Barbieri's meaning*. Indeed, the concept of nominable entity matches the attempted definition of *an* information outlined above (Sects. 2.1 and 2.2), since its representative, the *information message*, is unique. Information theorists, and first of all Shannon himself, dealt with information in

[4] The progress of mathematics has refuted Zeno's paradoxes like that of Achilles and the tortoise.

general and, thanks to the postulated divide between information and semantics, did not need to precisely define an information. On the contrary, outlining the relationship of information with semantics is necessary for applying information theory outside communication engineering. It is why we try to do so in Sect. 4.3.5 below.

The length of possible symbolic sequences is not limited. Their number is thus infinite, and so the number of possible nominable entities is unlimited. Moreover, the number of sequences is a fast increasing function of their length $k$, namely the exponential $\alpha^k$, where $\alpha \geq 2$ is the alphabet size.

Before proceeding further, let us give a few examples borrowed from the daily experience. A sequence which uniquely labels a nominable entity may be, for instance, an address or an identification number. A telephone number is an example of such a sequence. I want to call Mr. Dupont whose telephone number is 0123456789, but if I erroneously dial 0123456788, I fail to be connected with him. The number I dialled differs from that of Mr. Dupont only in its less significant digit if I interpret his telephone number according to the usual numeration system, but the same negative issue occurs if I dial a very different number, e.g., 9012345678. It would make actually no sense to interpret a telephone number as an ordinary integer. In mathematical words, nominable entities cannot be ordered (no one is larger or smaller than another one) and they ignore any topology: a nominable entity has no neighbours. Performing operations of ordinary arithmetic on such 'numbers' would obviously be meaningless. Having discovered the role of nominable entities in the life processes, Barbieri is right when writing that 'they are a new kind of natural entities' (Barbieri 2007, p. 200).

However, just like a telephone number, any address or identification label can be written as a natural integer by means of a numeration system. Natural integers are endowed with a topology when they are intended to express the result of counting and then their status of nominable entity (of being uniquely expressed as a product of primes) is no longer relevant. Then, instead of a nominable entity, they represent a quantity. For instance if I am a fireman who intends to count the attendants in a theater in order to estimate how long it would take to evacuate it in case of emergency, 100 and 101 people make little difference, at variance with 100 and 999 people. Ordering these numbers and stating that they are more or less close to each other, i.e., endowing them with a topology, becomes meaningful. However, if I use numbers as labels for identifying the attendants, like the numbers worn by the competitors of a race, the one who bears number 100 is just as different from the bearer of 101 as he/she is from the bearer of 999. In other words, no degrees of difference are meaningful in such a case: identities ignore any topology. Then natural numbers used as labels act as nominable entities and are not endowed with any topology nor can be ordered: although the natural number 999 is larger than the natural number 101, such an inequality is meaningless when these numbers are used in order to label individuals, hence represent nominable entities.

As still another example of a nominable entity, a person is identified (in France) by a 13-digit number the successive decimal digits of which indicate his/her sex (1 digit), then the year, month and place of birth (2, 2, and 5 digits, respectively), plus the rank of the line in the register where the birth has been recorded (3 digits). As uniquely identifying someone, this sequence[5] is a nominable entity, too. But it is obtained by the *concatenation* of sequences of 1, 2, 2, 5, and 3 digits having the meanings indicated above, each of which being itself a nominable entity. In such a case, sequences are successively appended to each other so as to constitute some single string of symbols. It must be possible to separate each peculiar sequence from the other ones as an intrinsic nominable entity. This condition is fulfilled in the above example because the order of the component sequences as well as their lengths are known. In general, means should enable determining the precise places where a component sequence begins and ends. Sequences which can be separated from each other inside a symbol string which includes them are said to be *decipherable*, and how this can be achieved is dealt with in Sect. 3.2. The lengths of the concatenated sequences necessarily obey the Kraft inequality (4.32). Concatenating short sequences is a means for creating longer ones which is often useful. If the short concatenated sequences are decipherable, the longer sequence which results from their concatenation specifies an object as simultaneously belonging to different classes each identified by one of the shorter component sequences. We shall refer to the nominable entities corresponding to the component sequences as *nested*. Then the concatenation reflects a hierarchical taxonomy.

A musical theme is another example of a nominable entity, all the more interesting since no semantic content is generally associated with it (if we except Wagner's *Leitmotive*).

Nominable entities are rather foreign to the usual practice of science, at least if physics is taken as reference. A physical measurement, for instance, consists of assessing some physical quantity by means of an experimental apparatus. Except if it consists of counting objects and thus results in an integer, a measurement provides a number of a quite different kind, said 'real' in the mathematical meaning of the word, which however may be somewhat misleading (see Sect. B.2 in Appendix B below). A margin of uncertainty about the measured quantity *always* remains, due to the intrinsic limitations of the measurement apparatus and possibly to the lack of a sharp delimitation of the measured quantity itself. The result of a measurement should always be considered together with its uncertainty margin, so its actual mathematical status is that of a 'real' *random variable*. A single measurement result is merely a realization of this random variable, of *a priori* unknown mean and variance although they can often be estimated. Repeating many times the experiment, provided it is possible, enables as a consequence of the law of large numbers to make their estimation more an more accurate, and the mean of the random variable is referred to as the measurement result.

The relative precision of a measure, i.e., the ratio of the uncertainty margin to the measured quantity, widely depends on the measurement apparatus, especially on the quality of the tools available to build it and on the skillfulness of the craftsmen who did so. Nowadays, the apparatuses still need to be precisely manufactured but measurements furthermore rely on sophisticated data processings based on information theory and computer science. These processings are intended to diminish the

[5] It is redundant: fortunately, less than $10^{13}$ people live in France.

uncertainty margin, but risk of making the results dependent on more or less implicit underlying hypotheses. In any case, a nominable entity *cannot* result from the measurement of a physical quantity, regardless of its precision, as being intrinsically different from an integer (or 'natural number') as defined above.

The word 'real' was initially intended to be opposed to 'imaginary', both qualifying numbers. Imaginary numbers, now renamed 'complex numbers', revealed useful in the resolution of equations but are foreign to any counting process. However, it is likely that the slowly elaborated mathematical concept of *real number* has been developed in order to match a property that classical physics associates with the entities of the real world: *continuity*. It turns out that the mathematicians used to this end tools entirely foreign to the physical world like limits of infinite series. They did not take into account the necessarily limited precision of any physical measurement. The very progresses of physics have moreover shown that an apparent continuity often masks, by an effect of large numbers, the existence of very numerous discrete objects or events. The engineers use to express the divide between discrete objects and continuous (or seemingly continuous) ones the words 'digital' and 'analog'. This latter word is not assumed to refer to all the properties that mathematicians attribute to 'real' numbers since they are actually random variables, however sharply defined they may seem. It should just be intended as the contrary of 'digital'. The relevance of 'real' numbers in physics has been criticized by many mathematicians, e.g., Poincaré and, recently, (Chaitin 2005, p. 94) (also see Sect. B.2 below). The concept of real number has been fully elaborated only at the end of the XIX-th century. A few decades later, physicists dismissed continuity from the way they describe the world. They had to invent other mathematical tools like operators, which act through a discrete set of eigenvalues, in order to properly account for the discontinuous new vision of the world that quantum physics proposed.

It should be noticed that, according to our view of an information being necessarily inscribed onto a physical support, a real number in the mathematical meaning cannot represent an information since its symbolic representation would imply an infinite number of digits, hence need an infinitely large support. Even a single quantity measured by a real number in this meaning could not be stored in the physical world (we use this argument for exorcizing Laplace's demon in Sect. 6.3.3 below).

Nominable entities are nevertheless needed in physical sciences in order to designate singular objects, or sets of objects which share some singular property. For instance, astronomy, chemistry and the physics of particles need naming labels, just like biological taxonomy. The utility of these labels is restricted to the scientific literature, but they are not related to the phenomena themselves. What makes biology unique, radically different from any other science, is that certain *nominable entities* borne by genomes actually *participate* in the operation of life. As emphasized by Barbieri, far from being confined in the human culture, they are first rank actors of the living processes. Likening informations to nominable entities then legitimates considering as highly specific the relation of biology to information theory, a major thesis defended in this book (see Chap. 9).

### 2.4.4 Representatives of Nominable Entities Need to be Protected

Nominable entities are obviously of fundamental importance in culture as well as in any form of life. When the representative of an information, i.e., an information message, is a recorded sequence, the perturbing influences incurred by the recording medium threaten its integrity, hence the very existence of the information.

In the cultural field, we do not really perceive that nominable entities are so fragile, in part because the artificial memory devices we use to record sequences belong to the macroscopic world; they are thus made of a huge number of molecules so the states to be distinguished for identifying an alphabet symbol always involve myriads of atoms. When the sequences are borne by an acoustic or electromagnetic wave, its power widely exceeds that of the ambient noise. It is far from being so for natural memory devices or signals at the cellular or molecular scale. Especially, DNA and RNA sequences belong to the quantum world as molecular memories.

However, the physical robustness of our memory devices probably does not suffice to secure the needed integrity of cultural nominable entities. We do not realize that, owing to its enormous redundancy, our linguistic system acts as an error-correcting code at multiple levels (see Sect. 8.1.4). This error-correcting ability results in keeping the integrity of informations, as defined in Sect. 2.2 and now likened to nominable entities. We claim that similar error-correcting means must exist at the most basic levels of life. These means need to be even more powerful because, at the cellular and molecular levels, the risk of symbol errors is much larger than at the macroscopic scale: the smaller the spatio-temporal scale of the medium, the larger the risk. Moreover, genomes are conserved during the immense time intervals of geology.

The vulnerability of sequences to perturbations of any kind and the necessity of conserving nominable entities must be conciliated. The very existence of the nominable entities lies in their integrity, and the fragility of the recording media which are needed for their conservation demands that the sequences which bear them are highly protected (see Sect. 3.4). When sequences are communicated or recorded, each of their symbols incurs the risk of being replaced by another one with some non-zero probability, resulting in substituting another one for it. This event is referred to as a symbol error (for binary symbols, the error probability can be assumed to be less[6] than 1/2; a proper labelling of the symbols has always this result). In communication theory parlance, the channel is 'noisy'. If we consider the set of $n$-symbol sequences, the Hamming distance $d$ between two of them is just the number of symbols which need to be changed in order to transform one of them into the other one. The larger $d$, the less likely a pattern of symbol errors can perform this transformation. Keeping the integrity of sequences can thus exclusively be achieved by using sequences which belong to a set where a sufficiently large minimum Hamming distance $d$ exists between its elements. The original sequence can always be recovered if less than $d/2$ symbol errors occurred, since then the

[6] An error probability of 1/2 prevents any communication by means of the binary alphabet; see Sect. 5.2.2.

received word is closer to the actually sent codeword than to any other. This is the principle of error-correcting codes. The integrity of a sequence as a nominable entity can thus be ensured with high probability provided this sequence belongs to an error-correcting code, with a probability of failure the smaller, the larger the minimum Hamming distance $d$ of this code. *Error-correcting codes*, possibly in a meaning extended with respect to that of engineering, appear indeed as the only means for conserving nominable entities despite the fragility of the media. If some given medium on which a sequence is recorded is placed in given physical conditions, the probability of symbol error within some time interval does not vary with time. This error probability may be assumed constant for the need of analysis. The symbol errors are cumulative, so the error-correcting ability of any code, however large its minimum distance $d$, is exceeded after some time interval. The conservation of sequences thus needs the periodic or almost periodic *regeneration* of their recorded representatives, encoded by means of an error-correcting code, after a small enough delay which depends on both the frequency of errors and the minimum distance of the code.

The absolute necessity of protecting nominable entities is a major thesis of this book. Keeping the *integrity of nominable entities* may be thought of as a principle of extremely general reach and significance. Information theory states this necessity and, inspired by it, communication engineering developped the means of its implementation. As specifying living things, genomes are nominable entities which play a prominent role in the life processes. There is thus a necessarily close link between information theory and biology, similar to the link between information theory and communication engineering. The concepts of information theory must therefore have a prominent place in biology. They may be thought of as the bases of a future *theoretical biology*.

## 2.5 A Short History of Communication Engineering

Before we begin expounding information theory, we briefly discuss the historical development of communication engineering in order to help understanding its close relationship with it and why the application of information theory to other sciences remained as yet so limited.

Two events of capital importance for the future of communication engineering occurred almost simultaneously at the same place, the Bell Telephone Laboratories: in 1947, John Bardeen, William Shockley and Walter Brattain invented the *transistor*; and, in 1948, Claude Shannon published 'A mathematical theory of communication'. The technological developments based on the first event, i.e., the semi-conductor technology, provided means for implementing solutions to communication problems having their origin in the second one. In his seminal paper (Shannon 1948), which is more precisely a theory of communication *means*, Shannon introduced a quantitatively measurable entity referred to as information and developed its mathematical theory. Neither the word nor the concept of information were new, of course, but considering information as a *measurable scientific entity* was so.

Shannon's paper (Shannon 1948) gave rise to a new science. This paper constitutes a complete treatise of the nascient science, an event almost unique in history. It has actually been the starting point of a vast and fruitful stream of researches. Some further theoretical developments were needed to confirm Shannon's statements, especially for more rigourously proving some of his theorems, but little was left to Shannon's successors for deepening and expanding the core of information theory. Shannon introduced the main quantities needed for measuring information. He showed that communication is possible only within precise limits which he expressed in terms of the quantities thus introduced (see Appendix A). How to reach these limits, namely the needed processings as well as the means for their physical implementation, remained however entirely to be invented when Shannon's paper was published. Moreover, the promises of the theory exceeded by far the performance of devices which could be implemented when they were formulated, and even what was believed to be possible. The most paradoxical result in this respect is the theoretically proven possibility of *reliable* communication over an *unreliable* channel. This unexpected and very promising result has been a strong incentive to researchers, putting out a difficult challenge. The parallel progress of researches in error-correcting codes and in semi-conductor technology did not result in practically implemented means for closely approaching the theoretical limit that information theory sets for error-free communication, namely the *channel capacity*, before the invention of turbocodes, in 1993.

On the theoretical side, one of the most important events after the publication of (Shannon 1948) has been the introduction of the *algorithmic information theory*, by Solomonoff (Gàcs and Vitànyi 2011), Kolmogorov (Kolmogorov 1965, 1968) and Chaitin (Chaitin 2005) around 1965, which, at variance with Shannon's, does not rely on probabilities. It is inspired by computer science instead of communication engineering and provides a new insight about information (see Sect. 6.1).

If Shannon left comparatively little to be done as regards theory, his papers prompted countless and entirely unexpected applications in the fields of *source-* and *channel coding*. Source coding consists of replacing an initial message by a *shorter* but fully equivalent one. Channel coding aims at protecting a message against transmission errors, which demands introducing redundancy, i.e., replacing the original message by a *longer* but equivalent one. Then within certain limits symbol errors in the encoded message do not prevent recovering the original one.

As regards source coding, information theory established that the message delivered by a source can be encoded so as to reduce its length, but only up to a lower limit which depends on a quantity specific to the source referred to as its *entropy*. The Huffman algorithm asymptotically achieved this limit as early as 1952, for simple source models (see Sect. 4.3.4). Other efficient source coding algorithms were found later (arithmetic coding, Lempel-Ziv algorithm, ...). In sharp contrast, while information theory also stated the limit beyond which channel coding can no longer achieve errorless communication, namely, the *channel capacity*, no practical means to closely approach it were found during decades although it has been perceived as a challenge by thousands of mathematicians and engineers and thus prompted intense researches. This goal was not achieved earlier than 1993 when the invention of turbocodes by Berrou and Glavieux (Berrou et al. 1993; Berrou and Glavieux 1996; Guizzo 2004) provided means to communicate at information rates close to the channel capacity, hence experimentally proving that the limit set by Shannon's channel coding theorem is practically attainable. Analyzing how turbocodes achieve this result further revealed that a much older code family, Gallager's low-density parity-check codes (Gallager 1962, 1963) could also closely approach the channel capacity. When they were invented, the performance of these codes could not be assessed for lack of adequate computation or simulation means and, moreover, their decoding was too complex for being implemented by the technological means then available. Besides providing useful results, the availability of codes practically reaching the channel capacity can be thought of as a remarkable experimental confirmation of the theory itself. Codes able to reach the channel capacity have even been recently invented (Arikan 2009). Far from the purely mathematical discipline it was at its beginning when no means were available for its implementation, information theory can indeed no longer be separated from communication technology. The innumerable engineering achievements it enabled may be thought of as experimental proofs of it. Information theory now acquired a full operational validation.

It turns out indeed that the information-theoretic solutions to communication problems are the more efficient, the more complex. The progress of semi-conductor technology resulted in devices becoming at the same time more and more complex, more and more tiny and less and less expensive. By now, more than 60 years after the transistor was invented, a silicon chip of a few square centimetres can bear about a billion transistors. The tremendous evolution of semi-conductor technology towards increasing complexity and small size perfectly fitted the needs of communication engineering for implementing solutions inspired by information theory. It especially led to the development of very sophisticated *error-correcting codes* which reliable and inexpensive devices can by now implement. They actually invaded our daily life: computer memories, mobile phones, CD, DVD, digital television ... However, they remain invisible and very few people are aware of the high complexity which subtends electronic objects of daily use. With their trend towards complexity and small size, electronic devices tend to mimic biological devices. Just like we are unaware of the physiological processes which keep us alive, we are less and less conscious of the complexity of the electronic objects which we routinely use. Most of us completely ignore how they work and, moreover, explaining their operation often needs advanced mathematical concepts borrowed from information theory.

Although in its early years many researchers attempted to apply information theory to a number of scientific and even philosophical problems, in fact to almost everything, this has not been a fruitful stream of research, but rather a fad which left no significant results (see Sect. 3.2.1 below for an example). Shannon mocked this trend in an editorial written in 1956 where he very lucidly wrote (Shannon 1956):

> A thorough understanding of the mathematical foundation [of information theory] and its communication applications is surely a prerequisite to other applications. I personally believe that many of the concepts of information theory will prove useful in these other fields—and, indeed, some results are already quite promising—but the establishing of such applications is not the trivial matter of translating words to a new domain, but rather the slow tedious process of hypothesis and experimental verification.

The failure of the early attempts shed doubts on the possibility of applying information theory elsewhere than its native field of engineering, resulting in lasting prejudices. More than half a century later, however, we may think that a much improved understanding of the foundation of information theory and its communication applications has been acquired through both theory and experience, so applying it outside communication engineering can reasonably be contemplated. We dare doing so in this book. The examples dealt with in what follows are hopefully not so trivial and not too tedious.

Besides the fad mocked by Shannon, we must mention that some resarchers, mostly coming from physics, were deeply interested in information theory and in its possible applications to natural sciences. Léon Brillouin, to be quoted in several places below, early investigated the relationship of information with thermodynamics (Brillouin 1956), but also asked deep questions about life and its possible connection with information (Brillouin 1959). We should also mention Henry Quastler, and Hubert P. Yockey who tirelessly tried to convince biologists of the interest and potential usefulness of information theory in their discipline (Yockey 1974, 1992, 2005).

## 2.6 Communication Over Space or Over Time

Information theory originated in a reflection about communication techniques, intended to deliver to a recipient a message emitted by a source. The source and the recipient are distinct, hence distant. Communicating the message is possible through the agency of *propagation phenomena* which involve a *wave* generated at the source location which propagates up to the recipient: a periodic time variation of some field occurring at the source entails that, at spatially distant locations, the same variation is reproduced (up to a propagation delay and an attenuation of its amplitude). Then some intentional modification of a parameter of the wave, intended to represent the message, a process referred to as *modulation*, is reproduced at the recipient location where it can be received. Emphasis will be laid on the case where this message is symbolic as defined above, i.e., consists of a sequence of elements from a finite alphabet. The wave is said the 'carrier' of the sequence of symbols, or message, which modulates it.

Although it was not initially intended to this case, communication theory is also relevant when the source and the recipient are separated by a *time* interval instead of a spatial distance. Then the message is recorded on some enduring medium and read later. A major phenomenon of life, heredity, may be thought of as relevant to this case. The recording techniques are different from those of telecommunication, but the basic formalism is the same in both cases so information theory applies to both. We consider communication over space in order to introduce information theory since this case has been the most studied and some concepts are easier to explain in this framework, but many problems relevant to biology, heredity being the first and most important one, concern communication over time. We indicate when needed how the concepts and vocabulary which pertain to communication over space can be transposed so as to fit communication over time. We shall often refer to communication over time as *conservation*, and the conservation of an object is a necessary condition for its present *existence*.

These two modes of communication are, moreover, not entirely foreign to each other. On the one hand, the propagation of the wave which carries a message lasts some finite time and thus actually performs communication over both space and time, even when its only purpose is communication over space and if the delay unavoidably incurred may be detrimental. On the other hand, writing a message on some medium which is later transported up to its destination is an ancestral means for ensuring human communication at a distance. As a biological example of this kind of communication, sexual reproduction involves the transportation of the genetic messages borne by male and female gametes up to the point they meet.

## References


Arikan, E. (2009). Channel polarization: A method for constructing capacity-achieving codes for symmetric binary-input memoryless channels. *IEEE Transactions on Information Theory, 55*(7), 3051–3073. doi:10.1109/TIT.2009.2037044. July 2009.

Barbieri, M. (2007). Is the cell a semiotic system? In M. Barbieri, (Ed.), *Introduction to Biosemiotics* (pp. 179–207). Dordrecht: Springer.

Berrou, C., & Glavieux, A. (1996). Near optimum error-correcting coding and decoding: turbo-codes. *IEEE Transactions on Computers, 44*(10), 1261–1271.

Berrou, C., Glavieux, A., & Thitimajshima, P. (1993). Near Shannon limit error-correcting coding and decoding: turbo-codes. *Proceeding of ICC'93*, Geneva, Switzerland, pp. 1064–1070.

Brillouin, L. (1956). *Science and information theory*. New York: Academic Press.

Brillouin, L. (1959). *Vie, matiere et observation*. Paris: Albin Michel.

Chaitin, G. J. (2005). *Meta Math!*. New York: Pantheon Books.

Gàcs, P., & Vitànyi, P. M. B. (2011). Raymond J. Solomonoff 1926–2009. *IEEE Information Theory Society Newsletter, 61*(1), 11–16.

Gallager, R. G. (1962). Low-density parity-check codes. *IRE Transactions on Information Theory, IT-8*, 21–28.

Gallager, R. G. (1963). *Low-density parity-check codes*. Cambridge: MIT Press.

Guizzo, E. (2004). Closing in on the perfect code. *IEEE Spectrum, 41*(3 INT), 28–34.

Kolmogorov, A. N. (1965). Three approaches to the quantitative definition of information. *Problems of Information Transmission, 1*, 4–7.

Kolmogorov, A. N. (1968). Logical basis for information theory and probability theory. *IEEE Transactions on Information Theory, IT-14*(5), 662–664.

Landauer, R. (1996). The physical nature of information. *Physics Letters A, 217*, 188–193. (Reprinted in Leff and Rex 2003, pp. 335–340).

Leff, H. S., & Rex, A. F. (2003). *Maxwell's demon 2, 2nd edition*. Bristol: IoP.

Poincaré, H. (1902). *La science et l'hypothèse*. Paris: Flammarion.

Ricard, M., & Trinh, X. Th. (2002). *L'infini dans la paume de la main*. Paris: Pocket.

Rovelli, C., (2004). Relational quantum theory. In N. Kolenda & A. Elitzur (Eds.), *Quo Vadis Quantum Mechanics?* Dordrecht: Springer.

Shannon, C. E. (1948). A mathematical theory of communication. *The Bell System Technical Journal, 27*, 379–457, 623–656. (Reprinted in Shannon and Weaver 1949, Sloane and Wyner 1993, pp. 5–83 and in Slepian 1974, pp. 5–29).

Shannon, C. E. (1956). The bandwagon. *IRE Transactions on Information Theory*. (Reprinted in Sloane and Wyner 1993, p. 462).

Shannon, C. E., & Weaver, W. (1949). *The mathematical theory of communication*. Urbana: University of Illinois Press.

Slepian, D. (Ed.). (1974). *Key papers in the development of information theory*. Piscataway: IEEE Press.

Sloane, N. J. A., & Wyner, A. D. (Eds.). (1993). *Claude Elwood Shannon, collected papers*. Piscataway: IEEE Press.

Trinh, X. Th. (2011). Un non-début de l'univers? *Nouvel Observateur Hors-Série* No. 77, *L'origine du monde*.

Yockey, H. P. (1974). An application of information theory to the central Dogma and the sequence hypothesis. *Journal of Theoretical Biology, 46*, 369–406.

Yockey, H. P. (1992). *Information theory and molecular biology*. Cambridge: Cambridge University Press.

Yockey, H. P. (2005). *Information theory, evolution, and the origin of life*. Cambridge: Cambridge University Press.

# Chapter 3
# Basic Principles of Communication Engineering

**Abstract** Chapter 3 briefly examines the technical means used for representing the symbols of an alphabet by physical signals, i.e., the basis of any communication. The reception of a binary signal in the presence of noise is given special emphasis. It consists of comparing hypotheses, hence it is wrong with some finite probability. As relevant to probability theory, literal communication is thus by no means a trivial matter. Sequences of symbols are represented by sequences of such signals, and it is examined how constraints linking the symbols, thus introducing redundancy, can endow a sequence with resilience to the errors which may individually affect its symbols. The significant parameter in this respect is the ratio between the signal and noise powers, or signal-to-noise ratio, which relates the symbol error rate of a sequence with the power used for communicating in the presence of noise.

Communication engineering is presumably foreign to many readers. It is why we briefly expound its fundamentals in the present chapter before dealing with the science which is founded on it. Why probabilities are so important in information theory will become clear. We also lay emphasis on the role of redundancy which, besides its engineering usefulness, is the key for understanding the very concept of information.

## 3.1 Physical Inscription of a Single Symbol

We consider here communication over time. A first and mandatory step of a literal communication consists of representing the symbols of the alphabet as distinct *signals*, a signal being defined as the variation in time of some physical quantity within a finite interval. The shape of its variation is particular to each signal, and a one-to-one correspondence of signals with the alphabet symbols is established. As functions of time, the signals are endowed with properties required for their propagation. Their association with the symbols is just a matter of convention. As a very simple introductory example, we first assume that we intend to represent a single symbol of the binary alphabet, referred to as *bit*, by the variation in time of some physical quantity, say the voltage $V$ at a point of an electrical circuit, during a time interval $T$ to be referred to as 'bit duration'. The simplest means for doing so consists of representing the digits 0 and 1 of this alphabet by constant values of the voltage during the time interval $T$. We may for instance choose $V$ volts for representing one of the bits and

G. Battail, *Information and Life*, DOI 10.1007/978-94-007-7040-9_3,

$-V$ volts for representing the other one. If the range of possible voltages is limited to $\{-V, +V\}$, this choice results in the largest possible difference between the two cases, hence provides the best way for discriminating between them in the presence of outer perturbations. Moreover, since what differentiates the values 0 and 1 of the bit is a sign, the signal still represents the same bit if it is multiplied by some positive constant, a property which can be referred to as *scale invariance*[1]. This property is useful since signals are attenuated when they are propagated and are amplified in receivers, hence it is an advantage that they are insensitive to multiplication by a positive constant. The association of the signs + and − with the symbols of the binary alphabet is just a matter of convention. We agree, for instance, that a voltage of $V$ volts within the time interval $T$ represents the bit 0 while $-V$ volts represents 1: this choice results in $(-1)^b$ being the sign which affects $V$, where $b$ denotes the bit 0 or 1.

Another equally efficient means of representing binary digits would consist of associating the signal made of $+V$ volts during the first half of the time interval $T$ and $-V$ during its second half with bit 0, and the opposite signal ($-V$ volts during the first half of the time interval $T$ and $+V$ during its second half) with bit 1. Other signal shapes obtained by cutting the given time interval into segments during which the voltage remains constant, equal to $+V$ or $-V$, an such that the signs associated with bit 0 and those associated with bit 1 are opposite would obviously be equivalent. Signals used for representing a bit which are opposite to each other are often referred to as *antipodal*.

We see here that we may represent a bit by a sign, associated with it according to some arbitrary convention, which affects some fixed time function. This function itself describes the variation of the chosen physical quantity, and can in turn be thought of as the product of a physical quantity, here a voltage $V$, by a function devoid of physical dimension which represents the shape of the signal. The value of $V$ determines the *energy* which is common to the signals associated with the bits, namely, $E = TV^2$ (up to a constant factor depending on the chosen units). The two independent factors of signal shape and energy are conveniently separated if the signal shape is represented by a signal $s(t)$ of unit norm. The norm $\|s\|$ of a signal $s(t)$ is defined[2] as the positive square root of

$$\|s\|^2 \triangleq \int s^2(t)\mathrm{d}t,$$

where the integration interval is the interval on which $s(t)$ is defined, namely, the bit duration $T$ in our examples. A normalized signal $s(t)$ is thus such that

$$\int s^2(t)\mathrm{d}t = 1. \tag{3.1}$$

[1] When the considered physical quantity is intrinsically positive, or when it is impossible to measure its sign, it is necessary to choose non-negative values, say $\{0, +V\}$. The property of scale invariance is then lost.

[2] Here and below, $\triangleq$ means 'equals by definition'.

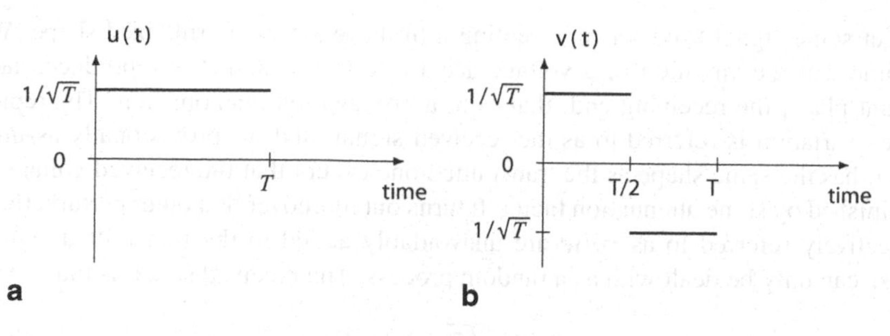


**Fig. 3.1** The two orthonormal functions $u(t)$ and $v(t)$

Let $u(t)$ denote the normalized signal of the first example above (a constant during the given time interval), namely:

$$\begin{aligned} u(t) &= 1/\sqrt{T},\ t \in (0, T) \\ &= 0, \text{ elsewhere} \end{aligned} \tag{3.2}$$

and $v(t)$ that of the second one:

$$\begin{aligned} v(t) &= 1/\sqrt{T},\ t \in (0, T/2), \\ &= -1/\sqrt{T},\ t \in (T/2, T), \\ &= 0, \text{ elsewhere.} \end{aligned} \tag{3.3}$$

The normalized signals $u(t)$ and $v(t)$ are depicted in Fig. 3.1.

Given two functions $f(t)$ and $g(t)$ over the same time interval, we define their 'scalar' (or 'correlation') product, denoted by $(f \cdot g)$, as

$$(f \cdot g) \stackrel{\Delta}{=} \int f(t)g(t)\mathrm{d}t, \tag{3.4}$$

the integration being effected on the whole interval where the functions $f(t)$ and $g(t)$ are defined. The squared norm of a signal as defined above is thus its scalar product by itself.

The two signal shapes $u(t)$ and $v(t)$ defined above have been chosen mutually 'orthogonal', meaning that

$$(u \cdot v) = \int u(t)v(t)\mathrm{d}t = 0. \tag{3.5}$$

Being both normalized and orthogonal, such functions are often referred to as *orthonormal*.

The signals contemplated up to now are generally not directly transmitted. They are most often used in order to *modulate* some 'carrier' wave which possesses the propagation properties necessary for transporting a sequence of symbols up to the intended destination (see Sect. 3.2.2).

Let some signal used for representing a bit have $s(t)$ as normalized shape. We assume that the variation of a voltage according to this signal is reproduced at a distant place, the receiving end, thanks to a propagation phenomenon. The reproduced variation is referred to as the received signal, and we provisionally assume that it has the same shape as the transmitted one except that the received voltage is diminished by some attenuation factor. It turns out moreover that outer perturbations collectively referred to as *noise* are unavoidably added to the transmitted signal. Noise can only be dealt with as a random process. The received signal is thus

$$r(t) = \pm\sqrt{E}s(t) + n(t), \qquad (3.6)$$

where the sign $\pm$ represents the transmitted bit, the normalized signal $s(t)$ indicates the shape of the time function which is used, $E$ is the received signal energy and $n(t)$ denotes a realization of the noise. The received signal is most often amplified in the receiver, but the signal and the noise are amplified by the same factor which at best, if the amplifier noise is ignorable, leaves unchanged the signal-to-noise ratio, the fundamental parameter of a communication to be more formally defined in Sect. 3.3.

## 3.2 Physical Inscription of a Sequence

### *3.2.1 Symbols and Sequences*

Before we contemplate the physical inscription of a sequence of symbols, we must warn the reader that there is no basic difference between symbols and finite sequences, except that lengthening a sequence can always be contemplated, while a symbol is assumed to belong to an alphabet of given finite size. An $n$-symbol sequence from an alphabet $\mathcal{A}$ of size $\alpha$ may as well be considered as a sequence of $n/k$ groups of $k$ successive symbols of $\mathcal{A}$ (assuming that $k$ divides $n$). A group (sequence) of $k$ symbols from $\mathcal{A}$ can be interpreted as a single symbol from the alphabet $\mathcal{A}^k$ of size $\alpha^k$. For instance, the 15-bit sequence 100001101100010 is equivalent to the 5-digit sequence 41542, since it can be written 100-001-101-100-010, where each of the 3-bit groups can be interpreted as representing in natural numeration a symbol of the octal alphabet (of size $2^3 = 8$). The sequence in the alphabet $\mathcal{A}^k$ is referred to as the $k$-th order extension of the original sequence. It is merely an alternative way of representing this sequence. The statistical properties of a sequence result from the probability of its symbols taken separately but also from their mutual dependency. Separately considering the symbol probabilities amounts to ignore this mutual dependency. Considering longer and longer extensions of the original sequences results on the contrary in taking more and more into account the mutual dependency of the symbols of the original sequence, which is then 'integrated' into the probabilities of the extended alphabet symbols. Considering longer extensions is actually a standard means in source coding algorithms for better exploiting the mutual dependency of the symbols in the original sequence (see Sect. 4.2.5).

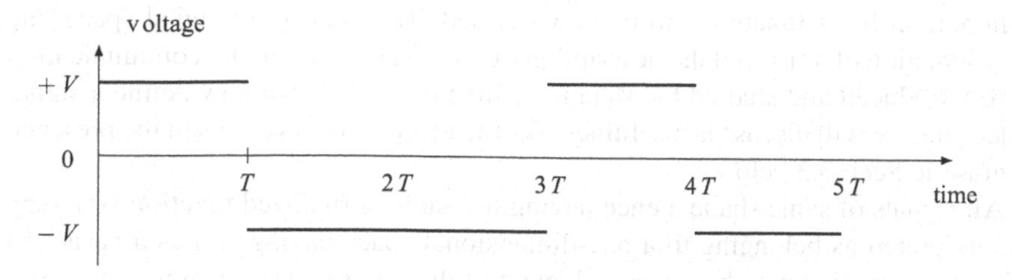


**Fig. 3.2** Example of a binary representation of the message 01101

In some early attempts to apply information theory to biology, e.g., by Lila Gatlin, the symbols of the alphabets found in nature, like the 4 nucleic bases of DNA, were dealt with separately (Gatlin 1972). The higher order extensions of the molecular sequences were not considered, resulting in the mutual dependency of their symbols being ignored. Applying information theory in such a truncated way provided rather trivial results of little benefit to biology. It turns out that the mutual dependency of symbols in a sequence, which was ignored, is extremely important: for instance, error-correcting codes owe their performance to it (see Sect. 5.5). Since the intended readers were not as experts in information theory as to have a critical look at the method, they concluded that 'information theory is of no benefits to biology'. The failure of such early attempts of applying information theory to biology hampered it for decades since this kind of prejudice is long-lived (all the more it prompts laziness).

### 3.2.2 *Representing a Sequence of Symbols by a Sequence of Signals*

We saw above how to represent a *single* bit. We intend however to communicate a *sequence*. An obvious means for obtaining this result consists of using a sequence of signals with normalized shapes like $u(t)$ or $v(t)$ above in successive disconnected time intervals. For instance, if $u(t)$ as in Eq. (3.2) is used as normalized signal, a voltage of $V$ volts represents 0 while $-V$ volts represents 1. Then a message is represented by a succession of such signals assuming the constant values $\pm V$ during each time interval of duration $T$. As an example, Fig. 3.2 represents the signal associated with the message 01101.

Then, only the noise present in the particular time interval where some signal is defined will perturb receiving the corresponding bit. Disconnected time slots as in Fig. 3.2 are not necessary for properly representing a sequence. It suffices, if $s(t)$ is the common normalized shape of the bits, that the functions $s(t)$ and $s(t \pm kT)$ are orthogonal for any integer $k$. Then, as will be stated in Sect. 3.3, the noise independently affects the signals which represent the bits.

There is however a less obvious way of representing more than a bit by a signal. It consists of using a *signal space* defined over a given time interval by mutually orthonormal basis functions, to be now defined. The concept of signal space, the representation of noise and the description of its effects on symbolic communication were introduced and studied by Shannon (Shannon 1949). We now define a signal space, and we will discuss its usefulness for the reception of symbols in the presence of noise in Sect. 3.3 below.

All signals of same shape, hence having the same normalized function $s(t)$, may be interpreted as belonging to a one-dimensional space having $s(t)$ as a basis. An axis being associated with $s(t)$, any element of this set, say $f(t)$, is represented as a point on this axis. The abscissa of this point is the correlation product $(s \cdot f)$. Two normalized functions, say $u(t)$ and $v(t)$, similarly define each a one-dimensional space. If moreover $u(t)$ and $v(t)$ are orthogonal, they define together the orthonormal basis of a 2-dimensional space, or plane. Then any function $f(t) = \lambda u(t) + \mu v(t)$, i.e., expressed as a linear combination of $u(t)$ and $v(t)$, has $\lambda$ and $\mu$ as coordinates, with $\lambda = (f \cdot u)$ and $\mu = (f \cdot v)$ (where $\lambda$ and $\mu$ are real numbers). A geometrical interpretation results if we think of $f(t)$ as a point of the plane, and $\lambda$ and $\mu$ as its projections on the axes, or coordinates. The projection operator onto an axis is then the correlation product of $f(t)$ by the basis function which defines this axis. Of course, there is no reason to limit the number of dimensions to 2: $n$ orthonormal functions (i.e., mutually orthogonal functions of unit norm) similarly define an $n$-dimensional signal space.

The 2-dimensional case is of particular importance because such a signal space can be associated with a sinusoidal wave. Such a wave acts as a 'carrier' of information in modulation systems when it is multiplied by signals representing sequences like that of Fig. 3.2. Indeed, for an integration interval of an integer number of half periods (or, approximately, for any integration interval much longer than the period), the functions $\sin(2\pi f t)$ and $\cos(2\pi f t)$ obey Eq. (3.5) and are thus orthogonal. After being normalized, they constitute the 'natural' orthonormal basis of a 2-dimensional signal space for describing modulation systems.

The signals represented by points of a signal space having $\pm V$ as coordinates specify as many bits as dimensions. For instance, two bits are simultaneously represented by the 4 combinations of signs in the signal $\pm u(t) \pm v(t)$ where $u(t)$ and $v(t)$ are the orthogonal normalized signals defined by Eqs. (3.2) and (3.3). It turns out that, due to the *theorem of irrelevance*, the noise components according to each of the dimensions defined by $u(t)$ and $v(t)$ separately affect receiving the corresponding bits. Figs. 3.3a and b represent patterns of points, or 'constellations', in the one- and two-dimensional cases, respectively, for binary data.

Another way of representing several bits by a single signal consists of distinguishing more than 2 values in each dimension. For instance, 4 distinct values can represent 2 bits at a time. A more concise representation results, but at the expense of diminishing the distance between the points hence making receiving the corresponding bits more vulnerable to noise. In Fig. 3.3c, 4-point constellations were used in each of the 2 dimensions of a signal space defined by an orthonormal basis, thus resulting in a 16-point, 2-dimensional constellation simultaneously representing $\log_2(16) = 4$ bits. Any set of $n$ orthonormal functions similarly defines an $n$-dimensional *signal space* and $2^k$ distinct points in it can be used for simultaneously representing $k$ bits.

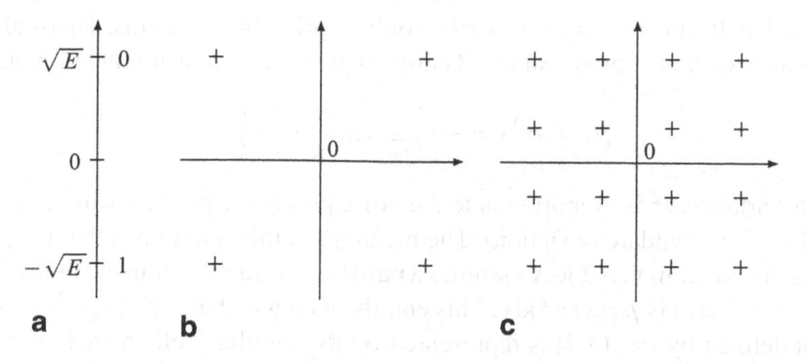


**Fig. 3.3** A one-dimensional constellation **(a)** and two two-dimensional constellations, **(b)** and **(c)**.

Actually, the propagation of signals with sharp transitions like $u(t)$ and $v(t)$ above is not possible without distorsion since such signals have significant frequency components far beyond $1/T$ whereas circuits and antennas are very generally selective, so only signals having a spectrum in a comparatively narrow frequency range can actually propagate. Any signal, say $s(t)$, can be decomposed into a sum of (infinitely many) sinusoidal functions, or single-frequency components, by the Fourier transformation. Besides its representation as a function of time $s(t)$, a signal can equivalently be thought of as a function $\tilde{s}(f)$ of the frequency $f$ and the range of frequencies where $\tilde{s}(f)$ differs from zero is referred to as its *spectrum*. Discontinuities as exhibited by $u(t)$ or $v(t)$ result in an infinitely broad spectrum, so smoother functions than $u(t)$ and $v(t)$ must be used. Of course, multiplying normalized smoother functions by $\pm$ can still be used in order to represent a binary digit. Slight shape differences between the transmitted and received signals (apart from the addition of noise) cannot entirely be avoided but have a negligible impact on the performance, which in properly designed systems almost solely depends on the noise.

## 3.3 Receiving a Binary Symbol in the Presence of Noise

The presence of the noise term $n(t)$ in Eq. (3.6) entails that receiving a signal does not provide any certainty as regards which of the bits 0 and 1 has been transmitted. What can be done is only to *assess the probability* Pr(0) (for instance) that the bit 0 has been transmitted. Because the probability Pr(1) that the bit 1 has been transmitted equals $1 - \Pr(0)$, assessing Pr(0) simultaneously assesses the probability Pr(1). Then we can only choose as received the most probable bit, a process referred to as *decision*. We must thus be able to describe the noise as random and to define the best means for assessing these probabilities.

The most fundamental perturbation which affects any received signal is thermal noise. Its power per dimension of the physical space at an absolute temperature of $T_{\text{abs}}$ is $k_{\text{B}}T_{\text{abs}}/2$, where $k_{\text{B}}$ is Boltzmann's constant. No precise shape can be assigned to thermal noise. It can properly be described only in probabilistic terms. Thermal noise has with an excellent approximation a Gaussian probability density function, namely

$$p_X(x;\sigma^2) = \frac{1}{\sigma\sqrt{2\pi}} \exp\left(\frac{x^2}{2\sigma^2}\right), \tag{3.7}$$

where its variance $\sigma^2$ is proportional to the noise power (its positive square root $\sigma$ is referred to as its standard deviation). The meaning of this function is that the probability that the random variable $X$ assumes a particular value $x$ within the infinitesimal interval $[x, x+\mathrm{d}x]$ is $p_X(x;\sigma^2)\mathrm{d}x$. This entails of course that $\int p_X(x;\sigma^2) = 1$. The function defined by Eq. (3.7) is represented by the familiar bell-shaped curve. According to the *central limit theorem*, any function which results from the addition of many independent elementary random functions of arbitrary probability distribution and similar amplitude has a nearly Gaussian probability distribution and it is why it is met in the many instances where a macroscopic phenomenon results from the joint effect of many independent microscopic events, like thermal noise.

In many practical situations, the noise has a much broader frequency spectrum than the signal, so it may be considered as having a constant power spectral density $N_0$ within the frequency range where the signal is defined, meaning that the noise power in an infinitesimal frequency interval $[f, f+\mathrm{d}f]$ is $N_0\mathrm{d}f$. Then the noise is said to be *white*. Although a power spectral density cannot be strictly constant within the whole frequency range for both mathematical and physical reasons (this range is infinite), white noise is a very convenient concept for modelling thermal noise. Its validity actually relies on the *theorem of irrelevance* which tells that a given signal is perturbed only by the noise components within its own signal space. All other noise components may thus be ignored. A standard assumption in communication theory, especially for comparing the performance of systems, is that the signal is received in the presence of *additive white Gaussian noise*.

The noise power is obtained by integrating the noise spectral density within the signal spectrum. The most important property of additive white Gaussian noise is that its components in *any* signal space have *all* the same variance $\sigma^2 = N_0/2$, regardless of the orthonormal basis functions which define it[3]. We may thus think of the white Gaussian noise as virtually 'containing' any possible signal shape. (I cannot refrain from quoting here a delightful anecdote, freely translated from (Pignon-Ernest 1999): At the beginning of the summer holidays, a sculptor usually working in a backyard where children play receives a block of rough marble. Coming back after the holidays and looking at the completed sculpture, a little girl asks: 'How did you know that there was a horse in this stone?'.) As a consequence, the noise variance per coordinate remains constant, equal to $N_0/2$, after any change of coordinates. Noise thus equally affects signals according to any dimension, in any coordinate system.

It turns out moreover that the probability of an erroneous decision in favour of one of the bits exclusively depends on the *Euclidean distance* between their representative

[3] The factor 1/2 results from the power spectral density $N_0$ being defined in terms of essentially positive frequencies, while the theory of the Fourier integral involves negative as well as positive frequencies, which halves the spectral density which is then referred to as 'bilateral'.

points in the signal space. Designing constellations so as to optimize the signal reception in the presence of noise thus reduces to a mere problem of Euclidean geometry become *independent* from the choice of the signal space basis for a given number of dimensions. The orthonormal functions of this basis can then be chosen in terms of their properties as regards signal propagation, essentially their spectral characteristics. The two problems of signal design and reception in the presence of noise of this kind become unrelated and thus can be dealt with separately. Since then the noise performance of a constellation only depends on the distances between its points, it is invariant with respect to transformations which preserve distances, especially rotations. For instance, a rotation of $\pi/4$ of the 4-point constellation of Fig. 3.3b results in the 2-point constellation of Fig. 3.3a on each of the coordinate axes, showing that the systems represented by these constellations are equivalent as regards their noise performance if the total received energy is the same for both (this result can be directly proved).

Let us go back to the binary case and consider how a signal $\pm\sqrt{E}s(t)$ is received in the presence of a realization $n(t)$ of additive white Gaussian noise, where $s(t)$ denotes a normalized signal, i.e., obeying Eq. (3.1). The choice of the sign made at the transmitting end is unknown at the receiving end so it can only be guessed in terms of the received signal $r(t) = \pm\sqrt{E}s(t) + n(t)$. The most likely estimate of this sign is that of the correlation product $(r \cdot s) = \int r(t)s(t)\mathrm{d}t = \pm\sqrt{E} + \mathcal{N}$, where $\mathcal{N} = \int s(t)n(t)\mathrm{d}t$ is a random Gaussian variable representing the noise. The correlation product of $r(t)$ and $s(t)$ is easily obtained as the response of a *filter matched* to the input signal $r(t)$. A filter matched to a signal $s(t)$ of duration $T$ is an electrical circuit which responds to an input very short with respect to $T$ by delivering an output proportional to $s(T - t)$; then its response to $r(t)$ at the instant $T$ (the input being assumed to begin at instant 0) is proportional to the correlation product $(s \cdot r)$. The noise term $\mathcal{N}$ in it is a centred (i.e., zero-mean) random Gaussian variable having as variance $\sigma^2 = N_0/2$, *regardless* of the normalized function $s(t)$, a very important result.

Let $\hat{b}$ denote the most likely hypothesis as regards the transmitted bit. The two possible outcomes of the correlation product of $r(t)$ and $s(t)$ are $\sqrt{E} + n_+$ and $-\sqrt{E} + n_-$, where $n_+$ and $n_-$ are two realizations of the random variable $\mathcal{N}$. If $n_+$ is smaller than $-\sqrt{E}$ when the correct sign is +, or if $n_-$ is larger than $\sqrt{E}$ when the correct sign is −, an error occurs, which means that the noise resulted in the receiver making the wrong decision. The logarithm of the ratio of the probabilities $\Pr(\hat{b} = 0)$ and $\Pr(\hat{b} = 1) = 1 - \Pr(\hat{b} = 0)$ to be assessed, referred to as the *log-likelihood ratio* of the received binary variable, is easily computed using the Gaussian probability distribution of the noise given by Eq. (3.7), namely,

$$\ell(x) = \ln\left[\frac{\Pr(\hat{b} = 0)}{\Pr(\hat{b} = 1)}\right] = \ln\left[\frac{\Pr(\hat{b} = 0)}{1 - \Pr(\hat{b} = 0)}\right] = 2x\sqrt{E}/\sigma^2 = 4x\sqrt{E}/N_0. \quad (3.8)$$

It measures the ratio of the probabilities of the two possible outcomes, and it is proportional to the received quantity $x$. Therefore the received quantity, say, the response of the matched filter to the incoming signal, directly measures the probabilities of the possible outcomes. Their comparison just relies on the sign of $x$: if it is positive the most probable one is 0; it is 1 if $x$ is negative. The best binary decision $\hat{b}$ is thus $[1 - \text{sign}(x)]/2$, where $\text{sign}(x)$ equals $\pm 1$ and has the same sign as $x$. Moreover, the reliability of the decision is measured by the magnitude $|x|$ which is proportional to the logarithm of the ratio of the probability of the best decision to that of the other one. (For instance if $x = 0$ the two probabilities are equal and no reliable decision can be taken. The corresponding received symbol can then rightfully be dealt with as *erased*.) The binary decision $\hat{b}$, referred to as *hard decision* in the parlance of communication engineering, is wrong with probability

$$p_{\text{su}} = \int_{\sqrt{E}}^{\infty} g(x;\sigma^2)\text{d}x = \frac{1}{\sigma\sqrt{2\pi}} \int_{\sqrt{E}}^{\infty} \exp{(x^2/2\sigma^2)}\text{d}x$$

where the subscript 'su' is intended to mean that the error consists of substituting a wrong symbol for the correct one. Introducing the 'error function'

$$Q(x) \triangleq \frac{1}{\sqrt{2\pi}} \int_{x}^{\infty} \exp{(t^2/2)}\text{d}t, \tag{3.9}$$

enables writing the probability of an error as

$$p_{\text{su}} = Q(\sqrt{E}/\sigma) = Q(\sqrt{2E/N_0}). \tag{3.10}$$

The error function $Q(x)$ equals 1/2 for $x = 0$, decreases when $x$ increases and vanishes as $x$ approaches infinity. A hard decision is wrong with the probability $p_{\text{su}}$ given by Eq. (3.10), which only depends on the ratio $2E/N_0$, a fundamental parameter in communication engineering referred to as the *signal-to-noise ratio*, often abbreviated as SNR. This parameter is extremely important because it relates the physical energy to information. It actually sets a limit to the information quantity that a channel can transmit (see Sect. 5.2.3).

We assume in the following that the log-likelihood ratio of the most likely decision $\hat{b}$ about any received bit is available according to Eq. (3.8). The binary decision $\hat{b}$ is wrong with a probability given by Eq. (3.10). The occurrence of such an error is (easily!) avoided by *not* taking a hard decision (unless of course doing so is mandatory), i.e., by keeping as far as possible the log-likelihood ratio $\ell(x)$ given by Eq. (3.8) as it results from the receiving process. This absence of decision is (a bit paradoxically) named *soft decision*. Then real numbers (as defined in mathematics; we already criticized in Sect. 2.4.3 the use of the word 'real') are dealt with at the receiving end instead of bits, a process referred to as *analog*. The importance of soft decisions for decoding error-correcting codes will become clear in Sect. 5.5.7.

Interestingly, the log-likelihood ratio $\ell(x)$ of a received binary variable, defined by (3.8), equals the derivative of the binary entropy function $\mathcal{H}_2(p_{\text{su}}) \triangleq -p_{\text{su}} \log_2{(p_{\text{su}})} - (1 - p_{\text{su}}) \log_2{(1 - p_{\text{su}})}$ (Eq. (4.11) below), with respect to the error probability $p_{su}$:

$$\ell(x) \triangleq \ln\left[(1 - p_{su})/p_{su}\right] = \frac{\mathrm{d}\mathcal{H}_2(p_{su})}{\mathrm{d}p_{su}}.$$

The binary entropy function $\mathcal{H}_2(p_{su})$ measures the quantity of information borne by a random binary variable assuming one of its values with probability $p_{su}$ (see Sect. 4.2.3).

## 3.4 Communicating Sequences in the Presence of Noise: Channel Coding

### *3.4.1 Channel Coding is Needed*

The *literal* communication of a sequence cannot be secured with certainty by separately transmitting each of its symbols as described above because a received symbol incurs an error with probability $p_{su}$ given by Eq. (3.10). The probability that an $n$-bit sequence is correctly received (i.e., without any erroneous symbol) is $(1 - p_{su})^n$, a probability which approaches 0 however small is $p_{su}$ when $n$ sufficiently increases. The probability that the sequence as a whole is incorrectly received is its complement to 1, namely,

$$P_{se}(n) = 1 - (1 - p_{su})^n, \tag{3.11}$$

which tends to 1 as $n$ approaches infinity.

The development of $(1 - p_{su})^n$ limited to its first two terms is $1 - np_{su}$, which entails that $P_{se}(n) \approx np_{su}$ if $p_{su}$ is small enough. The probability of a sequence error is according to Eq. (3.11) an increasing function of its length $n$: however small is the bit error probability $p_{su}$, hence however large is the signal-to-noise ratio, the correct literal communication of a sequence is not secured and, moreover, is the more unlikely, the larger is the sequence. Indeed, $(1 - p_{su})^{1/p_{su}}$ approaches $1/\mathrm{e}$ when $p_{su}$ tends to 0, where $\mathrm{e} = 2 \cdot 718\ldots$ is the base of the natural logarithms, showing that the error probability of a sequence becomes very high when $n$ becomes of the order of $1/p_{su}$. This seems to prevent the reliable communication of arbitrarily long symbolic sequences since thermal noise is ubiquitous and the energy available for representing sequences is necessarily limited.

Yet, extremely large sequences can be communicated or conserved, not only as the product of human activities like written texts or registered data, but also in nature, the most striking case being that of genomes billions of nucleotides long. Moreover, genomes are conserved at the geological timescale, and the number of occurring symbol errors is an increasing function of time. Clearly, the conservation of such very large sequences over large time intervals is not compatible with the expression (3.11) of the sequence error probability. *Information theory* fortunately states that long sequences can be communicated or conserved with an arbitrarily small probability of error despite symbol errors and thus appears as the sole framework which can account for the reliable communication of long sequences. One of its main results, the *fundamental theorem of channel coding*, is that *reliable* communication is possible within a precisely defined limit referred to as *channel capacity* through an *unreliable* channel, however paradoxical it may look. Firstly a mere theoretical result proven without exhibiting any means for its practical implementation, this statement became a communication engineering reality with the development of *error-correcting codes*.

Error-correcting codes owe their efficiency to their *redundancy*. For securing the communication of some sequence of length $k$, it is first transformed into a *longer* one of length $n > k$ which is transmitted instead of the initial sequence. Then, the dependency that this encoding establishes between the symbols entails that errors in certain symbols can be corrected thanks to the errorless ones. To this end, the received version of the encoded sequence, hence affected by symbol errors, is processed so as to recover the encoded sequence, hence the original one. Referred to as *decoding*, this process fails with a non-zero probability which can however be kept ignorable by simultaneously increasing $k$ and $n$ while maintaining the information rate $R = k/n$ approximately constant, provided $R$ does not exceed an impassable theoretical limit stated by the fundamental theorem of channel coding, referred to as the *channel capacity*.

Replacing the given initial sequence by a longer one seems at first sight to go against its conservation in the presence of errors. First of all, even if the probability of symbol error is kept constant, more symbol errors occur in the longer encoded sequence than in the original one. Moreover, if a constant energy per symbol of the original message is available and results in a received energy $E$, the total received energy $kE$ must be shared among the $n > k$ symbols of the encoded message, leaving only $kE/n < E$ for each of them. For a constant noise power density, this results in an increase of the symbol error probability according to Eq. (3.10). Lengthening the sequence and sharing the available energy among more symbols seem both to be detrimental to the probability of correctly receiving the sequence. However, information theory paradoxically tells that such codes, provided they are properly decoded, can make the error probability arbitrarily small up to the channel capacity. This theorem will be expounded in Sect. 5.4 below and an overview on error-correcting codes will be given in Sect. 5.5.

It must be emphasized that the existence of a drastic limit that noise imposes to literal communication as well as the possibility of escaping the effects of noise within this limit were *entirely novel results* brought by information theory and communication engineering. Prior to Shannon's statement of the channel coding theorem (Shannon 1948), it was accepted as obvious that an unreliable channel would result in unreliable literal communication. That it is not so, but that noise only sets a limit to the rate at which information quantity can be communicated, is a major result of information theory. The channel noise thus acquired a crucial importance.

The results of information theory and communication engineering about the conservation of sequences should be used in many domains other than engineering. They should be of paramount importance in sciences devoted to symbolic sequences, especially *linguistics* and *semiotics*. Unfortunately, these sciences were founded much before Shannon's paper was published. They ignored that literal communication is by no means a trivial problem and implicitly considered it as secured. Nowadays, their contemporary upholders still ignore information theory and communication engineering, mainly because they misunderstand or do not know them for lack of proper popularization. The results of information theory and communication engineering remain a dead letter in these sciences (Battail 2009, 2012). In them as well as in biology, error-correcting codes appear as the *necessary* solution to the *crucial* problem of sequence conservation, but both the problem and its solution passed unnoticed as pertaining to information theory, which remained an almost cryptic science. If the problem is not perceived, how could its solution be understood and accepted?

### 3.4.2 Redundancy Enables Channel Coding

No concept of information theory is more useful than *redundancy* in engineering as well as biology because of its prominent role in making sequences distinct, thereby enabling their conservation in the presence of symbol errors. Unfortunately, no one is more misunderstood. It is often perceived as a mere repetition which is often spurious and at best can provide spares. In sharp contrast, it plays the essential role of ensuring the integrity of an information. Contrary to the usual presentation of information theory which conforms more or less to the framework of Shannon's seminal article (Shannon 1948), we anticipate here on our following discussion of information theory by giving an example of channel coding, the most important application of information theory to communication engineering. Then the meaning of redundancy and its necessity will become clear, together with the meaning of the very concept of information.

Let us assume that we have to transmit a sequence of time-successive binary symbols. A binary symbol is referred to here as a *bit*, an acronym for 'binary digit'[4]. The bits of this sequence are mutually independent. They are arbitrary as well as their number $k$. They are referred to as the 'information bits' and this sequence itself as the *information message*. It is unique and cannot suffer any alteration without losing its identity. It is thus a nominable entity in Barbieri's meaning (Barbieri 2007), as defined in Sect. 2.4.3.

The channel available to communicate this information message is unreliable, in the sense that a received bit has some non-zero probability of being different from the transmitted one. The physical process which results in channel errors escapes any control. For securing the faithful transmission of the information message, hence to ensure its integrity, we wish to endow it with resilience to channel symbol errors. To

[4] Contrary to Shannon and most engineers and information theorists, we *do not use* 'bit' for naming the binary unit of information quantity because doing so risks to hide redundancy to hasty or lay readers. In order to avoid such a confusion, we name this unit 'shannon'; see Sect. 4.2.1

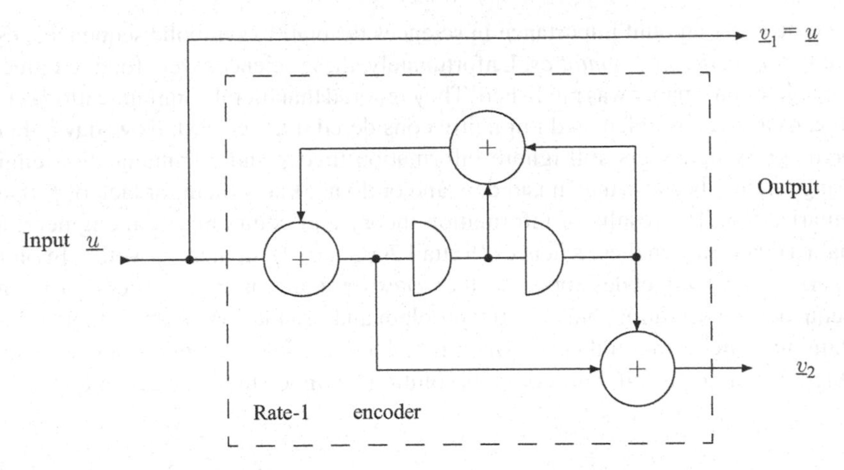


**Fig. 3.4** Channel coding: example of a convolutional recursive encoder. The information sequence $\underline{u}$ enters at *left* and also appears at the upper output at right as $\underline{v}_1$. The D-shaped devices are delay elements of 1-bit duration which together constitute a 2-bit memory, also referred to as a *shift register*. A combination of the input bit and of the 2 bits which precede it (available in the 2-bit memory) according to the indicated connections (+ denotes addition modulo 2) appears as the lower output, and the successive bits thus calculated constitute the redundancy sequence $\underline{v}_2$. The bits of sequences $\underline{v}_1$ and $\underline{v}_2$ are transmitted in alternance. The dashed box which calculates the redundancy bit is a 'rate-1 encoder' which delivers a single bit in response to each input bit

this end we transform it, according to a process referred to as *channel coding*, into an *encoded* sequence which is actually transmitted over the available channel.

As an example, the encoded sequence may be generated by the device[5] of Fig. 3.4. The successive bits of the information message, coming from the left, enter the dotted box labelled *rate-1 encoder*. Simultaneously, they leave the device through the upper right output. The rate-1 encoder consists of a memory of length $\mu = 2$ and contains modulo-2 adders which combine the input information bit and the bits contained in the memory according to the indicated connections (addition modulo 2 of two bits is the same as ordinary addition except that $1 + 1 = 0$). The output of the rate-1 encoder is connected to the lower output of the whole device. Every time an information bit enters the encoder, two bits leave it: the information bit itself and the one calculated by the rate-1 encoder, to be referred to as *redundancy* or *check* bit. The sequence of redundancy bits will be referred to as the redundancy or *check sequence*. The bits of the upper and lower output, i.e., the information and redundancy bits, are sent alternately (this operation, referred to as 'parallel-to-serial conversion', is not shown in the figure).

A set of $n$-symbol sequences can be endowed with the *Hamming metric*, which is relevant when errors possibly affect their symbols. The Hamming distance between

[5] This encoder is referred to as 'convolutional recursive systematic'; when combined with an interleaver, two such encoders generate a turbocode like that which first succeeded in closely approaching the channel capacity (Berrou et al. 1993; Berrou and Glavieux 1996; Guizzo 2004) (see Sect. 5.5.5).

two sequences of same length is defined as the number of positions where their symbols are different. The larger their Hamming distance, the more unlikely an error pattern can change one of them into the other. An error-correcting code like the one generated by the device of Fig. 3.4 is a set of $n$-symbol sequences which are made distinct from each other as having a minimum Hamming distance between any two of them larger than 1.

Examples of sequences generated by the encoder of Fig. 3.4 are now given. It is easily shown that in response to the sequence $\underline{u} = 10\ldots$ made of a single 1 followed by infinitely many bits 0, it generates $\underline{v}_2 = 111011011\ldots$, consisting of 111 followed with the motif 011 repeated infinitely many times. Then the encoded sequence is at an infinite Hamming distance from the all-0 sequence generated by the all-0 input sequence. If the input sequence is $\underline{u} = 1110\ldots$ (where $\ldots$ stands for infinitely many zeros), however, the response $\underline{v}_2 = 1010\ldots$ (similarly followed with infinitely many zeros) has a finite number of ones, so the generated sequence where the bits of the sequences $\underline{u} = \underline{v}_1$ and $\underline{v}_2$ alternate, namely 110111, is at a Hamming distance of 5 from the all-0 sequence. The number of bits 1 in a sequence, referred to as its *weight*, is the Hamming distance from the all-0 sequence. The code generated by the encoder of Fig. 3.4 is referred to as *linear*, meaning that its words are defined by only using addition and multiplication of the $q$-ary field, $q$ denoting the alphabet size ($q = 2$ above). The vast majority of currently used codes are linear. Only primes or primes raised to an integer power can be endowed with the mathematical structure of a field. In the binary field, for instance, the addition rule is modulo 2. Linear codes are such that the smallest non-zero weight of its codewords equals the minimum distance of the code, i.e., the main parameter used to assess its performance as regards error correction.

At the receiving end, both the information and redundancy sequences possibly contain erroneous bits due to channel symbol errors. The connections in the rate-1 encoder have been chosen such that the encoded sequence where the information and redundancy bits alternate is at a large enough minimum Hamming distance $d$ from any other sequence that the encoder can generate. If less than $d/2$ bit errors occurred, the sequence actually emitted can be recovered with certainty as being closer for the Hamming metric to the received sequence than any other encoded sequence. In the example of Fig. 3.4 the minimum distance secured by the encoding is $d = 5$, so the encoded sequence, hence the information message, can be exactly recovered in the presence of any pattern of 1 or 2 bit errors. Using a longer memory size $\mu$ can result in larger values of the minimum distance $d$, hence in more numerous bit errors being corrected by appropriate decoding means. The device of Fig. 3.4 is *redundant* as delivering 2 bits every time it receives an information bit. It is referred to as a rate-(1/2) encoder. The *rate* of a binary code having $n$-bit words with $k$ information bits is defined in general as the ratio $R = k/n$. Any rational rate less than 1 can actually be obtained by the use of more complicated but similar encoding devices.

Notice that the encoded sequence delivered by an encoder like that of Fig. 3.4 obeys precise constraints which result in keeping a minimum distance with respect to all other sequences that this encoder can generate. The set of sequences, or *words*, that an encoder can generate is referred to as an $(n, k)$ binary *code* if it is made of

$n$-bit words with $k$ information bits, with $n > k$ since the code is redundant. Only $2^k$ $n$-bit words belong to it while $2^n$ different $n$-bit sequences exist. The generated code is thus a minority subset of all possible binary sequences of length $n$, and it owes its success in correcting errors to this property. It is only a fraction $2^{-(n-k)} = 2^{-n(1-R)/R}$ of the total number of $n$-bit sequences, which vanishes if $n - k$ approaches infinity. The probability of a decoding error can thus be made vanishingly small as the length $n$ of the codewords approaches infinity.

Each encoded bit, whether of information or of redundancy, actually contains an information quantity of $k/n$ binary units (referred to here as 'shannons'; see Sect. 4.2.1 below), equal in the binary case to the code rate $R$: it turns out that the encoding actually 'diluted' the information among all symbols. Indeed, at the receiving end, the whole encoded sequence, hence the information message, can be exactly recovered given the correct values of at least $k$ of its symbols arbitrarily chosen (the other symbols being ignored[6]), regardless whether they are information or redundancy symbols[7]. The same holds true for a code using an alphabet with $\alpha > 2$ symbols, referred to as $\alpha$-ary; then each encoded symbol bears an information quantity of $(k/n) \log_2(\alpha) = R \log_2(\alpha)$ shannons.

The sequence generated by the encoder of Fig. 3.4 is said 'systematic' to mean that the information message is explicitly present in it at a set of given positions (it is here the sequence of bits at odd positions, due to the parallel-to-serial conversion which alternates the information and redundancy bits). A further transformation of the output sequence leaving the rate unchanged, like scrambling or rate-1 encoding in the above meaning, makes the information message no longer explicitly readable in the encoded sequence. Yet the sequence thus obtained still represents the original information message. We may define *an information*[8] as the equivalence class of *all* the encoded sequences which can be associated with a same information message. Since there are infinitely many possible encodings, this equivalence class contains an infinite number of elements. All the possible encodings are redundant except leaving the message uncoded or transforming it using a rate-1 encoder, so the length $k$ of the original information message appears as the length of the smallest sequences belonging to this class, which can thus be taken as a quantitative measure of the information. The information message may be likened to the information itself, *provided* it is kept in mind that it is merely the *representative* of a wide equivalence class. It has already been said that the information message is a *nominable entity* in Barbieri's meaning (Barbieri 2007) since it suffers no change. One may also notice

[6] In engineering words, ignoring a symbol is referred to as its 'erasure': when the receiver cannot take any decision about a symbol, it does not take it into account. An erasure must be distinguished from an error which consists of taking a wrong decision.

[7] This is a theoretical result. Actually correcting up to $n - k$ erased bits within an $n$-bit word can always be performed by comparing each of the codewords with the the received sequence, a very complex (brute force) process that a specific algorithm possibly alleviates.

[8] Notice that we refer here to *an* information, not to information in general. We make here more precise a statement of Sect. 2.2.

that the binary numeration associates with it a natural integer which is *unique* as regards its arithmetic properties.

## References


Barbieri, M. (2007). Is the cell a semiotic system? In M. Barbieri (Ed.), *Introduction to Biosemiotics* (pp. 179–207). Dordrecht: Springer.

Battail, G. (2009). Living versus inanimate: The information border. *Biosemiotics, 2*(3), 321–341. doi:10.1007/s12304-009-9059-z.

Battail, G. (2012), Biology needs information theory. *Biosemiotics, 6*,(1), 77–103. doi:10.1007/-s12304-012-9152-6.

Berrou, C., Glavieux, A., & Thitimajshima, P. (1993). Near Shannon limit error-correcting coding and decoding: Turbo-codes. *Proceedings of ICC'93*, Geneva, Switzerland, pp. 1064–1070.

Berrou, C., & Glavieux, A. (1996). Near optimum error-correcting coding and decoding: turbo-codes. *IEEE Transactions on Communications, 44*(10), 1261–1271.

Gatlin, L. (1972). *Information theory and the living system.* New York: Columbia University Press.

Guizzo, E. (2004). Closing in on the perfect code. *IEEE Spectrum, 41*(3) (INT), 28–34.

Pignon-Ernest, E. (1999). *Ousmane Sow, le soleil en face*. Paris: Le P'tit Jardin.

Shannon, C. E. (1948). A mathematical theory of communication. *The Bell System Technical Journal, 27,* 379–457, 623–656. (Reprinted in Shannon and Weaver 1949, Sloane and Wyner 1993, pp. 5–83 and in Slepian 1974, pp. 5–29).

Shannon, C. E. (1949). Communication in the presence of noise. Proc. IRE, pp. 10–21. (Reprinted in Sloane and Wyner 1993, pp. 160–172 and in Slepian 1974, pp. 30–41).

Slepian, D. (Ed.), (1974). *Key papers in the development of Information Theory*. Piscataway: IEEE Press.

Sloane, N. J. A., & Wyner, A. D. (Eds.), (1993). *Claude Elwood Shannon, Collected papers.* Piscataway: IEEE Press.

# Chapter 4
# Information Theory as the Science of Literal Communication

**Abstract** Chapter 4 is devoted to information theory as the science of literal communication. It begins with describing Shannon's paradigm, which identifies the actors of any communication: (1) the *source*, which generates some message; (2) the *channel* which propagates the message; and (3) the *destination* which receives it. The matching of these entities to each others needs using devices which transform the *message* by *coding*, of two main types. Source coding is intended to shorten the message that the source delivers. Channel coding is intended to protect it against the symbol errors which occur in the channel, which demands lengthening the message. Both are assumed to be exactly reversible. Quantitative measures of information are defined, based on the improbability of symbols and messages. The source entropy measures the average information quantity borne by each of the symbols of the message it delivers. The channel capacity measures the largest information quantity that it can transfer. Two fundamental theorems state that source coding can reduce the message length up to a limit set by the source entropy, and that errorless communication is possible in the presence of symbol errors, but only provided the source entropy is less than the channel capacity. A normalized version of Shannon's paradigm assumes that the message is transformed by source coding followed by channel coding, both achieving their theoretical limit. A simple proof of the fundamental source coding theorem is presented and the Huffman source coding algorithm is described. Comments about source coding help understanding the very concept of information and its relationship with semantics.

Chapter 3 briefly described the fundamentals of communication engineering and has shown that messages and signals must be dealt with as random. First of all, as stated in Shannon's quotation of Sect. 2.2 because any communication system must accept a broad variety of messages, so the one to be actually communicated is not known in advance. Second, as emphasized in Chap. 3, because noise is the most important factor which limits the performance of communication systems and can only be described as a random process. It is why Shannon's information theory is entirely based on probabilities. The present chapter is devoted to it. Many treatises have been written on the subject besides Shannon's seminal work (Shannon 1948), e.g., (Gallager 1968; Cover and Thomas 1991; Battail 1997), but they are not aimed at popularization. Space lacks here for summarizing their content. We now present information theory in a rather informal manner mainly intended to give an insight on the matter. For simplicity's sake, and because this case suffices for introducing the main concepts needed in what follows, this presentation is mainly restricted to

G. Battail, *Information and Life*, DOI 10.1007/978-94-007-7040-9_4,

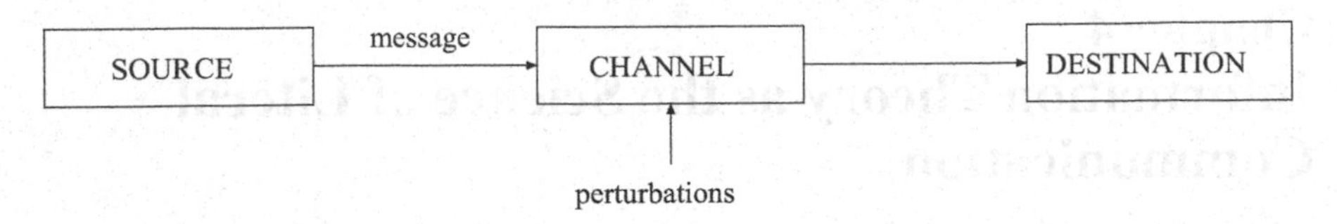


**Fig. 4.1** Basic scheme of a communication: Shannon paradigm

*discrete* information. Appendix A also contains in Sect. A.3 an overview of Shannon's information theory.

Shannon's papers have been collected in (Sloane and Wyner 1993). Papers of historical significance in the development of information theory have been gathered by D. Slepian (1974), and we will refer to this work to cite them.

## 4.1 Shannon's Paradigm and its Variants

### *4.1.1 Basic Paradigm*

The general scheme of a communication, referred to as *Shannon paradigm*, is represented in Fig. 4.1. A *source* generates a *message* intended to a *destination*. The source and the destination are distinct entities, hence spatially separated, but there exists between them a *channel* which, on the one hand, propagates its input up to the destination where the corresponding output response can be observed; and, on the other hand, suffers perturbations, collectively referred to as *noise*, which entail that the input to the channel does not suffice to determine its output with certainty. The destination has no other means to access the transmitted message than observing the channel output.

The source is most often assumed to permanently generate messages. As already stated in Sect. 3.4, the perturbations are very important. It is why information theory integrates them in its channel representations. Contrary to Fig. 4.1 where the perturbations are represented outside the box labelled 'CHANNEL', the channel models in use actually describe how they act on the channel input and determine its output (see Sect. 5.1 below). From now on, thus, we consider the perturbations as intrinsic to the channel.

The scheme of Fig. 4.1 or its variants is still relevant if the source and the destination are separated in time rather than in space, a case already mentioned in Sect. 2.6. Communication then consists of recording a message which is read later. The channel is no longer a propagation medium, but consists of some physical support which saves long-lasting modifications specified by writing signals. Receiving consists of reading the written signals. This interpretation of the recording and reading process as a channel in the information theoretic sense is important as showing that information theory is relevant to this case as well as to communications over space.

Giving some examples, the source may be somebody who speaks and the destination somebody who listens: then, the channel is ambient air; or the source is a person who writes and the channel a sheet of paper, or maybe the writer and the reader are connected through a proper electrical or electronic medium. The scheme of Fig. 4.1 clearly applies to a wide variety of sources, channels and destinations. The word 'paradigm' designates the general model of some structure, independently of the interchangeable objects the relations of which it describes (for instance, in grammar). This scheme was introduced by Shannon in 1948, in a slightly different form, at the beginning of his seminal paper (Shannon 1948). It may now look trivial, but this simple identification of the partners of a communication was a necessary prerequisite for developing the theory.

As recalled above, the main property of the channel in the information-theoretic meaning is the presence of perturbations which affect the transmitted message. If one wonders at the importance given to phenomena which often pass unnoticed in daily life, one should keep in mind that observing the channel output, which is necessary to perceive the message, is a physical measurement which can be performed only with finite precision. The reasons which limit the measurement precision are many and provisions can be made to improve it. However, the pervasive presence of *thermal noise*, which was dramatically confirmed by the discovery of the 'fossil noise' of the universe by Penzias and Wilson in 1965, suffices to justify the central role given by the theory to perturbations. One of the most important conclusions of information theory is indeed to identify noise as the factor that ultimately limits the communication possibilities.

### *4.1.2 Variants of Shannon's Paradigm*

If we assume that the source, the channel and the destination are arbitrary, nothing *a priori* guarantees the mutual compatibility of the source and the channel, on the one hand, and of the channel and the destination, on the other hand. For instance, in radiotelephony, the source and destination are humans, but the channel consists of propagating electromagnetic waves. A human being has no natural means for transmitting and receiving such waves, except the perception of light. We thus have to augment the scheme of Fig. 4.1 with blocks representing the devices intended to perform the technical functions of converting and matching. One then obtains the scheme of Fig. 4.2a. It is merely a variant of the scheme of Fig. 4.1, since the set comprising the source and the transmitting device on the one hand, that comprising the reception device and the destination on the other hand, may be interpreted as a new source–destination pair now matched to the initially given channel (Fig. 4.2b). One may as well consider the set comprising the transmitting device, the channel and the receiving device as being a new channel, matched to the original pair source–destination (Fig. 4.2c); thus, we considered in previous examples as a channel a telephonic or telegraphic circuit, made of the set of transmitting devices, a propagation medium, and a set of reception devices.

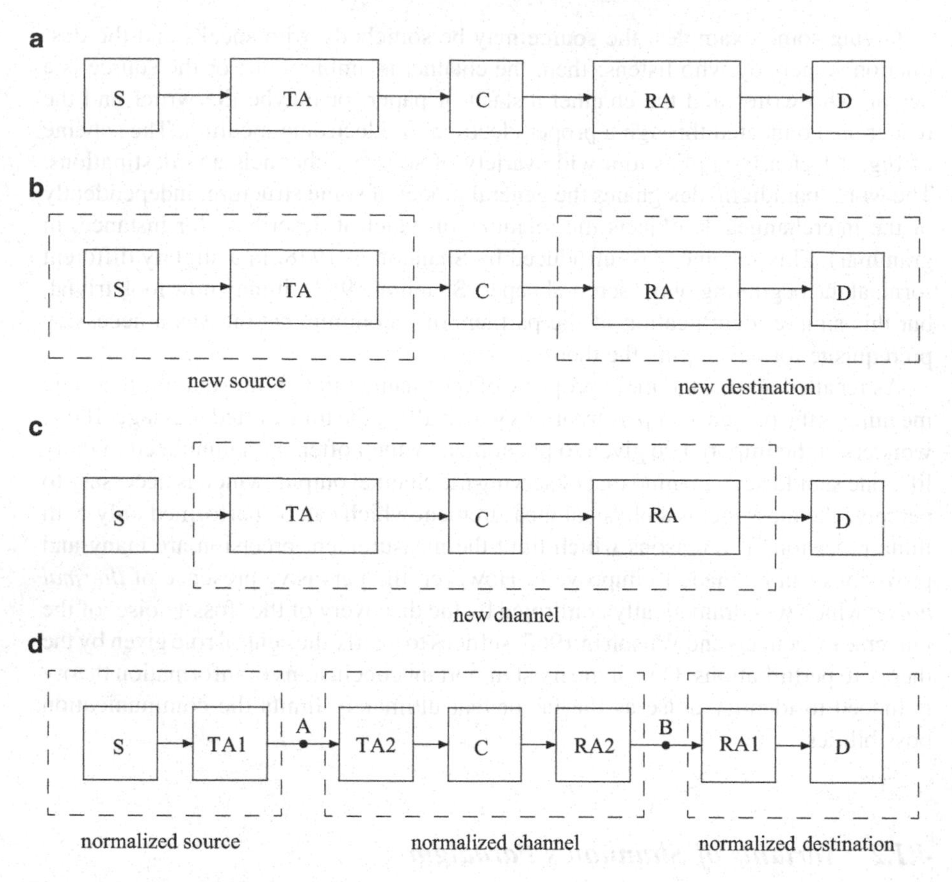


**Fig. 4.2** Variants of Shannon's paradigm. S stands for 'source', C for 'channel' and D for 'destination'. TA means 'transmission apparatus' and RA, 'reception apparatus'

A more fruitful point of view actually consists of splitting each of the transmission and reception apparatuses into two blocks, one matched to the source (or destination), the other one matched to the input (or output) of the channel. Interestingly, this scheme enables *normalizing* the characteristics of the blocks of Fig. 4.2, redefined as follows: new source before point A in Fig. 4.2d; new channel between points A and B; new destination beyond point B. The engineering problems then consist of *separately* designing the pairs of matching blocks denoted in the figure by TA1 and RA1 on the one hand, by TA2 and RA2 on the other hand. We differ to Sect. 4.1.3 defining what this announced normalization consists of because we first need to introduce the concepts of source and channel coding.

In general, one is free to redefine the borders of blocks in Fig. 4.2 for the sake of analysis; cutting any apparatus chain linking a source and a destination in two points such that the origin of the message useful to the destination is on the leftmost block, and *all* places where *perturbations* occur are located in the central block, defines a new triplet source–channel–destination.

Description of the perturbations and of the transmitted messages, and of their transformation into signals which can propagate, belongs to the field of *signal theory*. Messages and signals undergo the transformations necessary for their transmission, especially in the form of modulation and coding (the latter word having several meanings) but bear a more fundamental (and more difficult to define) entity which remains unchanged in such transformations: the *information*. Information theory provides a *quantitative measure* of it, at the expense indeed of important restrictions to be later expounded, studies its communication and the impairments it undergoes. The relevant quantities have a *statistical* meaning and the main theorems establish the existence of *limits*.

The invariance of information with respect to the messages and signals which bear it implies that one may choose, among the set of equivalent messages which represent the same information, those which possess some *a priori* favourable properties. In this way we shall introduce the *coding* operations, in the main two meanings of this word.

### 4.1.3 Functions and Limits of the Coding Processes

Since it is possible to associate with a given information several equivalent messages, transformations of an original message may be used in order to endow it with *a priori* favourable properties. What are these properties? Of what will consist these transformations, referred to as *coding* processes? How, more specifically, will be performed the normalization of the blocks 'source', 'channel' and 'destination' which has been announced above?

We now intend answering these questions. First of all, let us notice that the most important results of information theory concern the ultimate limits of the coding processes and that they are expressed in terms of the quantities which measure information, i.e., source entropy and channel capacity, to be introduced in Sect. 4.2.1. These quantities acquire therefore an operational significance, which confirms their adequacy to communication problems and enlightens their meaning.

One may *a priori* contemplate to transform a symbolic message by source coding and by channel coding. (Cryptography is a third commonly used kind of coding which is of no interest to us and will be left aside.)

**Source coding** Source coding aims at *maximum conciseness*. In engineering, using a channel costs the more, the longer the message, where 'to cost' should be understood in a very broad sense, that of needing the use of some limited resource like time, power or bandwidth. Coding may thus, for diminishing this cost, aim at substituting a message as short as possible for the original message. This transformation should be reversible in the sense that the exact recovering of the original message should be possible.

**Channel coding** The aim of *channel coding* is to protect the message against the channel perturbations. We laid emphasis above on the necessity of taking into account

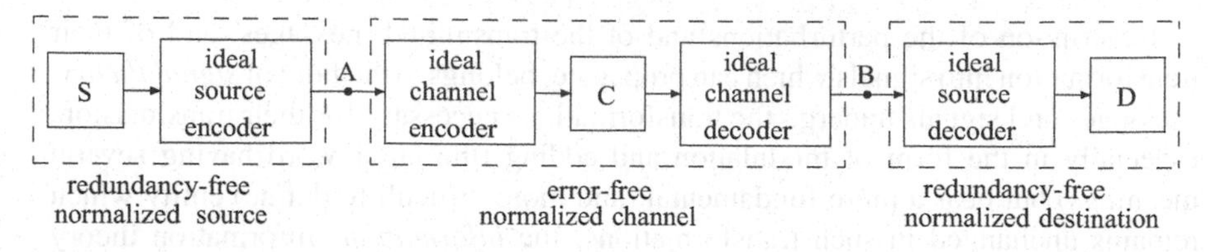


**Fig. 4.3** Normalization of the blocks 'source', 'channel' and 'destination' of Shannon's paradigm. S, C and D denote the original source, channel and destination, respectively

the channel perturbations, up to make of their existence the specific property of a channel. Channel coding is intended to improve the communication reliability over the given channel, despite the presence of noise.

**Normalizing the blocks of Shannon's paradigm** It now becomes possible to precisely define the announced normalization of the blocks of Fig. 4.2. The message generated by the source is first transformed by *source coding* into a message *devoid of redundancy*, i.e., where the successive symbols are mutually independent and where all the alphabet symbols occur with an equal probability. This encoding matches the source characteristics and its result is highly vulnerable to perturbations since any of its symbols is essential to its integrity. *Channel coding* thus becomes mandatory so as to protect the message generated by the source encoder against the channel noise, which necessarily demands reintroducing redundancy.

We may assume that source coding has been performed ideally. Then, channel coding just protects a redundancy-free message against the channel perturbations. If the message to be encoded is not completely devoid of redundancy, the protection achieved is at least that for a redundancy-free message: exploiting some remaining redundancy may be but beneficial as regards protection against noise. We may similarly assume that ideal channel coding has been performed, resulting in error-free communication. One may then redraw Fig. 4.2d as in Fig. 4.3, where the normalized source generates a redundancy-free message and where the normalized channel is error-free.

Suppressing the redundancy of the original message by source coding and then reintroducing redundancy by channel coding may look self-contradictory, but the redundancy in the original source is not necessarily well fitted to the properties of the channel to which it is connected. Moreover, rather than designing a coding system in order to match a particular source to a particular channel, the normalization as just defined enables dealing with the source and the channel separately, which is simpler and more flexible. Any source can then be connected to any channel without needing that this particular source be specifically fitted to this particular channel. This normalization has also an auxiliary advantage: the message alphabet at points A and B of Fig. 4.3 is arbitrary. We may thus assume that it is the simplest possible, i.e., binary, without significantly restricting generality.

**Fundamental theorems** We based the proposed normalization on coding processes resulting in:

- a redundancy-free message, although the original message is generated by a redundant source, as concerns source coding;
- an error-free message after decoding, although the coded message is received through a perturbed channel, as concerns channel coding.

The possibility that coding processes have these results is established by the *fundamental theorems* of information theory, under conditions they precisely state. Their proofs do not demand the explicit knowledge of the means which enable obtaining these results and, in many instances, these means remain unknown.

The fundamental theorem of *source coding* tells that it is possible to eliminate all redundancy from a stationary source. Optimized source coding results in a message with an *average length* per symbol of the original source equal to its *entropy* expressed using the alphabet size $\alpha$ for defining the information unit, i.e., using logarithms to the base $\alpha$. Entropy is defined in Sect. 4.2.2.

Algorithms which can achieve this result are known, especially the Huffman algorithm, as we shall see in Sect. 4.3.4. Very schematically, such algorithms aim at constructing a set of words (finite sequences of symbols) for representing the symbols of the $\alpha^k$-ary alphabet, where $k$ is an integer. It should first be possible to identify these words without ambiguity, which one expresses by referring to such a code as decipherable. These words should be the shortest, the more frequent the $\alpha^k$-ary symbols they represent. If optimum encoding can be reached for finite $k$, then this length is proportional to the opposite of the logarithm of the occurrence probability of the corresponding symbol. If this is not the case, increasing $k$ enables improving the relative precision of the approximation of real numbers by integer word lengths. Besides and more important, increasing the number of words entails that the distribution of their lengths can better and better match the probability distribution of the corresponding messages. (More subtle means exist which avoid the need of using a source extension, e.g., arithmetic source coding algorithms.)

The fundamental theorem of *channel coding* is asymptotic as the length $n$ of the words associated with blocks of $k$ source symbols approaches infinity. Keeping constant the code rate $R = k/n$ implies that $k$ increases proportionally to $n$. (This statement is not the most general possible, especially because it is assumed here that the channel input and output alphabets are the same.) The necessary condition for the decoding error probability to approach 0 as $n$ increases can be stated as follows: the source *entropy* should be *less* than the channel *capacity* (source and channel are assumed to satisfy regularity conditions to be indicated in Sect. 5.4, which are mandatory to guarantee the existence of these quantities). Thus, the presence of channel perturbations does not limit the reliability of the message communication, as measured for instance by the decoding error probability, but only the possible information rate through this channel. The highly desirable result of an arbitrarily small error probability can however be obtained only when using an appropriate code. Before we can more precisely deal with the fundamental theorems of source- and channel coding, we must introduce the quantities which measure information, in terms of which the limits of what is possible in both cases can be expressed.

## 4.2 Quantitative Measures of Information

### *4.2.1 Principle of Information Measurement*

The *quantitative measurement* of information was a preliminary condition for elaborating the theory. The obvious remark that

*communicating a message is of no use if its destination already knows it*

led Shannon

a) to deal with an information source as generating *random* events, the sequence of which constitutes the transmitted message;
b) to define the information quantity borne by this message as a measure of its *unpredictability*.

When the *source* is *stationary*, i.e., if its operation does not depend on the chosen time origin, one may define an average information quantity generated by this source and borne by the transmitted message: its *entropy*.

One should notice that the distinction between a useful message and perturbations entirely depends on the *destination's aim*. As an example, the sun generates a radiation which is spurious in a satellite communication system. For a radio-astronomer who studies the solar radiation, on the contrary, the signal of the satellite is obviously a perturbation. Actually, one *agrees* to locate in the block 'source' of Shannon's scheme the events interesting the destination and in the block 'channel' the perturbing ones (still according to the destination's interest). Noise is a purely *subjective* concept. This important remark could be thought of as defining a relativity principle of information. It should moreover be noticed that the quantitative measures of information are not intrinsic to an information, but depend only on the probability distribution which is associated with it.

Another information measure concerns the information quantity that the knowledge of the output of a channel provides about its input. It turns out that this quantity is symmetric, in the sense that it also measures the information quantity provided by the channel input as regards its output (as we shall see), so it is called *mean mutual information* (often shortened into 'mutual information', where 'information' is itself a shortening of 'information quantity'). This quantity is different from the entropy of the output message and smaller than it. Indeed, far from providing more information, random perturbations of the channel can but impair the transmitted message.

The mean mutual information does not characterize the channel alone, since it also depends on the source. In order to measure the ability of a channel to communicate information, information theory defines its *capacity* as the maximum of its mean mutual information (the existence of which is proved provided some regularity conditions are met) with respect to all possible *stationary and ergodic* sources connected to its input. Ergodicity, a concept distinct from stationarity, implies the homogeneity of the set of messages that the source is likely to transmit. For an ergodic source, indefinitely observing a single transmitted message is almost surely sufficient to statistically characterize the set of all possible transmitted messages.

Information theory thus enables associating an average quantity with the first two blocks of the scheme of Fig. 4.2: the entropy with the source and the capacity with the channel. When the events which bear information are choices of elements, called symbols, in some predetermined set called alphabet, which we assume to be not only discrete but finite (in order to make the symbols practically distinguishable from each other), the message consists of a sequence of symbols and is referred to as *symbolic*. One may obviously replace the transmitted message by any other one which results from the original message by deterministic (i.e., non-random) and *reversible* transformations. There is no creation and, at best, no destruction of information in such encoding processes, so information appears as invariant with respect to the set of messages which may be used in order to communicate it (we defined it as an equivalence class with respect to these messages in Sect. 2.2).

The blocks of Fig. 4.2 must often be drastically simplified in order to make computations tractable. The conclusions drawn will nevertheless be general enough to pertain to many concrete situations. The simplifications will most often be necessary only in order to make computable some fundamental quantities, the existence of which remains however guaranteed under fairly broad assumptions. Moreover, even if these assumptions are not exactly satisfied (it is often difficult or impossible to acquire the experimental certainty they are so), the solutions to communication problems provided by the theory, consisting of device structures or algorithms, will generally remain usable, maybe at the expense of loosing the exact optimality asserted by the theory when the relevant assumptions are actually satisfied.

### 4.2.2 *Proper and Mutual Information*

First of all, is a *quantitative* measure of information just conceivable? As we have its real-life experience, information obviously has *qualitative* aspects (semantic, affective or aesthetic, for instance) which by hypothesis escape a quantitative measure. Quantitative measures of information and the science which is based on it, i.e., information theory, are thus *a priori* reductionist. They only retain a very partial facet of the concept of information but, as it will clearly appear in the following, precisely the one which is relevant to communication techniques. We may not too much insist on this restriction: the theory was conceived in a modest framework, out of which it is imprudent to extrapolate its conclusions.

The messages delivered by the source are sequences of symbols, i.e., of elements of some predetermined set called alphabet. In order to introduce the quantitative measure of information (and, more generally, the quantities and concepts of information theory) we assume that this alphabet is discrete, and even finite. For instance, such a source generates a text made of letters from the Latin alphabet, or a sequence of binary digits. The point of view of information theory leads us to consider *random* sources and channels. We saw indeed that one of the essential properties of a channel is the presence of perturbations, which cannot generally be described but in probabilistic terms. Therefore the destination can only make *hypotheses* as regards the transmitted message and choose the most likely of them, given the received message. The *source* output is thus dealt with as *random*. Of what use would indeed be to transmit a message if it were already known by its destination?

As unpredictability appears as an essential attribute of information, one is led to equate the quantitative measure of information to the measure of *unexpectedness*, likened itself to that of *improbability*. Thus, the information quantity $h(x)$ brought by the occurrence of an event $x$ of probability $\Pr(x)$ will be measured by some increasing function $f(\cdot)$ of its improbability $1/\Pr(x)$:

$$h(x) = f[1/\Pr(x)]. \tag{4.1}$$

If the occurrence of some event $x$ is certain, $\Pr(x) = 1$. This event brings no information at all, so the function $f(\cdot)$ should be such that $f(1) = 0$.

It is reasonable to assume that the joint occurrence of two *independent* events $x$ and $y$ brings the *sum* of their individual information quantities, hence that we should have:

$$h(x, y) = h(x) + h(y) \tag{4.2}$$

hence

$$f[1/\Pr(x, y)] = f[1/\Pr(x)\Pr(y)] = f[1/\Pr(x)] + f[1/\Pr(y)] \tag{4.3}$$

since, for independent events, $\Pr(x, y) = \Pr(x)\Pr(y)$.

Using for $f(\cdot)$ the *logarithmic function* fulfills this requirement, as well as that to cancel when its argument is unity since $\log(1) = 0$. The logarithmic function has thus been chosen for $f(\cdot)$ in Eqs (4.1) and (4.3). The choice of the logarithmic base defines the information unit. Shannon proposed to take this base equal to 2, so as to make the information unit equal to the information borne by the choice of one among two equally probable issues. He named this information unit 'bit' as an acronym of *bin*ary digi*t* (Shannon 1948). It turns out that 'bit' has since that time very often be used as an abridgement for 'binary digit'. A unit and a digit are very different entities, so it makes generally no sense to associate an information quantity with a binary digit, for lack in general of an associated probability; even if it bears a meaningful probability, the corresponding information quantity equals the binary unit in the sole case where this probability equals 1/2. A binary digit thus bears *at most* an information quantity of one binary unit. It bears a smaller, often *much smaller*, information quantity, if any, in a redundant binary sequence as met in practical situations. The ambiguity of the acronym 'bit' does not mislead communication engineers who are aware of it and use the context for determining if 'bit' refers to the information unit or to a binary digit. That the same word designates such different entities is, however, puzzling for the layman who excusably tends to equate a binary symbol with an information unit since both are named 'bit', thereby denying redundancy. We lay much emphasis is the sequel on the crucial importance of *redundancy* so we cannot accept this meaning of the acronym 'bit'.

To avoid this confusion, the International Standards Organization (ISO) proposed the name of *shannon*, abbreviated as Sh, for the binary information unit

(Roubine 1970). We use it in the following, at variance with most engineers and information theorists who still designate by 'bit' the information unit. We exclusively use the acronym 'bit' in order to designate a binary digit or a binary symbol. Thus, it will make sense to express an information quantity in shannons per bit, which will be specially useful when studying the coding processes.

Binary logarithms will be used throughout in the sequel, denoted by $\log_2(\cdot)$ and more often by $\log(\cdot)$. Natural logarithms will be occasionally useful and will be denoted by $\ln(\cdot)$.

One thus associates with the occurrence of an event $x$ the information quantity

$$h(x) = \log[1/\Pr(x)] = -\log[\Pr(x)]. \qquad (4.4)$$

It is a positive (more precisely, non-negative) quantity, since a probability is a positive number smaller than 1. The logarithm of such a number is negative so the minus sign makes the right hand side positive.

Let now $x$ and $y$ be two events, possibly mutually dependent. One may associate with the pair of events $(x, y)$, as a straightforward extension of Eq. (4.4), the information quantity

$$h(x, y) = -\log[\Pr(x, y)], \qquad (4.5)$$

where $\Pr(x, y)$ designates the joint probability of the two events. We may also define the information quantity associated with $x$ (for example) *conditioned* on the occurrence of $y$, $h(x|y)$, as:

$$h(x|y) = -\log[\Pr(x|y)], \qquad (4.6)$$

where $\Pr(x|y)$ is the probability of $x$ conditioned on $y$, i.e., the probability that $x$ occurs given that $y$ occurred.

From *Bayes' rule*

$$\Pr(x, y) = \Pr(x|y)\Pr(y) = \Pr(y|x)\Pr(x), \qquad (4.7)$$

and from Eqs. (4.5) and (4.6), it follows that

$$h(x, y) = h(x|y) + h(y) = h(y|x) + h(x). \qquad (4.8)$$

Notice that, if $x$ and $y$ are independent, $h(x|y) = h(x)$ so then Eq. (4.2) obtains.

One may also wish to measure the information quantity that the occurrence of one of the events, say $y$, provides as regards the other one, $x$. This information quantity is especially relevant when one identifies $x$ to the choice of a signal as a channel input and $y$ to the corresponding output. $\Pr(x)$ is then the *a priori* probability that $x$ is transmitted and $\Pr(x|y)$ the *a posteriori* probability that $x$ has been transmitted, $y$ being received. A measure of this information quantity is:

$$i(x; y) = \log[\Pr(x|y)/\Pr(x)], \qquad (4.9)$$

a logarithmic measure of the probability increase which results from $x$ being conditioned on $y$. It actually reduces, as expected, to $h(x)$ if $\Pr(x|y) = 1$, since then the occurrence of $y$ is equivalent to that of $x$; it is actually 0 if $x$ and $y$ are independent, since in this case $\Pr(x|y) = \Pr(x)$, so $y$ brings indeed no information about $x$.

Applying Bayes' rule (4.7) shows that $i(x; y)$ is actually symmetrical:

$$i(x; y) = \log [\Pr(x, y)/\Pr(x)\Pr(y)] = h(x) + h(y) - h(x, y) = i(y; x).$$

This quantity is referred to as *mutual information* (the quantity defined by Eq. (4.4) is then said proper information). At variance with the proper information $h(x)$ which is obviously non-negative, the mutual information $i(x; y)$ may be negative. Rather than this quantity, its average is intrinsically positive and much more useful, as shown in Sect. 4.2.4. Positiveness is indeed an expected property of an information measure.

### 4.2.3 *Entropy and Average Mutual Information*

**Entropy of a discrete random variable** Individual events are generally less important than averages in a probabilistic context. Thus, we consider an information source as generating *repetitive* events (for instance, but not necessarily, periodic), obeying a known probability distribution which will be assumed to be invariant with respect to time. Such a source will be referred to as *stationary*. We shall moreover restrict ourselves to the case where the source is *discrete* and even *finite*. The relevant events will then be interpreted as choices of symbols in a given alphabet. We shall moreover provisionally assume that the successive choices are *mutually independent*. In other words, the symbols of this alphabet being denoted by $x_1, x_2, \ldots, x_n$, each event occurring in the source is described by the random variable $X$ equal to each of the alphabet symbols, say $x_i$, with probability:

$$p_i = \Pr(X = x_i), \quad i = 1, 2, \ldots, n,$$

with, of course,

$$\sum_{i=1}^{n} p_i = 1.$$

This probability distribution does not depend on previous symbol choices since they are assumed to be mutually independent, neither on time since the source is assumed to be stationary. It thus completely describes the source. The set of values that $X$ can assume, which is here the source alphabet, is refered to as its *sample space*.

The information quantity associated in the average with each symbol generated by this source is the mean (or expectation, denoted by $\mathrm{E}[\cdot]$) of the proper information of the events $X = x_i$, i.e.,

$$H(X) = \mathrm{E}[h(X)] = \sum_{i=1}^{n} p_i \log(1/p_i) = -\sum_{i=1}^{n} p_i \log(p_i), \qquad (4.10)$$

which is referred to as the *entropy* of the source (or of $X$, or of its probability distribution). This word comes from the similarity, at least formal, of this quantity with the expression of thermodynamic entropy given by statistical mechanics. More precisely, this expression defines the entropy per symbol; one may also define an entropy rate (or entropy per second) $H'$ by multiplying the entropy per symbol by the (average) frequency at which symbols are generated by the source.

For instance, the entropy of a random binary variable which equals 1 with probability $p$ and thus 0 with probability $1 - p$, expressed in binary units equals:

$$\mathcal{H}_2(p) = \mathcal{H}_2(1-p) \triangleq \begin{cases} -p\log_2(p) - (1-p)\log_2(1-p), \ 0 < p < 1, \\ 0, \ p = 0 \text{ or } p = 1. \end{cases} \tag{4.11}$$

This function is symmetric with respect to $p = 1/2$, where it achieves its maximum equal to 1. It is represented in terms of $p$ in Fig. 4.4 below.

We may yet make the following two remarks.

1. The entropy function Eq. (4.11) achieves its maximum for $p = 1/2$, i.e., when the two possible events are equally probable. This may look at first sight contradictory with the fact that, according to Eq. (4.4), an event brings the more information, the more it is improbable. There is actually no contradiction since the information quantity $-\log[\Pr(x)]$ brought by one of the possible events is multiplied by $\Pr(x)$ when the mean is computed. If $\Pr(x)$ approaches 0, $\Pr(x)\log[\Pr(x)]$ itself approaches 0 and the contribution to the entropy of an event $x$ of very small probability is small, although the information quantity produced by this event is large *when* it occurs.
2. Although the notation of Eq. (4.4) seems to define $h(x)$ as a function of $x$, it does not actually depend on the value assumed by $x$ but on the probability which is associated with its occurrence. Thus, writing $H(X)$ as we did, for instance, is an abuse of notation: the random variable $X$ is not truly an argument. This notation is just used in order to identify the probability distribution of which $H$ is the entropy. In the discrete finite case, any transformation which merely permutes the probabilities associated with the values $X$ may assume leaves $H$ unchanged, as well as the substitution of symbols for others provided that the set of associated probabilities is saved. If there is a need to make explicit the dependency of the entropy (4.10) in terms of probabilities $p_1, p_2, \ldots, p_n$, we use the notation $H(p_1, p_2, \ldots, p_n)$ where the probabilities are now true arguments of the function. This remark applies as well to all the quantitative measures of information to be contemplated.

Let us now consider two random variables $X$ and $Y$ taking their values in two finite sample spaces $\{x_1, x_2, \ldots, x_n\}$ and $\{y_1, y_2, \ldots, y_m\}$. Applied to the elements $(x_i, y_j)$ of the set of all their pairs (also referred to as their Cartesian product), the definition of entropy leads to the *joint entropy*, which is the mean of (4.5):

$$H(X, Y) = \mathrm{E}[h(X, Y)] = -\sum_{i=1}^{n}\sum_{j=1}^{m} \Pr(x_i, y_j)\log[\Pr(x_i, y_j)]. \tag{4.12}$$

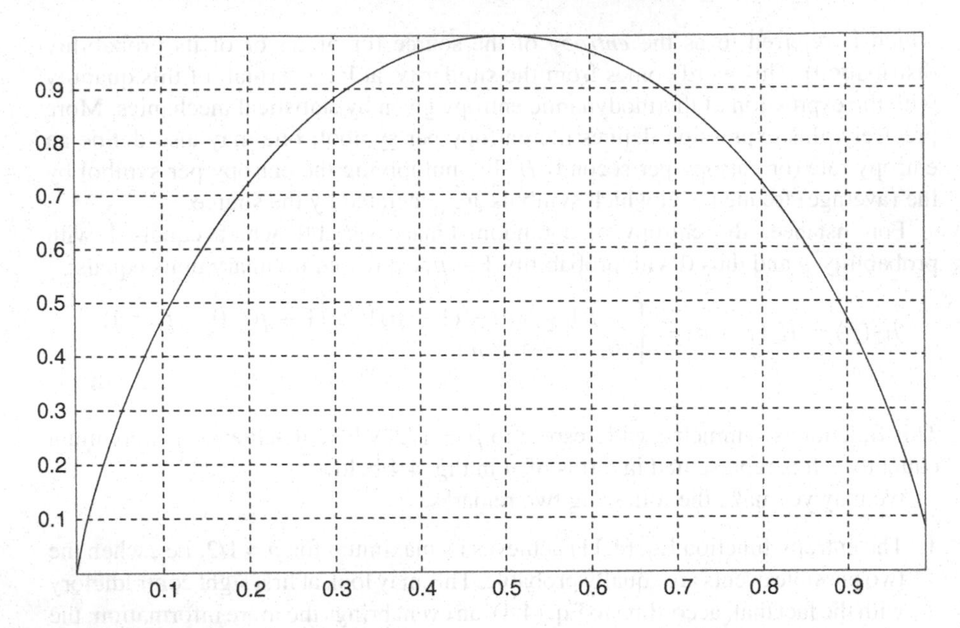


**Fig. 4.4** Binary entropy function $\mathcal{H}_2(p)$

Analogously, one may define *conditional entropies*; for instance, taking the mean of (4.6) results in:

$$H(X|Y) = \mathrm{E}[h(X|Y)] = -\sum_{i=1}^{n}\sum_{j=1}^{m} \Pr(x_i, y_j) \log[\Pr(x_i|y_j)]. \tag{4.13}$$

Notice that, at variance with the previous entropy expressions, the probability which factors the logarithm is different from that in its argument: the mean is still effected with respect to the Cartesian product of the variables $X$ and $Y$ so the coefficient of the logarithm is the *joint probability* $\Pr(x_i, y_j)$ while the argument of the logarithm is a *conditional* one, i.e., $\Pr(x_i|y_j) = \Pr(x_i, y_j)/\Pr(y_j)$, where $\Pr(y_j) = \sum_{i=1}^{n} \Pr(x_i, y_j)$ is the *marginal probability* of $y_j$.

This definition of entropy can be shown to be the only continuous function of its arguments (i.e., the probabilities $\{p_i\}$) which satisfies a system of axioms accounting for the expected properties of an average information measure. Khinchin used as axioms the following three properties (Khinchin 1957):

1. For $n$ given and $\sum_{i=1}^{n} p_i = 1$, the function $H(p_1, p_2, \ldots, p_n)$ is maximum for $p_i = 1/n$ for any value of $i$ (a property of the entropy to be stated in Sect. 4.2.4).
2. $H(X,Y) = H(X) + H(Y|X)$ (equality (4.21), below).
3. $H(p_1, p_2, \ldots, p_n, 0) = H(p_1, p_2, \ldots, p_n)$, i.e., appending an $(n+1)$-th event of probability 0 to a set of $n$ possible events leaves the entropy unchanged.

Other systems of axioms were introduced, especially that of Fadeev which may be considered as "minimum" (Roubine 1970):

1. $H(p_1, p_2, \dots, p_n)$ is a symmetric function of its variables $p_1, p_2, \dots, p_n$;
2. $H(p, 1-p)$ is a continuous function of $p$ in the interval [0,1];
3. $H(p_1', p_1'', p_2, \dots, p_n) = H(p_1, p_2, \dots, p_n) + p_1 H(\frac{p_1'}{p_1}, \frac{p_1''}{p_1})$, with $p_1' + p_1'' = p_1$.

One readily checks that the equality set as axiom 3 is a direct consequence of the definition (4.10) of entropy.

We already stated that we do not deal here with the information measure associated with a continuous random variable $X$, say of probability density function $p_X(x)$, meaning that the probability that $X$ belongs to the infinitesimal interval $[x, x + \mathrm{d}x]$ equals $p_X(x)\mathrm{d}x$. Indeed, replacing the summation by an integral in the entropy expression (4.10) is impossible because the differential $\mathrm{d}x$ would appear in the argument of the logarithm. Instead, one formally defines the 'differential entropy' of $X$ as

$$H_{\mathrm{d}}(X) \triangleq -\int p_X(x) \log [p_X(x)] \mathrm{d}x$$

which has some of the properties expected from an entropy but not all. We actually do not need this quantity in what follows so we do not develop this topic.

When a continuous random variable is known through a physical measurement, the difference between two measured values is meaningless if it is less than some positive quantity $\varepsilon$ which measures the measurement uncertainty. It is then possible to define its $\varepsilon$-entropy (see Sect. 5.2.4), which has the same properties as the entropy of a discrete random variable.

**Mutual information** Another very important quantity, now associated with a pair of random variables, is the *mean mutual information* deriving from (4.9) by taking its mean, i.e.:

$$I(X;Y) = \mathrm{E}[i(X;Y)] = \sum_{i=1}^{n}\sum_{j=1}^{m} \Pr(x_i, y_j) \log \left[\frac{\Pr(x_i, y_j)}{\Pr(x_i)\Pr(y_j)}\right], \tag{4.14}$$

with $\Pr(x_i) = \sum_{j=1}^{m} \Pr(x_i, y_j)$ and $\Pr(y_j) = \sum_{i=1}^{n} \Pr(x_i, y_j)$.

Contrary to the entropy, extending the mean mutual information to continuous random variables $X$ and $Y$, or to the important case where $X$ is discrete but $Y$ is continuous because the channel noise is continuous, is straightforward. Indeed, the same differentials appear in both the numerator and denominator in the argument of the logarithm in Eq. (4.14) so they are eliminated.

The mean mutual information is at least as important as the entropy in the communication field. It is possible, analogously, to give an axiomatic definition of it and to show its uniqueness. Once the mean mutual information has been introduced, the entropy may be derived from it since:

$$I(X;X) = H(X)$$

and more generally, from (4.25):

$$I(X;Y) = H(X) \text{ if } H(X|Y) = 0,$$

i.e., if the occurrence of $Y$ implies with certainty that of $X$.

We just defined quantities intended to measure the mean information: the entropy which measures the average information quantity associated with a discrete random variable (or with a discrete stationary source where successively transmitted symbols are independent, which is thus completely described by such a random variable), and the mean mutual information which measures the average quantity of information which is provided by one of the variables of a pair as regards the other one. We now examine their main mathematical properties and their relations to each other.

### 4.2.4 *Properties of the Entropy and of the Mean Mutual Information*

The reader unfamiliar with mathematics may jump over the proofs given in this section, but should accept its conclusions.

**Positiveness of entropy** The entropy is obviously non-negative, i.e.:

$$H(p_1, p_2, \dots, p_n) \geq 0\,;$$

it is zero if, and only if, one of the probabilities in its argument equals 1 (so the other ones equal 0). Then, one of the $n$ events occurs with certainty so its outcome brings no information.

**Maximum entropy** $H(p_1, p_2, \dots, p_n)$ reaches a maximum, for $n$ fixed, when $p_i = 1/n$ for any $i$. This property is easily shown to result from the convexity of the logarithmic function (i.e., the curve which represents it is below its tangent at any of its points), thanks to the Gibbs inequality. Given two probability distributions over the same finite set, say $(p_1, p_2, \dots, p_n)$ and $(q_1, q_2, \dots, q_n)$, with $\sum_{i=1}^n p_i = \sum_{i=1}^n q_i = 1$, Gibbs inequality reads:

$$\sum_{i=1}^{n} p_i \log(q_i/p_i) \leq 0, \tag{4.15}$$

with equality if, and only if, $p_i = q_i$, hence $p_i = 1/n$, for any $i$.

**Convexity of entropy** Replacing $p_1, p_2, \dots, p_n$ by averages

$$q_i = \sum_{j=1}^{n} a_{ij} p_j, \;\; a_{ij} \geq 0, \tag{4.16}$$

with

$$\sum_{j=1}^{n} a_{ij} = 1, \tag{4.17}$$

for any value of $i$, and

$$\sum_{i=1}^{n} a_{ij} = 1, \tag{4.18}$$

for any value of $j$, results in increasing entropy, i.e.:

$$H(q_1, q_2, \ldots, q_n) \geq H(p_1, p_2, \ldots, p_n). \tag{4.19}$$

In other words, $H(p_1, p_2, \ldots, p_n) = H(\underline{p})$ is a *convex function*[1] ∩ of its arguments.

Let us consider the function $f(x) \triangleq -x \log(x)$. Then $H(p_1, p_2, \ldots, p_n) = \sum_i f(p_i)$. The convexity ∩ of $f$ results, for coefficients $a_{ij}$ satisfying (4.17), in:

$$f\left(\sum_{j=1}^{n} a_{ij} p_j\right) \geq \sum_{j=1}^{n} a_{ij} f(p_j), \quad i = 1, 2, \ldots, n; \tag{4.20}$$

thus, summing with respect to $i$:

$$\sum_{i=1}^{n} f\left(\sum_{j=1}^{n} a_{ij} p_j\right) \geq \sum_{i=1}^{n} \sum_{j=1}^{n} a_{ij} f(p_j) = \sum_{j=1}^{n} f(p_j) \sum_{i=1}^{n} a_{ij},$$

hence, taking (4.16) and (4.18) into account:

$$\sum_{i=1}^{n} f(q_i) \geq \sum_{j=1}^{n} f(p_j),$$

an inequality equivalent to (4.19).

**Joint and conditional entropies of two variables** Let us now consider two random variables $X$ and $Y$. We already defined their joint entropy, denoted by $H(X, Y)$, by Eq. (4.12). It directly results from this definifion and from Bayes' rule (4.7) (see 4.8) that:

$$H(X, Y) = H(X) + H(Y|X) = H(Y) + H(X|Y). \tag{4.21}$$

The information quantity jointly brought by $X$ and $Y$ is thus larger than (or equal to) the one brought by $X$ or $Y$ separately, since the entropy is essentially non-negative. For the same reason:

$$H(X, Y) \geq H(X) \quad \text{or} \quad H(Y). \tag{4.22}$$

[1] Such a function is generally referred to as 'concave' in the mathematical literature. We prefer using the single word 'convex', the shape of its representative curve being indicated by ∩ or ∪.

As regards the conditional entropy, let us notice that

$$H(X|Y) \leq H(X), \tag{4.23}$$

with equality if, and only if, $X$ and $Y$ are independent.

In effect, inequality (4.20) expressing the convexity $\cap$ of the function $f(x) = -x \log(x)$ results in:

$$f\left[\sum_i \Pr(x_j|y_i)\Pr(y_i)\right] \geq \sum_i \Pr(y_i) f[\Pr(x_j|y_i)],$$

$$-\sum_i \Pr(x_j|y_i)\Pr(y_i)\log\left[\sum_i \Pr(x_j|y_i)\Pr(y_i)\right] \geq$$

$$-\sum_i \Pr(x_j|y_i)\Pr(y_i)\log[\Pr(x_j|y_i)],$$

$$-\Pr(x_j)\log[\Pr(x_j)] \geq -\sum_i \Pr(x_j, y_i)\log[\Pr(x_j|y_i)]$$

and, after summing with respect to $j$:

$$H(X) \geq H(X|Y),$$

an inequality equivalent to (4.23).

Thus, the realization of a random variable which conditions another one can but decrease the information quantity brought by the latter.

An immediate consequence of (4.23), taking account of (4.19), is the double inequality:

$$H(X,Y) \leq H(X) + H(Y) \leq 2\, H(X,Y); \tag{4.24}$$

the second inequality merely results from (4.21) and of the positiveness of entropy.

**Mean mutual information of two variables** As regards the mean mutual information, its very definition results in:

$$I(X;Y) = H(X) - H(X|Y) = H(Y) - H(Y|X), \tag{4.25}$$

$$I(X;Y) = H(X) + H(Y) - H(X,Y). \tag{4.26}$$

Then, (4.23) shows that $I(X;Y)$ is *positive* or zero; it is zero if, and only if, the two variables $X$ and $Y$ are independent. Interpreting $X$ as the input symbol of a noisy channel and $Y$ as its output, one may comment the double equality (4.25) as follows. The mean mutual information equals:

> the transmitted information minus the indeterminacy as regards the transmitted symbol which remains when the received symbol is known;
> symmetrically, the received information minus the indeterminacy as regards the received symbol which remains when the transmitted symbol is known.

The mean mutual information is convex $\cap$ under conditions met in most practical situations.

### 4.2.5 *Information Rates; Extension of a Source*

The quantities introduced above in order to measure information have been defined per *symbol*: that generated by the source as regards the entropy, the pair of input and output symbols of a channel as regards mutual information. One should then precisely indicate the points in Shannon's scheme where these symbols are considered, i.e., well define the borders between the blocks.

It is uncecessary to provide such precisions when the source generates its symbols at a constant average rate (the symbol choices may be exactly periodic or effected at a constant average frequency) and if the remainder of the communication link works 'in real time', i.e., at the same pace as the source. In this case, it suffices to consider information rates with respect to time: the entropy or mutual information per second, product of the entropy or mutual information per symbol by the frequency (possibly the average frequency) of occurrence of these symbols at a given point. Then, the same information rate may correspond to equivalent messages with different symbol frequencies (as resulting for instance from a change of the alphabet size). An often useful change of alphabet consists of replacing a source by its $k$-th extension. Given a source generating a message the symbols of which belong to the $\alpha$-ary alphabet, its $k$-th extension $S^k$ consists of grouping its symbols by successive blocks of $k$, each block being interpreted as a symbol of the $\alpha^k$-ary alphabet. The frequency at which these symbols are generated by $S^k$ is the product by $1/k$ of that of source $S$. For instance, if a binary source generates the message

$$000\ 100\ 101\ 101\ 011\ 100\ 110\ldots$$

its third extension generates the message

$$0\ 4\ 5\ 5\ 3\ 4\ 6\ldots,$$

when the symbols now belong to the 8-element alphabet and are represented here by decimal digits. The $k$-th extension $S^k$ of a source $S$ is not a different source, but another means for describing it.

### 4.2.6 Cross-Entropy

We now introduce a quantity to be referred to as *cross-entropy* from which the main two measures of average information, namely, the entropy and the mutual information, can be derived. This quantity measures in a sense the distance between two probability distributions and possesses interesting properties. It has been introduced by Kullback and Leibler (Kullback 1959). Its importance comes from Kullback principle, to be stated in the sequel, as well as from its axiomatic proof given by Shore and Johnson (Shore and Johnson 1980). The reader unfamiliar with mathematics may find this section rather difficult and may jump over it. It has been included because of its importance in information theory, but it is of little direct use in the sequel.

**Definition and main properties of cross-entropy** Let us consider two probability distributions on a same finite sample space of $n$ elements, namely, $\{p_i\}$, denoted by $\mathbf{p}$, and $\{q_i\}$, denoted by $\mathbf{q}$. By definition,

$$\sum_{i=1}^{n} p_i = \sum_{i=1}^{n} q_i = 1.$$

It is moreover assumed that none of the probabilities $\{q_i\}$ is zero. The Gibbs inequality (4.15) implies that the quantity

$$H(\mathbf{p}\|\mathbf{q}) \triangleq \sum_{i=1}^{n} p_i \log\left(\frac{p_i}{q_i}\right) \tag{4.27}$$

is positive or zero. It is zero if, and only if, $p_i = q_i$ for any $i$, hence $\mathbf{p} = \mathbf{q}$. The logarithmic base is arbitrary, for example 2. The quantity $H(\mathbf{p}\|\mathbf{q})$ is useful for measuring the degree of proximity between the two distributions $\mathbf{p}$ and $\mathbf{q}$. Kullback named it *directed divergence* (Kullback 1959) and it is often referred to as the 'Kullback-Leibler divergence'. It bears also the names of *cross-entropy*, to be used here, and *relative entropy* in the book by (Cover and Thomas 1991). We shall refer here to $H(\mathbf{p}\|\mathbf{q})$ as 'cross-entropy' of $\mathbf{p}$ with respect to $\mathbf{q}$. As regards its notation, we adopted that used by Cover and Thomas, where the arguments are separated by the sign $\|$. Separating arguments by a comma generally refers to the variable jointly associated with its two arguments, which is not relevant here. On the other hand, the arguments $X$ and $Y$ of the mutual information $I(X;Y)$ are usually separated by a semi-colon in order to indicate that it depends on two arguments and not on the variable jointly associated with its two arguments. Mutual information is symmetrical with respect to $X$ and $Y$ since $I(X;Y) = I(Y;X)$. On the contrary, if one swaps $\mathbf{p}$ and $\mathbf{q}$ in Eq. (4.27), one formally obtains

$$H(\mathbf{q}\|\mathbf{p}) = \sum_{i=1}^{n} q_i \log\left(\frac{q_i}{p_i}\right),$$

which now is meaningful only if none of the probabilities $\{p_i\}$ is zero. If this condition is met, one generally has:

$$H(\mathbf{q}\|\mathbf{p}) \neq H(\mathbf{p}\|\mathbf{q}).$$

Using the special separation sign $\|$ is intended to recall the asymmetry of cross-entropy with respect to its arguments.

The additivity of cross-entropy with respect to the product of distributions immediately follows from definition (4.27). Let $\{p_i\}$, denoted by $\mathbf{p}$, and $\{p'_i\}$, denoted by $\mathbf{p}'$, be two probability distributions over a same finite sample space of $n$ elements, assumed to be *independent*. Let $\{q_i\}$, denoted by $\mathbf{q}$, and $\{q'_i\}$, denoted by $\mathbf{q}'$, be two other independent probability distributions over the same finite set. One denotes by $\mathbf{P}$ and $\mathbf{Q}$ the distributions associated with the Cartesian product of the sample spaces endowed with the product of the corresponding probabilities, i.e., the $n^2$ components of $\mathbf{P}$, for example, are of the form

$$P_k = p_i p'_j,\ k = n(j-1)+i,$$

where $i, j \in \{1, 2, \dots, n\}$. If no component of $\mathbf{Q}$ is zero, the cross-entropy of $\mathbf{P}$ with respect to $\mathbf{Q}$ is the sum of the cross-entropies of $\mathbf{p}$ with respect to $\mathbf{q}$ and of $\mathbf{p}'$ with respect to $\mathbf{q}'$:

$$H(\mathbf{P}\|\mathbf{Q}) = H(\mathbf{p}\|\mathbf{q}) + H(\mathbf{p}'\|\mathbf{q}').$$

The cross-entropy has a particularly simple expression when $\mathbf{q}$ is equal in Eq. (4.27) to the uniform distribution $\mathbf{u}$, such that $u_i = 1/n$ for each value of $i$, namely:

$$H(\mathbf{p}\|\mathbf{u}) = \log(n) - H(\mathbf{p}), \tag{4.28}$$

where $H(\mathbf{p})$ is the entropy associated with the probability distribution $\mathbf{p}$, as defined in (4.10). But $\log(n)$ is the maximum entropy associated with $n$ events. The cross-entropy of a probability distribution with respect to the uniform distribution thus measures the difference between the maximum entropy and that which corresponds to $\mathbf{p}$, i.e., the *redundancy*.

We now consider two non-independent random variables $X$ and $Y$. We take for $\mathbf{p}$ the joint distribution of $X$ and $Y$ ($\mathbf{p} \triangleq \{\Pr(X, Y)\}$) and for $\mathbf{q}$ the product of the marginal distributions of $X$ and of $Y$ ($\mathbf{q} \triangleq \{\Pr(X)\Pr(Y)\}$), which would be their joint distribution if, and only if, $X$ and $Y$ were independent. The cross-entropy of $\mathbf{p}$ with respect to $\mathbf{q}$ is nothing but the mean mutual information as defined in Eq. (4.14): $H(\mathbf{p}\|\mathbf{q}) = I(X;Y)$.

It is thus possible to first define the cross-entropy as (4.27), and then derive from it the main two mean information measures introduced in Sect. 4.2.2: the entropy as defined by (4.28) and the mean mutual information as we just have seen it.

Contrary to the entropy but similarly to the mutual information, the cross-entropy can easily be extended to the continuous case. The distributions $\mathbf{p}$ and $\mathbf{q}$ appear in argument of the logarithm in formula (4.27) which defines cross-entropy in terms of the ratios $p_i/q_i$, which enables extending its definition to the continuous case, provided only that distributions $\mathbf{p}$ and $\mathbf{q}$ possess probability density functions $p(x)$ and $q(x)$. Assuming their existence, the cross-entropy of $\mathbf{p}$ and $\mathbf{q}$ is expressed as:

$$H(\mathbf{p}\|\mathbf{q}) \triangleq \int_{-\infty}^{+\infty} p(x) \log\left[\frac{p(x)}{q(x)}\right] \mathrm{d}x \tag{4.29}$$

and the inequality $H(\mathbf{p}\|\mathbf{q}) \geq 0$ remains true. Equality (4.29) can even be extended to more general distributions.

The previous statements remain valid for cross-entropy as extended to the continuous case, except that now $q(x)$ should be non-zero almost everywhere. One has $H(p\|q) = 0$ if, and only if, $p(x) = q(x)$ almost everywhere, i.e., except maybe on a set of measure zero.

**Kullback principle** The positiveness of cross-entropy is for the moment the sole (and meager) reason which justifies its use as a proximity measure between two distributions $\mathbf{p}$ and $\mathbf{q}$. Besides the fact it is dissymmetric in $\mathbf{p}$ and $\mathbf{q}$, this quantity does not even obey the triangular inequality, which seems to poorly qualify it as a distance measure. The importance of cross-entropy as a proximity measure between two distributions, in spite of these reservations, actually relies on the Kullback principle (Kullback 1959) and on the axiomatic proof of it given by Shore and Johnson (Shore and Johnson 1980; Johnson and Shore 1983).

Let $\mathbf{p}$ be a probability distribution on the sample space of a certain random variable, referred to as *a priori*. Let us assume that this quantity is also submitted to constraints ignored in the expression of $\mathbf{p}$, in the form of either some linear combinations of probabilities in $\mathbf{p}$ being equated to zero (i.e., equating to zero the mathematical expectation of some random variable defined in terms of $\mathbf{p}$), or of an inequality concerning such a combination, specifying for instance that it should be non-negative. One intends to determine the probability distribution on the same sample space, to be referred to as *a posteriori*, which takes account of both the distribution $\mathbf{p}$ and these constraints. The answer to this question is given by Kullback principle:

> *among all probability distributions* $\mathbf{q}$ *compatible with the constraints, the best* a posteriori *distribution* $\mathbf{q}^*$ *is the one which minimizes the cross-entropy* $H(\mathbf{q}\|\mathbf{p})$.

In the discrete case, when the *a priori* distribution is uniform (or if, in the absence of any *a priori* data, it is assumed to be so), Kullback's principle reduces, according to relation (4.28), to that of maximum entropy previously stated by (Jaynes 1957).

In order to prove Kullback's principle, Shore and Johnson use simple axioms which aim at logical consistency for determining $\mathbf{q}^*$. They actually first define this quantity in the continuous case i.e., using (4.29), and put as axioms the following statements, written here in informal terms:

a) the result is unique;
b) it is invariant by scale change;
c) it is equivalent to take account of independent informations on independent quantities and to consider them together in the form of a joint distribution;
d) considering a distribution over a complete set of events has the same result as decomposing it into successively taken disjoint subsets. For each subset, one then replaces the initial probability distribution by the distribution conditioned on the membership to this subset.

In the discrete case, Shore and Johnson replace axiom b) by

b') the result is invariant by permutation;

and they moreover append to the set of axioms the following one (Shore and Johnson 1981):

e) taking into account of a constraint already satisfied by the *a priori* distribution leads to an *a posteriori* distribution identical to the *a priori* distribution.

The general idea which led to the choice of these axioms was to ensure the identity of the results when there exist several ways for obtaining them. In the case of axiom d), for example, one may either consider the complete set of events, or split it in subsets.

Many other proximity measures between probability distributions have been proposed, but *minimized cross-entropy* is the only one which satisfies the set of all the axioms written above and it is why it is important.

One finds applications of cross-entropy in all cases where some result may be obtained as a consequence of the Gibbs inequality (4.15). In source coding (to be considered in the next section), for instance, let us consider $\mathbf{p}$ as the probability of the $n$ messages that a source can generate and $\mathbf{q} \triangleq \{\alpha^{-n_i}\}$, where $\alpha$ is the source alphabet size, $n_i$ is the length of the word associated by some source encoding to the $i$-th message, the Kraft inequality $\sum_{i=1}^{n} \alpha^{-n_i} = 1$ (4.32) being moreover satisfied with equality (see Sect. 4.3.2). The probability distribution $\mathbf{q}$ is submitted to the constraint that each of the numbers $n_i$ is an integer. Then,

$$H(\mathbf{p}\|\mathbf{q}) = \overline{n} - H_\alpha(\mathbf{p})$$

where $\overline{n} \triangleq \sum_{i=1}^{n} p_i n_i$ is the average length of the codewords per source message and where $H_\alpha(\mathbf{p})$ is its entropy, expressed using logarithms to the base $\alpha$. Hence, $H(\mathbf{p}\|\mathbf{q})$ measures here the cross-entropy between the average length obtained after encoding and its lower limit as assigned by the fundamental theorem of source coding. In particular, this cross-entropy will be strictly positive even for the coding system which makes it minimum i.e., the best possible for a given value of $n$, if probabilities $p_i$ are not all of the form $\alpha^{-n_i}$ with $n_i$ integer. (It is known, as shown in Sect. 4.3.4, that it is possible to diminish it by performing the optimized encoding on the messages of the $k$-th extension of the original source, instead of the original messages of this source.)

Many other examples of applying Kullback's principle and of cross-entropy may be quoted in other branches of information theory, but also in the fields of signal processing and physics. In the latter case, Kullback's principle generally reduces to the maximum entropy principle of Jaynes. Kullback's principle has also been applied to the optimal decoding of channel codes (Moher 1993).

**Properties of minimized cross-entropy** We now assume that the distribution $\mathbf{q}$ results from an *a priori* distribution $\mathbf{p}$ and of a certain set $I$ of constraints by applying Kullback's principle. The cross-entropy $H(\mathbf{q}\|\mathbf{p})$ is then, by definition, minimum for the distribution $\mathbf{q}$ compatible with the constraints $I$.

This cross-entropy then verifies the *inverse triangular inequality*, namely:

$$H(\mathbf{q}\|\mathbf{p}) \geq H(\mathbf{q}\|\mathbf{r}) + H(\mathbf{r}\|\mathbf{p}), \tag{4.30}$$

where $\mathbf{r}$ is a probability distribution which minimizes the cross-entropy for a partial set of constraints $J \subset I$, $\mathbf{r}$ being then used as the *a priori* probability distribution for determining by the same way the probability distribution $\mathbf{q}$ now taking into account the set $I - J$ of the remaining constraints. Still better, if the constraints are in finite number and are all expressed by equalities, then (4.30) becomes an equality. This *triangular equality* fully justifies using the *minimized cross-entropy* as a proximity measure between *a priori* and *a posteriori* distributions. One will find in (Shore and Johnson 1981) a more precise and complete discussion of the properties of minimized cross-entropy.

### 4.2.7 Comments on the Measurement of Information

As it is used in the theory, the word 'information' has a *technical* meaning which is very restrictive with respect to the usual one. In order to define a quantitative measure of information, we just relied on an initial remark, without going further. We actually defined the information quantity as a statistical measure of *unexpectedness*, itself likened to *improbability*. The entropy adequately measures the average time or cost for communicating messages generated by a source. But the meaning of a message, its veracity, its documentary, affective or aesthetic value, its usefulness for the destination, the consequences it may have if it consists of a command, all notions relevant to *semantics*, are by hypothesis foreign to information theory. Confusing information in its technical meaning and information in its usual sense should be avoided. Extending information theory out of its technical validity domain (which is also the place of its origin), where excluding semantic was undoubtedly a necessary condition for elaborating the theory, demands a special attention for avoiding this confusion.

This situation is not very different from that encountered much earlier with physical quantities. For instance, any material object has many attributes like form, colour, texture, internal structure ... If one has to describe the movement of this object when driven by a mechanical force, however, a single quantity is relevant: its mass. Similarly, among the many attributes of a message, a single one is relevant as far as literal communication is concerned: the entropy of the source which generates it.

It is often asserted that the entropy measures uncertainty or disorder, and it is actually so that physics interprets it. As used in information theory, however, it actually measures a *resolved* uncertainty: the information associated with an event is defined in terms of its probability distribution *before* it occurs. That the entropy of information theory measures a resolved uncertainty also appears in the expression (4.25) of the mean mutual information $I(X, Y)$. If we interpret $X$ as the 'source' event i.e., the one the occurrence of which brings information, and $Y$ as its perturbed observation, then (4.25) shows that the uncertainty as regards $X$ is only partially resolved, since the entropy of $X$ is diminished by the conditional entropy $H(X|Y)$ which measures the residual uncertainty. Since the entropy of information theory measures the resolution of an uncertainty, the word *negentropy*, i.e., negative entropy, as used by (Schrödinger 1943) and (Brillouin 1956), would be more appropriate for referring to it. Thus the famous formula engraved on Boltzmann's tombstone in Vienna which expresses the thermodynamic entropy $S_{\text{th}}$ as

$$S_{\text{th}} = k_{\text{B}} \ln (W),$$

where $k_{\text{B}}$ is referred to as Boltzmann constant, measures the indeterminacy associated with $W$ equally probable configurations which are indistinguishable at the macroscopic scale although distinct at the microscopic level. We further examine the relationship of thermodynamic entropy with information in Sect. 6.3.1.

Even if one is interested as at the beginning of this section in a single event, the definition (4.4) of information quantity refers to a set of objects endowed with probabilities. Stating the obvious, this implies that the elements of this set are defined and that probabilities are assigned to them. Defining the elements implies a preliminary *agreement* of the source and the destination (an indissociable pair for the theory) as regards the events assumed as symbols (and more generally, as regards their meaning, but we ignore semantics). Assigning probabilities to future events is often impossible, but in the form of more or less arbitrary forecasts. The only case when it is meaningful is that of a *stationary and ergodic* situation, since then observing the frequency of occurrence of the past events enables reliably assessing probabilities to the events to come. But receiving a message results in some modification of the destination: it is quite obvious when it is a command for destroying a rocket, much less obvious in many cases, but of what use would be to communicate a message unless there is some lasting modification, e.g., its recording, at the receiving end? If one tries to measure information in terms of the probability estimates which the receiver can make, one has to take account of the destination modification due to its receiving the previously transmitted messages. This is a learning process, which is typically non-stationary.

At a very fundamental level, information is involved in the following vicious circle. I have a dice and, in the absence of further information, I assign to each of its 6 faces the same probability 1/6. If I actually know the position of the gravity center of the dice, thus exploiting a complementary information, I am able to revise the probability estimates associated with the faces: an information measure thus depends here on ... a preliminary information. Moreover the event I refer to, namely, the outcome of one of the faces, should be identifiable. In other words, each face of a dice should bear an ... information in order to enable distinguishing it from the others. This is not difficult at the macroscopic level, but it is not so at the microscopic one. Can we play dice with an atom or a molecule?

Does information exist prior to probabilities? We may deem that the algorithmic information theory to be briefly dealt with in Sect. 6.1, which is not based on probabilities, will eventually result in founding the theory of probabilities.

## 4.3 Source Coding

Source coding will not be directly useful to us for applying information theory to sciences of life. However, we devote this section to it for the sake of completeness, deeming that it is absolutely necessary for understanding information. Moreover, briefly dealing with source coding will be helpful to shed some light on the relationship between information and semantics. For instance, describing the Huffman algorithm in Sect. 4.3.4 will provide the opportunity to meet the concept of decipherable code, which is of fundamental importance as regards this relationship.

The aim of source coding, as already said, is to replace some sequence generated by a source with a shorter but entirely equivalent one. In other words, it is intended to remove its redundancy. We first introduce source models and discuss some of the properties they should possess in order to be tractable. Then, we shortly state the fundamental theorem of source coding with the help of the concept of source extension. We more lengthily describe the Huffman algorithm, an optimum source coding algorithm, which will give us the opportunity of introducing the representation of a binary sequence by a tree and of defining 'irreducible codes' as an example of decipherable codes. It will be shown that they obey the Kraft inequality which is of general relevance as satisfied by any decipherable code.

### *4.3.1 Source Models*

A written text will be used as an example of source. This source generates for a reader a sequence of signs which belong to the Latin alphabet, for instance, with a few more auxiliary signs like the space used in order to separate the words, punctuation signs, numerals. The letters themselves can take different forms, for instance capitals or italics. All these signs belong to some finite set which we may refer to as the 'extended alphabet' and its elements as 'symbols'. Words rather than separate letters have been chosen from some repertoire, the lexicon of the language which is used. This repertoire is very large and the grammatical rules according to which words are assembled are complicated. As a result, a thorough probabilistic description of such a source is impossible, and only models resulting from drastic simplifications can be contemplated. Moreover, some properties of the source models must be assumed so as to make them simply tractable.

The main two properties assumed to this end are that the sources are *stationary* and *ergodic*. 'Stationary' means that the probabilities which describe the source operation do not vary with time. 'Ergodic' means that observing the source output during a long enough time interval suffices to estimate its statistical properties; this implies in particular that the initial state of the source has no influence on its present operation, or that this influence can be ignored. Ergodicity and stationarity are unrelated properties, although often associated.

Now, assuming a source model to be stationary and ergodic, an important parameter of it is its *memory*, informally defined here as the number of consecutive

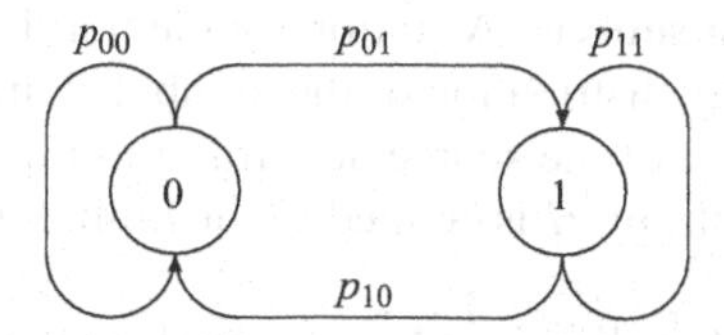


**Fig. 4.5** Two-state Markov chain

symbols which are mutually dependent. The simplest case, of course, is that of a *memoryless* source where each symbol is independent of others. Then this source model is entirely defined by the probability distribution of its symbol outcomes. For instance, the average frequency distribution of the letters in, say, English texts has been measured, so using this frequency distribution as the probability distribution of the source symbols provides a statistical model of a text written in this language. This very simple model is also very unrealistic since in almost all sources of any practical or theoretical interest the successively transmitted symbols are more or less mutually dependent, hence have memory according to our definition. A more realistic model, although still widely simplified with respect to most actual situations, is that of a *Markov chain*. Markov was a linguist who introduced the chains which bear his name for modelling languages (Roubine 1970). A Markov chain is some system which can assume $n$ distinct states. Changes of state regularly occur (e.g., periodically) at random between some of them. The source is described by the probabilities of the transition from each of its states to other ones, or transition probabilities. In a stationary Markov chain, the transition probabilities are constant. The specific property of a Markov chain is that the dependency with respect to the whole past is limited to that on the actual state, regardless of the succession of past states.

An $n$-state Markov chain is conveniently represented by a diagram where $n$ points correspond to the states. The transitions from one of the states, say $s_i$, to another one, say $s_j$, is represented by an arrow between the corresponding points, labelled with the corresponding transition probability, denoted by $p_{ij}$. Fig. 4.5 represents a very simple example of such a diagram, for a 2-state Markov chain,

In order to model a source, we assume that each transition in the chain entails the transmission of some definite symbol, to be identified in this very simple case with the state which is assumed after this transition occurred. Otherwise, whenever a transition occurs, some symbol similarly indicating the new state is transmitted. The chain reaches a steady-state operation if the probabilities of the states tend to constant values. If it occurs, and furthermore if these probabilities do not depend on the initial conditions, the chain is said to be *regular* or *completely ergodic*. Since they do not depend on time, the constant values eventually assumed by the probabilities of the states are referred to as 'stationary'. A sufficient condition for a Markov chain to be completely ergodic is that all states actually communicate in the sense that there exists a path without any zero-probability branch between any state and any other one, possibly through other intermediate states. In other words, all states should communicate.

**Defining the entropy of a source** A stationary memoryless source is completely described by the probability distribution of the symbols it transmits. The entropy of this distribution defines that of the source according to Eq. (4.10). For a stationary source with memory, its entropy $H$ is defined as the limit, for $k$ approaching infinity:

$$H \triangleq \lim_{k\to\infty} -\frac{1}{k}\sum_{s} p(s)\log[p(s)], \tag{4.31}$$

where $s$ denotes a block of length $k$, $p(s)$ its probability, the summation being effected for all blocks of length $k$ that the source can generate. In other words, it is the limit of the entropy of the $k$-th extension of the original source, divided by $k$. This definition is meaningful only if the source has a finite memory, i.e., if the dependency between two symbols $k$ symbols apart in the transmitted sequence tends to 0 as $k$ appraoches infinity.

In the case of an $n$-state Markov source, one easily shows that this entropy is equal to

$$H = -\sum_{i=1}^{n}\pi_i\sum_{j=1}^{n}p_{ij}\log(p_{ij}),$$

where $p_{ij}$ denotes the probability of the transition from state $i$ to state $j$ and $\pi_i$ denotes the stationary probability of state $i$ (which can itself be computed in terms of the transition probabilities).

## 4.3.2 Representation of a Code by a Tree, Kraft Inequality

We now consider how a set of words, i.e., a *code*, can be represented by a tree. We saw in Sect. 2.4.3, especially in the example of an identification number, that a sequence must often be separated into subsequences, referred to here as a *words*, having each its own semantic interpretation. We need also segmenting into words a sequence intended to represent the output of a source, especially if it results from source coding, i.e., aimed at representing this output by a shorter sequence. An *a priori* necessary property of such a code is that a sequence made of its words be *decipherable*, meaning that it can unambiguously be separated into its constituent words regardless of their order in the sequence. We first assume that a codeword can moreover be *immediately* recognized as such, i.e., as soon as it is received. A code having this property will be referred to as *instantaneous* or *irreducible*. We restrict ourselves here to a code using the binary alphabet, but the extension to an alphabet of arbitrary size is straightforward.

Let us first consider a $k$-bit word. It may be interpreted as uniquely specifying one of the $2^k$ possible paths in a 'complete' binary tree of length $k$. In Fig. 4.6, the ascending branches conventionally represent the bit '0' and the descending ones the bit '1'. A word is represented by a path starting from the root of the tree.

We obtain an instantaneous code if we impose to its words the prefix condition, namely, that no one is allowed to be a prefix of another one. This condition implies

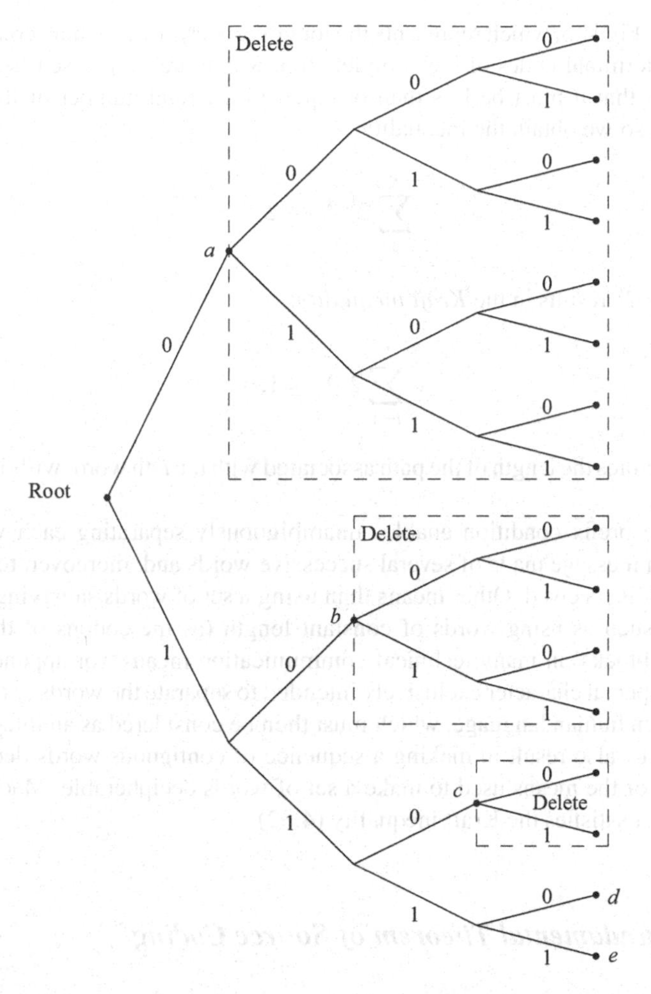


**Fig. 4.6** Complete binary tree of length $k = 4$ and, after deleting the content of the *dotted boxes*, tree representing an instantaneous code. Any word of length less than or equal to 4 is represented by a path in the tree starting from its root. When a word is chosen to belong to an instantaneous code (here a 5-word set), the prefix condition forbids using the branches of the complete tree beyond the extremity of its representative path so the partial trees contained in the dotted boxes of the figure must be deleted. *Black points* represent the root and the terminal extremities, or leaves, of the original complete tree on the one hand, and those of the incomplete tree obtained after deletion (labelled from $a$ to $e$), on the other hand. The number of leaves of the complete tree which are deleted to satisfy the prefix condition cannot exceed their total number, $2^k = 16$, which results in the Kraft inequality (here an equality)

that once a word is chosen as belonging to the code, the path which represents it in the tree is not further prolonged. The branches beyond this node have thus to be deleted so it becomes a terminal node of the tree which represents the code. Then, if the length of this path is $k_i$, $2^{k-k_i}$ terminal nodes of the complete tree are suppressed. For a $K$-word code, with $K < 2^k$ ($k$ now denoting the length of the longest path in the tree (see Fig. 4.6, which represents the longest codeword), we may count the total number of terminal nodes of the complete tree which are suppressed (see Fig. 4.6) and express that it must be less than or equal to the total number of these nodes, namely, $2^k$, so we obtain the inequality

$$\sum_{i=1}^{K} 2^{k-k_i} \leq 2^k.$$

Dividing by $2^k$ results in the *Kraft inequality*:

$$\sum_{i=1}^{K} 2^{-k_i} \leq 1, \tag{4.32}$$

where $k_i$ denotes the length of the path associated with the $i$-th word, with $1 \leq k_i \leq k$, $1 \leq i < K$.

Then the prefix condition enables unambiguously separating each word in an information message made of several successive words and, moreover, to identify it as soon as it is received. Other means than using a set of words satisfying the prefix condition, such as using words of constant length (as the codons of the 'genetic code' and 'blocks' in many technical communication means), or appending to the alphabet a special character exclusively intended to separate the words (as the 'space' in the written human language, which must then be considered as an integral part of the alphabet), also result in making a sequence of contiguous words decipherable. Regardless of the means used to make a set of words decipherable, MacMillan has shown that it satisfies the Kraft inequality (4.32).

### *4.3.3 Fundamental Theorem of Source Coding*

Let $H$ denote the entropy per symbol of a stationary, ergodic and finite-memory source, expressed in shannons, i.e., using logarithms to the base 2. The output of this source can be encoded into a decipherable binary sequence having an average length of at least $H$ bits per symbol of the given source, from which the original source output can be exactly recovered. No shorter encoded sequences enable an exact recovery.

By decipherable sequence, it is meant as above a sequence which can be segmented into words, such that each of them can unambiguously be separated from the others regardless of their order in the sequence. We assume here that the encoded sequence is binary, but the extension of the statement of the theorem to an arbitrary alphabet is straightforward. Then the logarithms used in order to express the entropy $H$ of the source should have as base the alphabet size of the encoded sequence.

For a sketchy proof of this theorem we first consider a memoryless source. Let $\alpha$ denote the size of its alphabet, assuming for the moment $\alpha > 2$, and $\overline{n}$ denote the shortest average length per source symbol of a decipherable encoded binary sequence which encodes it, expressed in bits, namely:

$$\overline{n} \triangleq \sum_{i=1}^{\alpha} p_i n_i, \tag{4.33}$$

where $p_i$ is the probability of occurrence of the $i$-th source symbol and $n_i$ the length of the word which represents it, in bits. Let us define $q_i \triangleq 2^{-n_i}, i = 1, 2, \ldots, \alpha$. Being assumed decipherable, the code should satisfy the Kraft inequality (4.32), namely, $\sum_{i=1}^{\alpha} 2^{-n_i} \leq 1$. We choose the word lengths so that the equality obtains, i.e.,

$$\sum_{i=1}^{\alpha} 2^{-n_i} = 1$$

and this choice actually leads to the smallest possible lower bound on $\overline{n}$. Then $\{q_i\}$ can be dealt with as a probability distribution. We may thus apply the Gibbs inequality (4.15) to $\{p_i\}$ and $\{q_i\}$, which results in

$$\sum_{i=1}^{\alpha} p_i \log_2 \left[\frac{2^{-n_i}}{p_i}\right] \leq 0.$$

Provisionally assume that the equality holds. This would mean that $p_i = 2^{-n_i}$ or, equivalently, $n_i = -\log_2(p_i)$, for any $i$. There is no reason why the probabilities $\{p_i\}$ should assume the form of 2 raised to some negative integer power, since they are parameters of the given source. However, it is always possible to find $\alpha$ integers $\{n_i\}$ such that

$$-\log_2(p_i) \leq n_i < -\log_2(p_i) + 1.$$

Multiplying by $p_i$ and summing with respect to $i$ results in:

$$-\sum_{i=1}^{\alpha} p_i \log_2(p_i) \leq \sum_{i=1}^{\alpha} p_i n_i < -\sum_{i=1}^{\alpha} p_i \log_2(p_i) + \sum_{i=1}^{\alpha} p_i.$$

Taking into account the definitions Eq. (4.10) of the entropy and Eq. (4.33) of the average codeword length, and since $\sum_{i=1}^{\alpha} p_i = 1$ because $\{p_i\}$ is a probability distribution, the following double inequality results:

$$H \leq \overline{n} < H + 1.$$

If the original source is replaced by its $k$-th extension, as already defined in Sec. (4.2.5), its entropy per symbol becomes $kH$ and the average length per symbol of the original source becomes $k\overline{n}$. The double inequality above still holds for the $k$-th extension, namely:

$$kH \leq k\overline{n} < kH + 1.$$

Dividing by $k$ shows that $\overline{n}$ can be made arbitrarily close to its lower bound $H$ by increasing the extension order $k$. The use of a large extension order $k$ entails moreover

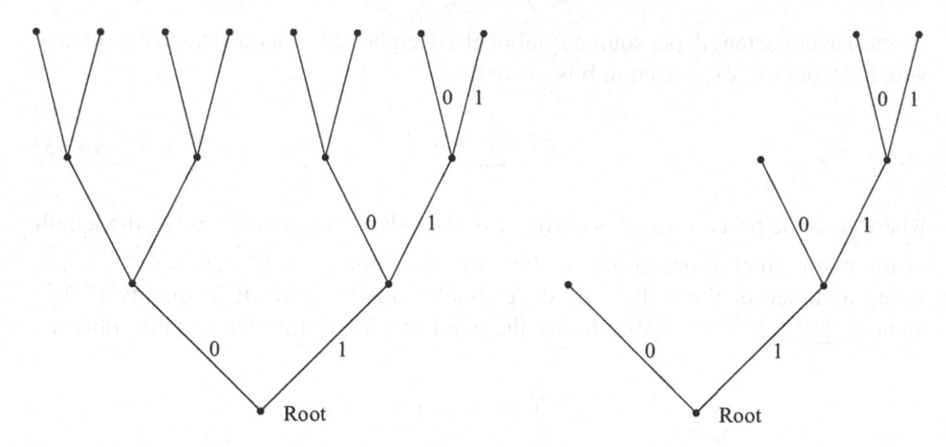


**Fig. 4.7** Trees representing sets of binary sequences. On the left, tree associated with the set of all binary sequences of length 3. On the right, tree depicting a 4-word code obeying the prefix condition, extracted from this set. The dots represent the nodes

that there is no need to assume that the initially given source is memoryless. It may be of finite memory since the mutual dependence of the symbols of the initial source is then taken into account within each of its $k$-symbol blocks.

### 4.3.4 Source Coding by the Huffman Algorithm

Let us first consider the problem of representing the successive symbols generated by a stationary memoryless $\alpha$-ary source $S_\alpha$ by a binary sequence as short as possible, in a fully reversible way. We assume that $\alpha$ is larger than 2; for instance $\alpha = 4$ will provide a simple example. Let $a, b, c, d$ denote the symbols of $S_4$ and $p_a, p_b, p_c, p_d$ their respective probabilities. Of course, $p_a + p_b + p_c + p_d = 1$. We assume that none of these probabilities is 0 (then 4 is the actual size of the source alphabet). We intend to represent each of the source symbols by a binary word according to a one-to-one correspondence, and we further demand that the codewords which represent the source symbols satisfy the *prefix condition* already defined in Sect. 4.3.2 above.

The set of all binary sequences of length $k = 3$ has been represented as a binary tree in the left part of Fig. 4.7, similar to the complete tree of Fig. 4.6 but drawn for $k = 3$. Any branch bears a binary label according to some arbitrary convention, here 0 for branches leaning to the left and 1 for those leaning to the right. Then the paths in the tree graphically represent all possible binary sequences of length $k = 3$. The tree representing a 4-word code obeying the prefix condition, drawn on the right of the same figure, is extracted from the tree at left by the same deletion process as in Fig. 4.6. The source symbols are associated with each of the terminal nodes of this tree. The word intended to represent one of these symbols is the sequence of binary labels read along the path from the root to the corresponding terminal node. The words of the code in the figure are thus 0, 10, 110 and 111.

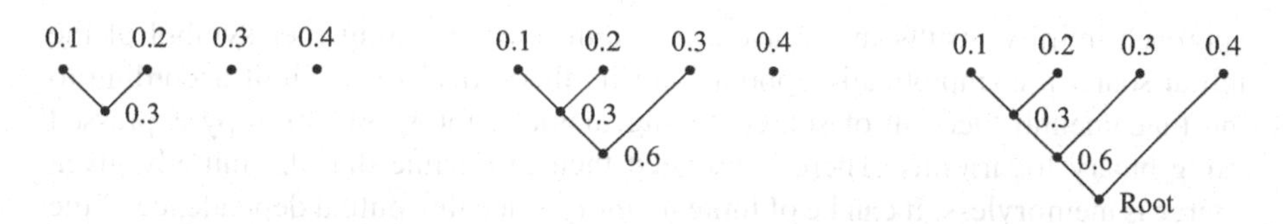


**Fig. 4.8** Successive steps in the encoding by the Huffman algorithm of a quaternary memoryless source having 0.1, 0.2, 0.3 and 0.4 as symbol probabilities. The nodes are labelled with their probabilities

Let $n_a, n_b, n_c$ and $n_d$ be the length of the words associated with the symbols $a, b, c$ and $d$, respectively. The average number of bits, or length, of the message which results from this encoding is thus

$$\overline{n} = p_a n_a + p_b n_b + p_c n_c + p_d n_d$$

and the code should be designed in order to minimize it. The Huffman algorithm is an optimal means for designing the corresponding tree (Huffman 1952). It consists of building node by node the binary tree which represents the code. The terminal nodes, each being associated with one of the source symbols and affected with its probability, are initially given. It is assumed that they are ranged in non-decreasing order of their probabilities, i.e., such that $0 < p_a \leq p_b \leq p_c \leq p_d < 1$. The algorithm builds the tree by creating new nodes as follows: it starts from the two terminal nodes having the smallest probability (the leftmost two in the figure) and connects them to a new node having as probability the sum of their probabilities. Then, a node connected to the two nodes of lowest probability among the remaining ones is further introduced. This process is continued until a node is given the probability 1: it is the root of the tree. The codeword associated with a source symbol is then represented by the path from the root to the terminal node which corresponds to this symbol.

Fig. 4.8 shows the successive steps of the Huffman encoding, assuming that the source symbols have as respective probabilities $p_a = 0.1$, $p_b = 0.2$, $p_c = 0.3$ and $p_d = 0.4$. The tree at right, obtained when the process is completed, is equivalent to the tree at right in Fig. 4.7. The resulting average length is $\overline{n} = 0.1 \times 3 + 0.2 \times 3 + 0.3 \times 2 + 0.4 \times 1 = 1.9$ bits, less than 2 bits as in the absence of source coding, but larger than the entropy 1.8464 ... Sh.

The tree constructed by means of the Huffman algorithm is optimal since when a new node is introduced, hence when 1 is added to the length of two paths at each step of the algorithm, it is the length of the most improbable two paths which is incremented. However, this tree is optimal only for a given alphabet size $\alpha$. The given source can be described as well by its higher order extensions, as already defined in Sect. 3.2.1 and 4.2.5, and the performance can be improved (the average length after encoding can be diminished) by replacing the original $\alpha$-ary source with a higher extension $k$ of it, of alphabet size $\alpha^k$. Exponentially increasing the number of symbols of the alphabet, the use of a higher extension makes the number $q_i = 2^{-n_i}$, where $n_i$ denotes the length of the word associated with the $i$-th alphabet symbol, better and better approximate its actual probability $p_i$. Letting the extension order approach infinity results in the average length after encoding (per symbol of the initial source) asymptotically approaching its theoretical lower limit according to the fundamental theorem of source coding, namely, the source entropy expressed using binary logarithms. There is no need then to assume that the initially given source is memoryless. It can be of finite memory since the mutual dependence of the symbols is then taken into account within each block of the initial source symbols which defines a symbol of the extension, of increasing size. Our initial assumption that the original source should have an alphabet of size $\alpha$ larger than 2 does not prevent encoding a binary source by the Huffman algorithm, which then has just to operate on an extension of this source.

Source coding ideally results in the information message which, according to the fundamental theorem, cannot have a length less than the source entropy. However, the source coding algorithms assume that the probabilities associated with the source symbols are perfectly known. An exact statistical description of the source is thus assumed to be known. Let us illustrate by an example the difficulty it may entail. It may seem that the frequency of the letters of a language, at the scale of a book (which contains some 50,000 letters), varies little from a book to another, and this is most often true. It is well known that the most frequent letter, in English as well as in French, is 'e'. There exists however a book written in English where this letter is *never* employed: "Gadsby", a novel by E.V. Wright published in 1939. The same literary *tour de force* was achieved in French by Georges Perec with his novel "La Disparition" (1969). A source coding system based on the average frequency of the letters in English or in French, operating on these books, would obviously give results much worse than expected.

This problem may be solved by the use of algorithms referred to as 'universal', which do not demand the preliminary knowledge of a statistical description of the source, or of 'adaptive' algorithms which evolve in order to optimally fit the actual source characteristics and are thus able to follow its possible variations. As an example of adaptive source coding algorithm, Gallager noticed that the operation of the Huffman algorithm results in sibling nodes (i.e., the extremities of branches originating in a same node) being neighbours in the list of nodes ranged by non-decreasing probability order. He used this property to design an adaptive version of the Huffman algorithm, where the tree is modified so as to eventually restore the sibling node property (using frequencies measured by the very operation of the algorithm instead of probabilities). Then it reaches optimality if the initial symbol probabilities are ill-known or even unknown, or restores optimality if the source symbol probabilities vary (Gallager 1978).

The recourse to a high-order extension which is generally needed in order to obtain good performance with the Huffman algorithm entails an exponentially increasing complexity. Other source coding means, especially arithmetic coding, approach the theoretical limit by less complex means (Rissanen 1976; Rissanen and Langdon 1979; Guazzo 1980). An adaptive version of the Guazzo algorithm has been proposed by us (Battail 1990). The Lempel-Ziv algorithm is still another asymptotically optimal algorithm which consists of determining the most frequent groups of symbols and use them as symbols of a new alphabet, instead of the original one (Ziv and Lempel 1978).

Describing the Huffman algorithm has shown that it is necessary to segment the encoded sequence into words, each intended to designate a symbol of the source (often of an extension of the original source). The necessity of distinguishing the words implies that the code must be decipherable. That a word refers to a source symbol, i.e., to something foreign to the code, may be interpreted as its *meaning*, just like the identification number given as an example in Sect. 2.4.3, a sequence of decimal digits which has to be segmented into several words telling a number of features which, taken together, uniquely characterize a person. The correspondence between the words and these features is semantic in the usual meaning of the word. In the case of source coding by the Huffman algorithm, however, the meaning of a word is internal to the encoding process of a sequence. We meet here a kind of 'internal semantics' at the heart of information theory, which may be helpful to better understand the relationship of information theory with semantics in its usual meaning. We more closely examine this question in the next section.

### 4.3.5 Some Comments About Source Coding

**On the relationship of information and semantics** The Huffman source coding algorithm described in Sect. 4.3.4 establishes a one-to-one correspondence between the symbols of a non-binary source (often an extension of an original binary source, as defined in Sect. 3.2.1) and the words of a binary code. How it works can be described as the construction of a tree as that of Fig. 4.7 where the terminal nodes correspond to the source symbols, any branch represents a bit and the successive branches of a path from the root of the tree to a terminal node represent a codeword. Let us further comment it.

We already noticed that the symbols transmitted by the source succeed to each other in any order, so the words which represent them should be unambiguously distinguished regardless of this order. The code should thus be *decipherable*, and the Huffman algorithm fulfills this requirement by satisfying the prefix condition which makes the code irreducible. We also noticed that the source symbol to which a codeword refers may be interpreted as its *meaning*. The Huffman algorithm may thus be thought of as defining *semantic rules* internal to the coding process. Thus, although semantics has been *a priori* excluded, it nevertheless appears within a classical topic of information theory. This remark will be helpful to better understand the relationship of information theory with semantics in its usual meaning.

Source symbols are objects of communication engineering, but similarly associating binary words with outer objects, instead of source symbols, suffices to define a semantic rule in its ordinary meaning. Besides objects of the physical world, words can be associated with relations between such objects, hence to abstract entities as well. The choice of the binary alphabet has been only a matter of convenience and binary words can be replaced by finite symbolic sequences using an arbitrary alphabet. Thus any conceivable object, physical, abstract or mental, can actually be represented by a finite symbolic sequence.

The requirement that a set of words be decipherable also holds for any communication system, although the means to achieve it are often different from an irreducible code, such as using words of constant length (as the codons of the 'genetic code' and 'blocks' in many technical communication means), or appending to the alphabet a special character exclusively intended to separate the words, as the 'space' in the written human language.

Besides being endowed with a meaning by semantic rules, any word can itself be split into smaller symbolic sequences specifying each some property. For instance, the individual bits of a word may be interpreted as answers to dichotomic questions which eventually enable uniquely identifying an object within a hierarchical taxonomic system. Similarly, the number used for identifying a person, as discussed in Sect. 2.4.3, can be split into subsequences such that the properties they indicate, taken together, correspond to a unique identity. Doing so can be interpreted as endowing the words with a morphology. Phonetics also constraints the sequence of letters used as a word since it is intended to be pronounced.

This remark moreover reveals a quantitative relationship between information and semantics, since the length in bits of an information message indicates how many independent dichotomic semantic distinctions, at most, the information it represents can provide. Information theory uses it for measuring the information quantity it bears. The quantitative measure of information is thus relevant to semantics, in the above meaning, although semantics is basically qualitative. We may thus think of an information as a *container* for semantics. Each of the bits of an information message can be endowed with a meaning, that of answering a dichotomic question. Before a correspondence is established between the bits and the questions, the information does not contain any semantics but it is able to receive some. Information is like an empty shell and semantics a hermit crab. Then the quantity of information, i.e., the shell volume, limits the size (but only the size) of the crab it can host. Moreover, this metaphor matches the protecting ability of the shell since an information cannot 'survive' without error-correction coding.

Besides establishing a correspondence between words and objects (i.e., naming objects), linguistic systems use the location of a word in a sequence made of successive words so as to express the relationship of the object that this word represents to the objects associated with the other words, thus resulting in a *syntax*.

Notice that both morphology and syntax necessarily imply that the symbols of a message are *ordered*, meaning that any two symbols being given, it can be stated unambiguously that one of them precedes the other one. Let us more closely consider the case where a message is associated with a 'recipe', understood as the sequence of instructions needed for constructing some object. For instance, the genome of a living being contains a sequence of instructions which in a proper environment eventually results in the assembly of a phenotype. Then a bit or a block of successive bits of the corresponding information message instructs a step of the construction process, hence its semantic content entirely depends on its *location* in the message.

The steps of the construction process must be executed successively, so their order in the message must match their order in time. The time axis is a unidimensional space, and so should be the medium which bears the message. A recipe, thus, cannot be borne by a medium having more than a single dimension. An object fabricated according to a recipe may have, and generally has, more than a single dimension and then cannot itself act as a recipe. Insofar as it has been entirely specified by the recipe, however, it bears at least the same amount of information as the recipe itself. The information borne by a multidimensional fabricated object will be referred to as *structural*, or Aristotelian, as opposed to the *symbolic* information borne by the sequence associated with its recipe.

Not any sequence of events can act as a recipe. For instance, Shannon's information theory associates an information quantity to any event in terms of its probability of occurrence. To have any impact on the physical world, an event or sequence of events must have a lasting action, e.g., be recorded as a sequence of discrete symbols, thus becoming symbolic information in the above meaning. If it does not, as often occurs when it is associated with an event at the microscopic level, we say that it bears only potential information.

The above remarks enable us to distinguish *potential*, *symbolic* and *structural* information (Battail 2009), a distinction we believe of crucial importance. The potential information does not interact with anything which can be observed. Only the symbolic information can be copied or used as a recipe, since the structural information lost unidimensionality which enables communication. The following figure depicts the relationship of potential, symbolic and structural information. The transformation of potential into symbolic information involves recording, i.e., inscription of symbol(s) on a macroscopic physical support; the transformation of symbolic into structural information needs semantics.

Both the inscription of symbols and the use of a symbolic sequence as a recipe for assembling a multidimensional object are irreversible. We may thus interpret this statement as extending the *central dogma* of molecular biology which tells that genes determine the composition of proteins, not the other way round. Instead of molecules, however, it concerns information as an abstract entity.

Let us elaborate on the properties of symbolic information and on its place in the extended central dogma just stated. In order to make it uniquely defined and indefinitely reproducible, it must be discrete and finite, i.e., its representative must be finite sets of symbols from a finite alphabet. This excludes any analog entity. Being a unidimensional set of symbols, i.e., being orderly, enables it to bear the largest possible information quantity, i.e., $k$ shannons in the binary case if the sequence length is $k$ (or $k \log_2(q)$ shannons for an alphabet of size $q$). Failing to use orderly sequences would greatly reduce this quantity, to $\log_2(k+1)$ in the binary case and, in the $q$-ary case, to a quantity much smaller than $k \log_2(q)$ which has a less simple expression (Fig. 4.9).

Any bit in the shortest binary representative of a symbolic information (referred to in Sect. 2.2 as its information message) has a well defined location relatively to the others. Such a 'relation of order' is needed for assigning meanings by means of the tree which represents the information message, and moreover for expressing the syntactic

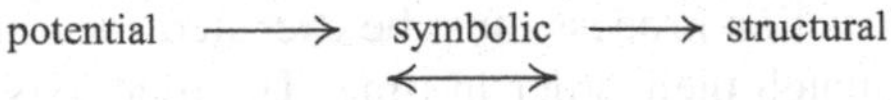

potential ⟶ symbolic ⟶ structural
⟷

**Fig. 4.9** Irreversibility of information transfer (extended central dogma). Only symbolic information can be copied, as indicated by the double arrow

rules which are themselves needed for any description or specification. Topology tells that only the elements of a unidimensional space are endowed with a relation of order. Symbolic information is borne by a unidimensional sequence and thus able to act as a recipe for constructing a multidimensional object. Being multidimensional, however, this object can no longer be used itself as a recipe[2]. This is the deep mathematical reason why the transition from symbolic to structural information is irreversible. We may think of it as *a posteriori* justifying that information theory is basically the science of sequences.

This situation is not foreign to biology since the genetic message borne by a gene instructs the assembly of a polypeptidic chain eventually becoming a protein, but a protein cannot itself be copied. Crick and Watson referred to this feature as the *central dogma* of molecular genetics. As a 3-dimensional object[3], a protein can be specified by a unidimensional message, i.e., a genomic sequence, not by an object having like itself more than one dimension. Yockey recognized that the central dogma is a constraint of mathematical character due to the genetic mapping being many-to-one, thus also valid outside molecular genetics for any mapping which takes a similar form (Yockey 1992).

This entails that a protein cannot be directly copied, and the same is true for a whole phenotype. Its symbolic description by a genomic sequence is needed to this end. Before the DNA replication mechanism was discovered, von Neumann showed that the existence of its symbolic description as a part of an object is a logical requirement for its self-reproduction (von Neumann 1966), and this result was applied by Howard Pattee to biology (Pattee 2005). A phenotype too has more than a single dimension; it may be thought of as 4-dimensional if we include time as a relevant dimension besides the spatial ones in order to account for its development. Then, the irreversibility of time is another reason why copying the structural information of a phenotype is impossible. This impossibility extends the reach of the central dogma far beyond molecular biology, confirming Yockey's statement.

**Some consequences of the fundamental theorem of source coding** Given a source of entropy per symbol $H$, the fundamental theorem of source coding entails that an $n$-bit sequence it generates cannot in the average represent an information message shorter than $k = nH$. A meaningful message thus cannot be arbitrarily short. This has been very cleverly illustrated by Molière in his comedy *Le Bourgeois Gentilhomme* (1670). Monsieur Jourdain, a bourgeois, is infatuated with nobility. Cléonte, a young man of common rank, is in love with his daughter. In order to be accepted as his

---

[2] Analog means for copying multidimensional objects exist, but they are only approximative so they are not reliable when they are repeatedly used.

[3] It may even be thought of as a 4-dimensional object since folding the assembled polypeptidic chain into a 3-dimensional molecule involves time.

son-in-law, he asserts he is the son of the Turkish sultan and speaks a fake Turkish language that his servant Covielle is assumed to translate. Here is a short excerpt of this play:

> Cléonte (in Turkish attire)
> *Bel-men.*
> Covielle
> He says that you should quickly go with him to prepare yourself for the ceremony, in order to later see your daughter, and to conclude the marriage. *Il dit que vous alliez vite avec lui vous préparer pour la cérémonie, afin de voir ensuite votre fille, et de conclure le mariage.*
> Monsieur Jourdain
> So much in two words? *Tant de choses en deux mots?*
> Covielle
> Yes. So is the Turkish language, it says much in few words. Go quickly where he wishes. *Oui. La langue turque est comme cela, elle dit beaucoup en peu de paroles. Allez vite où il souhaite.*

The audience intuitively perceives that Covielle's long sentence recounts too many circumstances to actually translate *Bel-men*, and laughs. (Were languages not redundant and would a meaning be associated with each possible combinations of letters, however, the set of six-letter words could designate no less than $26^6 = 308,915,776$ distinct instances.)

This example moreover shows that it is possible to represent a long message by a much shorter one provided it is explicitly agreed upon: for Monsieur Jourdain, the meaning of *Bel-men* is actually what Covielle told him! However, 'Bel-men' does not belong to any linguistic system and thus refers only to Covielle's sentence. A similar function is that of a title: when I say 'The Bible', I use a short message in order to evoke a long sequence. The title is not substituted for the work but designates it. It may be used as an address for retrieving the book in a catalog or in a library. Performing this function implies that it is a nominable entity, as discussed in Sect. 2.4.

As another example, the denizens of the French town of Marseilles are said to like jokes and to be somewhat lazy. Mr Durand travels in a train compartment with three of them and expects to hear pleasant jokes. One of the guys says 13, another says 8, the third says 24, and all three laugh after a number has been uttered. Mr Durand asks why they laugh. They reply that they all know the same set of jokes, so it costs less energy to endow each of them with a number and to pronounce this number than to actually tell the joke.

That it is possible to represent a long message by a much shorter one, its title or its address, does not actually contradict the fundamental theorem of source coding which tells that there is a lower bound to the length of the information message associated with any sequence generated by a source of non-zero entropy. Let us examine this case more closely. As contemplated by the algorithmic information theory to be considered in Sect. 6.1, the use of a universal computer enables generating *any* sequence. Designating a sequence by a title or an address, on the contrary, implies that some finite set of sequences, a corpus, has been defined prior to performing the encoding process. Addressing within a given corpus thus refers to a closed, preexisting set. The smaller the number of elements in the corpus, the shorter can be their titles or addresses, regardless of their intrinsic length. Source coding according to information theory refers to an open, potentially infinite, world of sequences. It does not prevent designating by its address any sequence available somewhere, instead of explicitly including the intended output sequence within the computer input as in Example d) of Sect. 6.1. Source coding in the information-theoretic meaning is thus far more general.

In human culture, references to a corpus actually play a major role, but languages give access to an open set of sequences. Whether references to some corpus are used in biological phenomena remains to be investigated.

**A source as a redundant encoder: the Shannon-McMillan theorem** We gave in Sect. 3.4.2 above an example of channel coding, which has been useful for understanding the concept of redundancy. A binary information message $\underline{u}$ of a well defined length $k$ was encoded into a sequence of length $n > k$. As belonging to a minority subset of all possible sequences of length $n$, this sequence could be reliably distinguished from others, thus becoming to a certain extent resilient to symbol errors and thus better fitted to be transmitted over the channel. It was then possible to define an information as an equivalence class among sequences, with respect to the possible encodings of some given information message. The information message is in itself a nominable entity, hence does not suffer any change, while its encoding depends on arbitrary choices and results in more or less efficient error correction.

The converse problem, namely *source coding*, consists, given some redundant sequence, of determining the information message of the equivalence class to which it belongs. The problem in its whole generality, referred to as 'universal source coding', has no explicit solution. However, when a statistical description of the source which generates the given sequence is available, i.e., when adequate source models are available, known algorithms like Huffman's optimally perform this task. Source coding ideally results in cancelling any redundancy, delivering messages made of independent equally probable bits, thus bearing each an information quantity of one shannon. Perfect source coding operating on the output of the channel encoder of Fig. 3.4 would thus result in the recovery of the sequence $\underline{u}$, the $k$-bit information message which was purposely encoded.

But what about sequences of 'natural' origin, i.e., not purposely generated by an encoder but found in human culture or in biology? In the general communication process described according to Shannon's paradigm (see Sect. 4.1) a source generates sequences intended to some distant, hence distinct, destination, to be transmitted over some channel. It turns out that the situation met in the case of purposeful encoding as in Sect. 3.4.2, and characterized by the rarity of the sequences which may actually be transmitted (as compared with all possible sequences), is also relevant to a rather general family of sources, provided the word 'almost' is used to weaken some statements by indicating that they are only approximate; they moreover become exact in the limit, as the length of the sequences tends to infinity. This is the content of the Shannon-McMillan theorem, to be now briefly stated without proof.

Consider a stationary and ergodic source, moreover assumed of finite memory. For simplicity's sake, assume that its alphabet is binary. Then the *Shannon-McMillan theorem* states that *almost* all sequences of length $n$ that the source generates belong to a set of *almost* $2^{nH}$ elements (referred to as 'standard' or 'typical' sequences), where $H$ is the source entropy as defined in Sect. 4.2.3 (assumed here to be expressed in binary units, i.e., in shannons) (McMillan 1953). If the word 'source' in this statement is replaced by 'encoder' and the word 'almost' is deleted, it describes indeed the case of deliberate redundant encoding as expounded in Sect. 3.4.2.

Notice that the actual information message often remains *hidden* although a redundant sequence of its equivalence class is known and used for processing the information. Perfect source coding of the redundant sequence should be performed in order to uncover its actual information message, but it is generally not needed since the available redundant sequence of its equivalence class adequately represents the information. Furthermore, the necessary source coding algorithm may well not be known.

## References


Battail, G. (1990). Codage de source adaptatif par l'algorithme de Guazzo. *Annales Télécommunic, 45*(11–12), 677–693.

Battail, G. (1997). *Théorie de l'information*. Paris: Masson.

Battail, G. (2009). Living versus inanimate: the information border. *Biosemiotics, 2*(3), 321–341. doi:10.1007/s12304-009-9059-z.

Brillouin, L. (1956). *Science and information theory*. New York: Academic Press.

Cover, T. M., & Thomas, J. A. (1991). *Elements of information theory*. New York: Wiley.

Gallager, R. G. (1968). *Information theory and reliable communication*. New York: Wiley.

Gallager, R. G. (1978). Variations on a theme by Huffman. *IEEE Transactions On Information Theory, IT-24*(6), 668–674.

Guazzo, M. (1980). A general minimum-redundancy source-coding algorithm. *IEEE Transactions On Information Theory, IT-26*(1), 15–25.

Huffman, D. A. (1952). A method for the construction of minimum redundancy codes. *Proceeding IRE, 40,* 1098–1101.

Jaynes, E. T. (1957). Information theory and statistical mechanics I & II. *Physical Review, 107/108,* 620–630/171–190.

Johnson, R. W., & J.E. Shore, J. E. (1983). Comments on and correction to axiomatic derivation of the principle of maximum entropy and the principle of minimum cross-entropy. *IEEE Transactions On Information Theory, IT-29*(6), 942–943.

Khinchin, A. I. (1957). Mathematical foundations of information theory. Dover.

Kullback, S. (1959). *Information theory and statistics*. New York: Wiley.

McMillan, B. (1953). The basic theorems of information theory. *Annals Of Mathematics Statistics 24,* 196–219. (Reprinted in Slepian 1974, 57–80).

Moher, M. (1993). Decoding via cross-entropy minimization. *Proceeding GLOBECOM'93*. 809–813, Houston, U.S.A.

Neumann, J. von. (1966). *Theory of self-reproducing automata, edited and completed by A.W. Burks*. Urbana and London: University of Illinois Press.

Pattee, H. (2005). The physics and metaphysics of biosemiotics. *Journal of biosemiotics, 1*(1) 281–301. (Reprinted in Favareau 2010, pp. 524–540).

Rissanen, J. J. (1976). Generalized Kraft inequality and arithmetic coding. *IBM Journal of Research & Development, 20*(3) 198–203.

Rissanen, J. J., & Langdon, G. G. Jr. (1979). Arithmetic coding. *IBM Journal of Research & Development 23*(2) 149–162.

Roubine, E. (1970). *Introduction à la théorie de la communication, tome III: théorie de l'information*. Paris: Masson.

Schrödinger, E. (1943). In *What is life?* and *mind and matter*. London: Cambridge University Press (1967).

Shannon, C. E. (1948). A mathematical theory of communication. *The Bell System Technical Journal, 27,* 379–457, 623–656. (Reprinted in Shannon and Weaver 1949, Sloane and Wyner 1993, pp. 5–83 and in Slepian 1974, pp. 5–29).

Shannon, C. E., & Weaver, W. (1949). *The mathematical theory of communication*. Urbana: University of Illinois Press.

Shore, J. E., & Johnson, R. W. (1980). Axiomatic derivation of the principle of maximum entropy and the principle of minimum cross-entropy. *IEEE Transactions. on Information Theory, IT-26*(1), 26–37.

Shore, J. E., & Johnson, R. W. (1981). Properties of cross-entropy minimization. *IEEE Transactions on Information Theory, IT-27*(4), 472–482.

Slepian, D. (Ed.). (1974). *Key papers in the development of information theory*. Piscataway: IEEE Press.

Sloane, N. J. A., & A.D. Wyner, A. D. (Eds.). (1993). *Claude Elwood Shannon, collected papers*. Piscataway: IEEE Press.

Yockey, H. P. (1992). *Information theory and molecular biology*. Cambridge: Cambridge University Press.

Ziv, J., Lempel, J. (1978). Compression of individual sequences via variable-rate coding. *IEEE Transactions on Information Theory, IT-24*(5), 530–536.

# Chapter 5
# Channel Capacity and Channel Coding

**Abstract** Chapter 5 continues the discussion of Shannon's information theory as regards channel capacity and channel coding. Simple channel models are introduced and their capacity is computed. It is shown that channel coding needs redundancy and the fundamental theorem of channel coding is stated. Its proof relies on Shannon's random coding, the principle of which is stated and illustrated. A geometrical picture of a code as a sparse set of points within the high-dimensional Hamming space which represents sequences is proposed. The practical implementation of channel coding uses *error-correcting codes*, which are briefly defined and illustrated by describing some code families: recursive convolutional codes, turbocodes and low-density parity-check codes. The last two families can be interpreted as approximately implementing random coding by deterministic means. Contrary to true random coding, their decoding is of moderate complexity and both achieve performance close to the theoretical limit. How their decoding is implemented is briefly described. The first and more important step of decoding enables regenerating an encoded sequence. Finally, it is stated that the constraints which endow error-correcting codes with resilience to errors can be of any kind (e.g., physical-chemical or linguistic), and not necessarily mathematical as in communication engineering.

The proof of the fundamental theorem of source coding given in Sect. 4.3.3 involves only a few lines of computation, moreover using easy mathematics. In sharp contrast, the proof of the fundamental theorem of channel coding demanded a very innovative method, namely Shannon's random coding. No proof of the theorem using simpler means has yet been found. The aim of source coding, although important, is not as vital as that of channel coding which concerns the very integrity of communicated messages, hence that of information. This chapter is thus capital and its topics will reveal of special importance in the second part of this book devoted to biological applications. We try in what follows to expound the main features of channel coding as simply as possible but without betraying it. We deliberately chose insight rather than formal rigour.

As the sources considered in the preceding chapter were mathematical models of message generators, the channels to be considered here are mathematical models of transmission means. They represent, in the form of a 'black box', i.e., of a description reduced to the relation between an input and an output, the set of transmitting devices

G. Battail, *Information and Life*, DOI 10.1007/978-94-007-7040-9_5,

and transmission channels or propagation media. Most physical channels are very complex and the models used in the theory are widely simplified (this is true also for sources). Even an assumption as necessary to the mathematical treatment as stationarity has often little rational or practical justification.

Since we limited ourselves as yet to finite-alphabet sources, it is logical to consider here channels with a finite input alphabet. It should be however underlined that the perturbing noise is often basically continuous (at least at the macroscopic scale), so restricting the channel *output* to a discrete alphabet in order to alleviate its further processing implies an *information loss* which is paradoxical in techniques intended to communicate information. Actually, studying channels with a continuous output is of real interest in communication engineering. It is shown indeed in the sequel that the most successful decoding algorithms deal with 'analog', as opposed to discrete, channel outputs.

## 5.1 Channel Models

As stated in Sect. 4.1, the effects of the perturbations are included in the channel models. Their presence results in the output channel depending only probabilistically on its input. In its simplest form, a channel can be described by a *transition diagram*. It is an oriented graph which represents how the channel transforms the symbols of its input alphabet into symbols of the output alphabet. Points at left represent all the input symbols, points at right all the output symbols, and arrows from left to right represent the possible transitions from input to output symbols that the channel operates. The arrows are labelled with the probability of the corresponding transitions. If all the transition probabilities do not vary with time the channel is said to be stationary. Figure 5.1 shows the transition diagrams of two stationary channels having both binary input symbols, as very simple and useful examples.

The input and output of the channel at left (Fig. 5.1a) are both binary. This channel is *symmetric*, meaning that the probability of the output '1' when the input is '0' equals the probability of the output '0' when the input is '1', their common value being denoted by $p_{su}$ (where the subscript 'su' stands for 'substitution', meaning that the wrong bit has been substituted for the correct one). This very simple model is frequently used and most of the error-correcting codes are designed for this channel. Being symmetric, it can be modelled as in Fig. 5.1b by the addition modulo 2 to the binary input variable $X$ of a random binary variable $Z$ equal to '1' with probability $p_{su}$, which may be referred to as an *error*.

The channel at right (Fig. 5.1c) has a binary symbol as input but its output alphabet is ternary, namely $\{0, 1, \epsilon\}$. At variance with the binary symmetric channel, no transition is allowed from the input '0' to the output '1' or from '1' to '0'. The only allowed transitions to an output symbol different from the input one lead to $\epsilon$, and their probabilities have been assumed to be the same, namely $p_{er}$. The subscript 'er' stands for *'erasure'*, the word used to refer to this case; it is a milder perturbation than the error in a binary symmetric channel since, if the output symbol is $\epsilon$, it is not

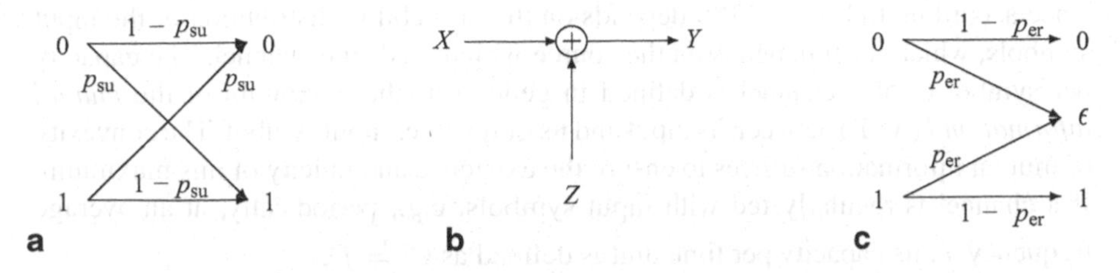


**Fig. 5.1** Models of channels with binary input alphabet. In **a**, transition diagram of a binary symmetric channel with error probability $p_{su}$. In **b**, alternative representation of the same channel by addition modulo 2 to the input $X$ of a random binary variable $Z$ assuming the value 1 with probability $p_{su}$. In **c**, transition diagram of a binary erasure channel, with erasure probability $p_{er}$

recognized as belonging to the binary alphabet and thus not taken into account, but the received symbol is not mistaken for its binary complement.

Another very important channel has a discrete input and a continuous output. If its input is binary, it models the case of a binary variable represented by antipodal signals (as defined in Sect. 3.1 above) perturbed by additive white Gaussian noise. The output of the channel is that of the matched filter used at the receiving end. It reads $\pm\sqrt{E}+\mathcal{N}$ where $\mathcal{N}$ is a realization of a real random Gaussian variable, while the sign $\pm$ represents the input binary variable (see Sect. 3.3).

It has been implicitly assumed as yet that an output symbol is determined by a single input symbol and thus does not depend on previous input symbols. Such a channel is referred to as *memoryless*. There are more complicated situations where it is not so. Models for such channels can be designed but their use is rather difficult. The usual means to practically use channels with memory consists of scrambling their input sequence by means of an interleaver, which makes contiguous bits in the scrambled sequence uncorrelated (although doing so suppresses a correlation which would be useful if it could be exploited).

Let us simply say that the usual requirements for making channel models theoretically tractable are that they should be causal (an output symbol cannot precede the corresponding input symbol) and should have finite memory, i.e., only a finite number of past input symbols besides the present one determine the channel output. We only consider in what follows stationary, causal, and memoryless channels.

## 5.2 Capacity of a Channel

### *5.2.1 Defining the Capacity of a Channel*

As defined by Eq. (4.14), the mutual information $I(X;Y)$ tells what information quantity knowing the output symbol $Y$ of a channel provides as regards its input symbol $X$ (or vice-versa since its expression has been shown to be symmetric with respect to them). However, it does not suffice to characterize the channel alone, since according to Eq. ( 4.25) it depends on the probability distribution of the input symbols, which itself depends on the source which feeds the channel. The capacity per symbol $C$ of a channel is defined in general as the maximum of the *mutual information* $I(X;Y)$ between its input and its output, per input symbol. The convexity of mutual information suffices to ensure the existence and unicity of this maximum. If a channel is regularly fed with input symbols, e.g., periodically, at an average frequency $f$, its capacity per time unit is defined as $C' \triangleq fC$.

In the absence of perturbations, the capacity per symbol of a memoryless channel is just $C = \log(\alpha)$ for an input alphabet of size $\alpha$. In the presence of noise, it is obviously upper bounded by the noiseless capacity $\log(\alpha)$, which is achieved only when the noise can be ignored. Else the capacity is less and depends on the transition probabilities.

The capacity per symbol of a memoryless channel has been defined as the maximum of the mean mutual information $I(X;Y)$, $X$ and $Y$ being the random variables associated with the input and output of the channel, respectively. For a finite-memory channel, the definition of $I(X;Y)$ should be extended in order to take into account the possible dependency of the channel output on successive input symbols. For the most general meaning of 'capacity', the maximum should be understood with respect to all sequences generated by *stationary and ergodic sources* having as alphabet that of the channel input. Then the channel capacity is said to be *ergodic*. If the channel is memoryless, the maximum should be understood with respect to all possible distributions of the input variable $X$. Properly extending the meaning of $X$ and $Y$, the channel capacity is defined in general as:

$$C = \max I(X;Y). \tag{5.1}$$

The capacity of a channel is thus the largest information quantity which it can transfer. Its maximum is defined with respect to the sole parameter which may be adjusted once the channel is given, i.e., the probability distribution of its input symbols in the memoryless case or, more generally, the choice of the source which is connected to its input because, according to Eq. ( 4.25), the mutual information $I(X;Y)$ depends on the channel, of course, but also on the source. Determining its maximum thus involves not only the computation of this mutual information, but of the source parameters which maximize it. The second step can be avoided in the important case of a symmetric channel, since it is easily shown that the mutual information then achieves it maximum for equally probable input symbols: $\Pr(X = x_i) = 1/\alpha$, where $x_i, i = 0, 1, \ldots, \alpha - 1$, is a symbol of the input alphabet of size $\alpha$. Then the source entropy $H(X)$ is maximized at the same time as $I(X;Y)$. A channel is said to be *symmetric* (1) if the set of transition probabilities from any of the input symbols does not depend on this input symbol and (2) if, moreover, the set of transition probabilities to any of the output symbols does not itself depend on this output symbol.

### 5.2.2 *Capacity of Simple Discrete Input Channels*

As regards the channels of Fig. 5.1, the binary symmetric channel obviously fulfills these conditions. The binary erasure channel does not fulfill condition (2) but one easily checks that the maximum of its mutual information is nevertheless achieved for equally probable input symbols. The mutual information of both channels is easily computed assuming the input bits to occur with probability 1/2. Using the second equality in (4.25), namely $I(X;Y) = H(Y) - H(Y|X)$, results for the binary symmetric channel in the capacity (expressed in shannons):

$$C_{\text{bsc}} = 1 - \mathcal{H}_2(p_{\text{su}}), \tag{5.2}$$

where the function $\mathcal{H}_2(\cdot)$ has been defined above by Eq. (4.11). One notices that $C_{\text{bsc}} = 0$ for $p_{\text{su}} = 1/2$, a case where the channel has actually become useless since its output is independent of its input. The capacity achieves its maximum value 1 for $p_{\text{su}} = 0$ and $p_{\text{su}} = 1$. If the first case is not surprising since then the channel is errorless, the second one is a bit more so, but it suffices to swap the labels '0' and '1' of the output symbols (which are in fact arbitrary) to revert to the case where $p_{\text{su}} = 0$. Taking account of the symmetry, no restriction of generality results from assuming that $p_{\text{su}} \leq 1/2$.

For the binary erasure channel, the computed capacity (again in shannons) is

$$C_{\text{bec}} = 1 - p_{\text{er}}, \tag{5.3}$$

a very simple expression which may be interpreted as follows: in the average, a fraction $p_{\text{er}}$ of the symbols are erased so the information they bear is lost. The remaining ones bring each an information quantity of one shannon, so the mean information quantity borne by an output symbol is $1 - p_{\text{er}}$ shannons per binary symbol.

As regards the channel with binary input and additive white Gaussian noise, its capacity $C_{\text{bg}}$ is well defined since the mutual information exists for a continuous output channel. This channel is symmetric so its capacity is achieved for the input variable assuming 0 or 1 with probability 1/2. It is a function of the signal-to-noise ratio $\rho \stackrel{\Delta}{=} 2E/N_0$, where $E$ is the energy of each received binary signal and $N_0/2$ the noise variance. The expression of this capacity is not simple so we do not write it out. Let us just say that it is a continuous increasing function of $\rho$. For very small values of $\rho$ the capacity $C_{\text{bg}}$ approximately equals $\rho/\ln(2)$ (as if no restriction were put on the input alphabet). When the signal-to-noise ratio $\rho$ approaches infinity, it tends to the horizontal asymptote $C_{\text{bg}} = 1$ shannon which obviously results from the restriction of the channel input to the binary alphabet.

### 5.2.3 *Capacity of the Additive White Gaussian Noise Channel*

The entropy defined in Sect. 4.2.2 above has been restricted to the case of a discrete random variable because it suffices to understand the entropy concept and because it is most often met in practical situations. Moreover, extending the entropy concept to a continuous random variable involves specific mathematical difficulties. It turns out that the differential entropy introduced in Sect. 4.2.3 lacks some of the useful properties of the discrete entropy. In the physical world, a continuous random variable is always known within some finite approximation so its $\varepsilon$-entropy, to be alluded to in Sect. 5.2.4 below, has the same desirable properties as the entropy of a discrete random variable.

We already noticed that, contrary to the entropy, the mutual information hence the channel capacity can be easily extended to continuous input and output, either separately or both. The channel having as input a continuous random function of time $f(t)$ of frequency range limited to the interval $(0, B)$ and of limited average power $S$, received in the presence of additive white Gaussian noise of spectral density $N_0$, is extremely important as expressing the largest information quantity which can be reliably communicated in the presence of thermal noise (as considered in Sect. 3.3) without any restriction to a finite alphabet since both the channel input and output are continuous. Shannon showed that the capacity of this channel is:

$$C = \frac{1}{2}\log_2(1 + 2E/N_0) = \frac{1}{2}\log_2(1+\rho) \tag{5.4}$$

shannons, where $E = 2BS$ denotes the average energy per sample and $\rho \triangleq 2E/N_0 = S/N_0B$ is the signal-to-noise ratio (Shannon 1948, 1949). This capacity is expressed per sample of $f(t)$ at the frequency $2B$. A sample is the value assumed by the function at a given instant, and sampling $f(t)$ at the frequency $2B$ means considering the set of samples $\{f(n/2B)\}$ for all integer values of $n$. The sampling theorem tells that, because the frequency range of the function $f(t)$ is limited to $(0, B)$, the successive samples thus obtained are independent so the discrete set of samples $\{f(n/2B)\}$ is fully equivalent to the continuous function $f(t)$. The capacity expressed by Eq. (5.4) is achieved when the input function has a Gaussian distribution. The curve which represents the capacity $C$ given by Eq. (5.4) is actually the envelope of those which represent the capacity in the case of a finite input alphabet. Regardless of the alphabet size, Eq. (5.4) thus expresses an *absolute limit* to the information quantity which can be reliably received when the available signal-to-noise ratio has some value $\rho$.

Considering the capacity $C'$ per time unit enables rewriting Eq. (5.4) as

$$C' = B\log_2(1 + S/N_0B) = B\log_2(1+\rho) \tag{5.5}$$

shannons. A very interesting geometrical interpretation of the formulas (5.4) or (5.5) has been given by Shannon in (Shannon 1949), and our development of Sect. 5.4.1 concerning the binary symmetric channel transposes Shannon's argument to this simpler case.

If the signal-to-noise ratio $\rho$ is very small, keeping only the first order term in the development of the logarithmic function with respect to $\rho$ results in the capacity being proportional to $\rho$, i.e., $C \approx \rho/2\ln(2)$ and $C' \approx B\rho/\ln(2)$. If on the contrary $\rho$ is very large, 1 can be ignored in (5.4), which results in $C \approx \ln(\rho)/2\ln(2)$ and $C' \approx B\ln(\rho)/\ln(2)$.

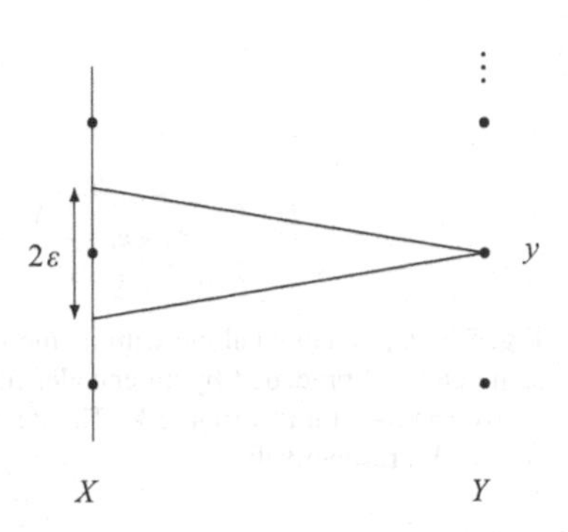


**Fig. 5.2** Channel representing the 'quantization' of an analog random variable $X$ into a discrete one $Y$. All the points of a $2\varepsilon$-long segment of the line where the values of $X$ are represented are mapped into a single point $y$ of $Y$

The equalities (5.4) and (5.5) are very important as establishing a relationship between the largest quantity of information which can be communicated over an additive white Gaussian noise channel and its signal-to-noise ratio $\rho$. In other words, they establish a link between the physical world of energy and the abstract world of information. We interpret in this spirit Boltzmann constant as a signal-to-noise ratio in Sect. 6.3.2 below.

### 5.2.4 Kolmogorov's $\varepsilon$-entropy

The intrinsic approximation of any physical measurement entails that the difference between two measured values is meaningless if these values are close enough. Shannon, in part V of his seminal paper (Shannon 1948), considered the case where a continuous quantity is approximately communicated while satisfying some 'fidelity criterion'. Kolmogorov defined the entropy of a continuous random variable from this point of view, and named it $\varepsilon$-entropy (Kolmogorov 1956). In any case, the $\varepsilon$-entropy has properties similar to those of the discrete entropy defined in Sect. 4.2.3. Although some fidelity criteria may lead to complicated expressions, we consider as an example the very simple case of what in engineering is referred to as 'quantization' of an analog quantity, which is widely used for approximately representing an analog quantity by a discrete one.

In order to define a finite entropy in this case, let us first consider the channel which has as input an analog value $X$ which results from a measurement. Let us divide the range of values assumed by $X$ into contiguous segments of length $2\varepsilon$. No value $x$ assumed by $X$ is at a distance from the centre $y$ of some segment larger than $\varepsilon$ which thus measures the largest possible absolute difference between the measurement result and its approximation which prevents to distinguish between them. We may thus represent $x$ by $y$. The set of segment centres constitutes a discrete random variable $Y$ and we may interpret the mapping of an element $x$ of $X$ by an element of $Y$ as a channel, as represented in Fig. 5.2 above. This kind of representation of an analog quantity is of common use in engineering where it is referred to as 'quantization'. Then the mutual information $I(X;Y)$, which remains meaningful for $X$ continuous as shown in Sect. 4.2.3, measures the information quantity of a

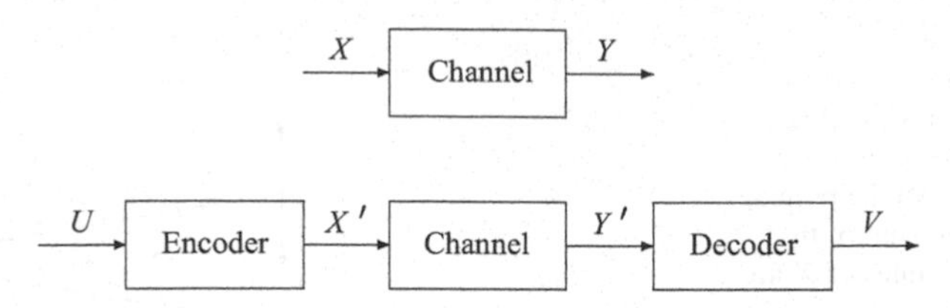


**Fig. 5.3** *Top*, channel alone with its input and output variables $X$ and $Y$, respectively. *Bottom*, the same channel preceded by an encoder receiving the input variable $U$ and followed by a decoder delivering the output variable $V$. The input and output variables of the channel are now denoted by $X'$ and $Y'$, respectively

random variable deriving from a continuous one by not distinguishing two values if they differ by less than $\varepsilon$.

## 5.3 Channel Coding Needs Redundancy

The mere possibility of communicating through a channel some information quantity measured by its capacity does not solve the problem of communicating through this channel a message generated by a source having this information quantity as entropy.

Let us consider the expression (4.25) of the mutual information $I(X;Y) = H(X) - H(X|Y)$. It equals $H(X)$, which measures the information quantity borne by the channel input $X$, coming from a source assumed to be stationary and memoryless and intended to the destination, minus a term which measures the uncertainty as regards $X$ which remains when $Y$ is given, measured by $H(X|Y)$, often named in this context ambiguity or *equivocation*. Clearly, the actual communication of a message demands that this term cancels: the messages are faithfully delivered to the destination only if a criterion of recovering quality which makes the equivocation negligible is satisfied. But $H(X|Y)$ solely depends on the channel once the probability distribution of $X$ is given. If the channel is noisy, $H(X|Y)$ cannot in general be neglected so directly connecting the source to the channel input does not ensure that messages are communicated as faithfully as desired. Intermediate devices, namely an encoder and a decoder, should be inserted between the source and the channel input on the one hand, between the channel output and the destination on the other hand. Any message generated by the source should be transformed by *channel coding* and the inverse transformation should be performed on the channel output so as to recover the message intended to the destination.

This encoding is necessarily *redundant*, as will be shown with the help of Fig. 5.3. The channel alone has been represented at the top. Its input and output variables are denoted by $X$ and $Y$, respectively. In the expression $H(X) - H(X|Y)$ of their mean mutual information, $H(X|Y)$ is strictly positive, depending on the channel. At the bottom, the channel is preceded by an encoder and followed by a decoder. We assume for simplicity's sake that the encoder output alphabet is the same as its input alphabet, which is itself identical to the input channel alphabet. Let $U$ be the input to the encoder and $V$ the output of the decoder, $U$ and $V$ being both random variables. Then the relation homologous to Eq. (4.25) is true:

$$I(U;V) = H(U) - H(U|V),$$

but the recovery criterion now demands that $H(U|V) < \varepsilon$, where $\varepsilon$ is a given positive constant smaller than $H(X|Y)$ which specifies how faithful should be the message recovery. Let $X'$ and $Y'$ denote the input and the output to the channel, respectively. The encoder and the decoder do not create, and at best do not destroy, information. For a well designed encoding system, no information is lost, so $I(U;V) = I(X';Y')$. It follows that $H(X') - H(U) = H(X'|Y') - H(U|V)$. The recovery criterion demands $H(U|V) < \epsilon$, which entails that $H(X'|Y')$ itself should be arbitrarily small. This is compatible with $H(X|Y)$ positive and possibly large because the source and the encoder taken together can be considered as a new source *with memory*, at variance with the original source, since the encoder acts on sequences of input symbols, not separately on each of them. Its successive output symbols $X'$ are thus mutually dependent. We may interpret $X$ as the *memoryless restriction* of the encoder output $X'$, i.e., ignoring the mutual dependency between the symbols due to the encoding operation, which entails that $H(X)$ is larger than $H(X')$ and $I(X;Y)$ larger than $I(X';Y')$.

## 5.4 On the Fundamental Theorem of Channel Coding

The fundamental theorem of channel coding states that an appropriate coding process enables satisfying a recovery criterion however stringent it is, hence performing errorless communication, provided the source entropy $H_s$ is smaller than the channel capacity $C$:

$$H_s < C, \tag{5.6}$$

regardless how close $H_s$ may be to $C$. Errorless communication is not possible if $H_s$ exceeds the channel capacity $C$.

This statement constitutes the most fruitful contribution of information theory to communication engineering. That noise does not prevent errorless communication but only sets an upper limit to the communicated information quantity was quite unexpected in 1948, when the theorem was first stated and is, anyway, paradoxical. Because the theory gave no explicit means for implementing errorless communication, this result prompted a vast research effort aimed at the design of codes and decoding algorithms, which eventually succeeded in obtaining an arbitrarily small error probability with an entropy $H_s$ (or code rate) close to the channel capacity $C$. This occurred no earlier than 1993, when the *turbocodes* were invented.

Since this book is not intended to offer a mathematical treatise of information theory but hopefully to provide intuitive insight, we just outline a proof of this theorem based on a geometrical interpretation, closer to Shannon's original proofs than the more general and rigourous ones which were later elaborated. This deliberate lack of rigour may be accepted since indisputable proofs are known to exist. Moreover, at variance with the early years of information theory when mathematical proofs were the only reasons for accepting the fundamental theorem as true, the progresses made since then in the field of error-correcting codes have provided a blatant experimental confirmation of its validity. All of us have in our pocket a proof of the theorem, I mean a mobile phone.

### 5.4.1 A Geometrical Interpretation of Channel Coding

We begin with defining a code intended to the binary symmetric channel and give a geometrical interpretation of it. A binary $(n, k)$ code $\mathcal{C}$ of length $n$ and dimension $k$, with $k < n$, is a subset of $M = 2^k$ sequences, referred to as *codewords*, among the $2^n$ $n$-symbol binary sequences. Encoding consists of establishing a one-to-one correspondence between each codeword and each of the $2^k$ possible information messages of length $k$. Transmitting a codeword then uniquely designates an information message. Since an $n$-bit codeword represents a $k$-bit information message, the inequality $k < n$ actually entails that this code is redundant.

Let us consider the finite space of $n$-bit words, which has been referred to as the Hamming space of dimension $n$ and denoted by $S_n$ in Sect. 3.4.2. Its $2^n$ elements are referred to as its points. The Hamming distance between any two points of this space is defined as the number of bits where the corresponding words differ. If a binary symmetric channel (as defined in Sect. 5.1) is used, the received word is represented by a point that errors make different from that which represents the actually transmitted codeword: if $t$ substitution errors occurred, the received point is at a Hamming distance of $t$ from the transmitted one. It is assumed that the distance between any two codewords of $\mathcal{C}$ has a minimum value $d$ which is larger than 1. If the minimum distance $d$ of the code is larger than twice the number $t$ of symbol errors which occurred, the received word enables unambiguously identifying the actually transmitted word since it is closer to it than to any other codeword. The occurring symbol errors thus do not prevent its correct identification, hence that of the corresponding information message. It by this means and in this meaning that the code $\mathcal{C}$ can be referred to as *error-correcting*.

The *optimum decoding rule*, which determines the most likely transmitted codeword, can thus be very simply stated:

*choose the codeword the closest to the received word for the Hamming distance.*

The set of points in the Hamming space which are closer to a codeword than to any other one is referred to as its Voronoi region. Optimally decoding a sequence represented by any point in this region thus results in this codeword.

The number of bit errors which occurred when a codeword has been transmitted over a binary symmetric channel of bit error probability $p_{\text{su}}$ is a random variable of mean $\bar{t} = np_{\text{su}}$. If $\bar{t}$ is smaller than $d/2$, this rule most often results in recovering the actually transmitted codeword. It fails to do so, an event referred to as 'decoding error', with a probability the smaller, the larger is $n$, which moreover vanishes as $n$ approaches infinity as a consequence of the law of large numbers. However, no reliable identification of the transmitted codeword is possible if $\bar{t} \geq d/2$: the above decoding rule then fails with a non-zero probability. When it fails, it results in an error pattern of at least $d$ bit errors.

In geometrical terms, the point which represents the received word is probably close to the surface of an $n$-dimensional hypersphere of $S_n$ centred on the transmitted word, the radius $r$ of which is the integral part of the expected number of errors $\bar{t}$. Asymptotically for $n$ approaching infinity, it tends to be *on* the surface of this hypersphere. If $\bar{t}$ is small enough with respect to the minimum distance $d$ of the code, the point which represents the received word is with high probability (surely when $n$ approaches infinity) closer to the transmitted word than to any other codeword. For the binary symmetric channel of bit error probability $p_{su}$, we may think of each codeword as surrounded by a 'noise (hyper)sphere' having as radius (for the Hamming metric) the integral part of $\bar{t} = np_{su}$, near the surface of which the point which represents the received word is almost surely located. For $n$ approaching infinity, Shannon wrote that the noise sphere becomes as well defined as a billiard ball (in a somewhat different context (Shannon 1949), but this metaphor is still valid here). Then, if all the noise spheres centred on the codewords become disjoint, the probability of a decoding error vanishes as $n$ approaches infinity and errorless communication is asymptotically possible. If the noise spheres intersect each others as $n$ approaches infinity, errorless communication is not possible.

The best code is intended to minimize the probability of a decoding error for the above optimum decoding rule. It can be thought of as distributing $M < 2^n$ points within $S_n$ so as to make them as far apart to each other as possible for the Hamming distance. For a given value of the symbol error probability $p_{su}$ of the binary symmetric channel, there is clearly a limit to the number $M$ of points which can be put in $S_n$ while keeping the distance between these points larger than or equal to $d = 2\bar{t} = 2np_{su}$. Let $M_{max}$ denote this number. The quantity

$$R_{max} = \lim_{n\to\infty} \frac{\log_2 (M_{max})}{n}$$

is the largest possible information rate per symbol which can be communicated through this channel: its capacity $C$. For an $(n, k)$ code, $2^k \leq M_{max}$ entails $k/n \leq C$. In geometrical terms, the design of the best code consists of *spreading* $M = 2^k$ points as evenly as possible, avoiding any local concentration of points which would diminish the largest minimum distance of the code. But how to spread as evenly as possible $M$ points within the Hamming space $S_n$? This problem has no known deterministic solution except for very few couples $(n, k)$. The solution given by Shannon is random coding, to be now examined.

### 5.4.2 Random Coding, its Geometrical Interpretation

Shannon's proof of the channel coding theorem relies on the extraordinary idea of *random coding* (Shannon 1948). Since no one knew (and no one still knows) how to design the best possible code, and *a fortiori* no one could compute the probability of erroneous decoding associated with it, Shannon considered, instead of a single code, a large set of randomly chosen codes. 'Randomly chosen' means here that each bit in a codeword is chosen at random independently of the other bits, each codeword in a code is chosen at random independently of the other codewords, and each code in the set of codes is chosen at random independently of the other codes.

The recourse to randomness may look strange since any random choice is by definition completely unexpected, hence a sequence of random choices is highly irregular. Do not forget however that when very many random choices occur with constant probabilities, a statistical regularity appears. Probability theory states precise laws which concern random choices successively effected very many times. The 'laws of large numbers' tell that the observed frequencies tend to the probabilities when the number of considered occurrences increases, and moreover tell how closely and how fast they do. Surprisingly, randomness actually results asymptotically in the most regular possible average distribution. If we think of a code, in geometrical terms, as a set of $2^k$ points among the $2^n$ points of the $n$-dimensional Hamming space, choosing them at random results in spreading them as evenly as possible in the average for $n$ and $k$ large enough.

Shannon computed the average probability of a decoding error for such a random ensemble of codes and showed that, provided the condition of the theorem is met (i.e., the source entropy is smaller than the channel capacity), then this average error probability can be made arbitrarily small by indefinitely increasing the codeword length. The ensemble of random codes contains at least a code as good as the average, which shows that 'errorless' communication is possible. This code, however, is not explicitly identified. Errorless should be understood in an asymptotic sense, meaning that the probability of error can be made as small as desired by increasing the codeword length and using codes optimally fitted to this increasing length.

The original proof given by Shannon was critized by pure mathematicians who pointed out its lack of formal rigour and even expressed doubts about the validity of the result itself. American mathematicians were especially sceptical, and the first mathematically sound proofs of the theorem came years later from the Soviet Union, especially that of (Khinchin 1957): Shannon's conclusions were fully confirmed but the conditions of validity of the theorem were more precisely stated. Among later proofs of the theorem, let us mention that given by Gallager. Using assumptions somewhat more restrictive but standard in channel coding theory, he gave a simplified proof of the theorem which moreover provides useful bounds on the best probability of a decoding error, for a code of finite length $n$, in terms of its code rate $R = k/n$ and of its length $n$ (Gallager 1965). However, his proof relies on mathematical computations which provide little intuitive insight on the problem so we do not discuss it here. All proofs of the theorem actually rely on the concept of random coding as Shannon's original one. In order to give an idea of the random coding method, we consider the simple case of a binary symmetric channel as defined in Sect. 5.1, using the simple geometrical interpretation of channel coding proposed in Sect. 5.4.1.

A random code results from drawing at random, independently of each other and once and for all, the $2^k$ words of the code among the $2^n$ sequences of $n$ bits. This code can thus be defined by the list of its $2^k$ $n$-bit words. It is comparatively easy to prove that errorless communication demands that the source entropy does not exceed the channel capacity, and it is what is done in the following. The positive statement of the theorem, i.e., that codes can be designed so as to perform arbitrarily close to this upper bound, is by far the most difficult to prove, at least from a formal point of view. However, we may accept as an axiom that no distribution of points in $S_n$ is more even in the average than that which results from random coding, asymptotically for $n$ approaching infinity. Then, no proof of the positive statement of the fundamental theorem is needed.

For some insight on the impossibility that the code rate $k/n$ exceeds the channel capacity, let us consider transmitting words of a given length $n$ over a binary symmetric channel where the number of errors per word is somehow kept constant: exactly $t$ bits, chosen at random among the $n$ bits of a word, are always wrong. This very unrealistic assumption is not necessary but will provide a useful simplification and will be justified below. The bit error probability of the channel is thus $p_{\text{su}} = t/n$. Thanks to this simplifying assumption, the total number of different erroneous words possibly received when a given codeword is sent is exactly the number of different possible choices of $t$ objects among $n$, or combinations, denoted by $\binom{n}{t}$. The channel errors can thus transform each of the possibly transmitted $n$-bit codewords into one among $\binom{n}{t}$ other $n$-bit words. No more than $M = 2^n/\binom{n}{t}$ distinguishable codewords can thus exist. But

$$\binom{n}{t} = \frac{n!}{t!(n-t)!},$$

where $n!$ denotes the factorial of the integer $n$, defined as the product of the $n$ first integers:

$$n! = 1 \times 2 \times \ldots \times n,$$

a fast varying function of $n$. A very useful approximation of $n!$ is provided by Stirling formula:

$$n! \approx \left(\frac{n}{\mathrm{e}}\right)^n \sqrt{2\pi n}\left(1 + \frac{1}{12n}\right) \tag{5.7}$$

which is very close to the actual value even for small values of $n$. Using Stirling's formula for $n!$, $t!$ and $(n-t)!$ results in the approximation

$$\binom{n}{t} \approx 2^{n\mathcal{H}_2(t/n)},$$

where the binary entropy function $\mathcal{H}_2(\cdot)$ has been defined by Eq. (4.11). The maximum number of distinguishable codewords $M$ has thus the approximate expression:

$$M \approx 2^{n[1-\mathcal{H}_2(t/n)]}.$$

A source connected to the channel input achieves errorless communication if its entropy $H$ is less than $(1/n)\log_2(M)$, thus approximately if:

$$H < 1 - \mathcal{H}_2(t/n).$$

The right hand side equals the channel capacity of the binary symmetric channel of error probability $p_{su}$ given by Eq. (5.2), so the equality

$$H < 1 - \mathcal{H}_2(p_{su}) \tag{5.8}$$

is asymptotically true for $n$ approaching infinity. The seemingly unrealistic assumption that exactly $t$ bits are in error in any received word becomes itself asymptotically true as $t$ and $n$ approach infinity. Indeed, when the errors occur at random with probability $p_{su}$, the number of bit errors tends to the constant value $\bar{t} = np_{su}$ when $n$ approaches infinity, as a mere consequence of the weak law of large numbers.

Although the calculations based on Stirling's formula (5.7) are not especially intuitive, the geometrical interpretation of the above argument is much more so. The Hamming space $S_n$ of binary words of length $n$ contains $2^n$ elements, or points. Assuming that exactly $t$ errors occur means that there is a distance of $t$ between the transmitted word and the received one. Let us define the 'volume' of some subset of this space as the number of words, or points, it contains. The volume of the whole space is $2^n$. The volume of the set of possible received words when some word $\underline{c}$ is transmitted is $\binom{n}{t}$. This is actually the 'surface' of an 'error sphere' of radius $t$ centred in $\underline{c}$ but, loosely speaking, when $n$ is large almost all the volume of a hypersphere in $S_n$ is concentrated on its surface, which is an $(n-1)$-dimensional volume. For $n$ becoming large enough, and if we relax the assumption that exactly $t$ symbol errors occur, the law of large numbers entails that a received codeword is with high probability on this surface. That this is true is checked by the fact that this grossly approximate assumption leads to the inequality (5.8) which expresses the fundamental theorem of channel coding, since the right hand side in it is the channel capacity as established by direct computation in Sect. 5.2.

This argument actually shows that it is impossible to achieve error-free communication if the source entropy is larger than the channel capacity. That it can be achieved if the entropy is smaller than the capacity, however close to it it may be, implies that the error spheres are as equally distributed in the whole space as possible. We accept as an axiom that choosing the codewords at random results in the average, asymptotically as $n$ approaches infinity, in the most regular possible configuration.

We already mentioned that Shannon wrote that the error spheres become as sharply defined as billiard balls, and that the number of points in the possible noise spheres is at most equal to the total number of points in the space. It should not be concluded that the noise spheres are strictly disjoint: they are only asymptotically disjoint, i.e., the volume that two of them have in common becomes negligible with respect to their own volume as the number of dimensions $n$ of the Hamming space tends to infinity. Shannon named this phenomenon 'sphere hardening', meaning that it is actually an asymptotic property.

### 5.4.3 Random Coding for the Binary Erasure Channel

As another simple example of the same method, the binary erasure channel considered above has the very simple capacity (5.3) which is quite different from that of the binary symmetric channel, Eq. (5.2). Let us try to understand why. First of all we may notice that, if we choose arbitrary binary symbols at the $n_{\text{er}}$ places where the original ones have been erased, only $n_{\text{er}}/2$ erroneous bits result in the average. This shows that, in some sense, an erasure is equivalent to half an error. The capacity of the erasure channel with erasure probability $p_{\text{er}} = 2p_{\text{su}}$ cannot thus be less than that of the binary symmetric channel of error probability $p_{\text{su}}$. It is even significantly larger since the location of the unidentified symbols is an information in itself which is destroyed when arbitrary binary symbols replace the signs $\epsilon$ which pinpoint the locations where the erasures occurred. Destroying an information symbol contained in the channel output necessarily reduces its capacity. Notice that the 'equivalence' of two erasures with a single error only holds in the binary case. If the alphabet size is $\alpha$, the number of erroneous symbols which result in the average from arbitrary choices becomes $(\alpha - 1)n_{\text{er}}/\alpha$ so an erasure is then 'equivalent' to $(\alpha - 1)/\alpha$ errors.

The simplest interpretation of this case is that the length $n$ of these words is reduced in the average by the factor $1 - p_{\text{er}}$. The largest information quantity that a word of length $n$ can bear, $n \log_2(\alpha)$ Sh, is thus reduced in the average by the occurring erasures to $(1 - p_{\text{er}})n \log_2(\alpha)$.

### 5.4.4 Largest Minimum Distance of Error-Correcting Codes

No explicit means for designing a code with the largest possible minimum distance $d$ is known in general. However, asymptotically for $n$ approaching infinity, random coding arguments show that the largest possible minimum distance of a code is at least equal to the Gilbert-Varshamov bound $d_{\text{GV}}$, defined in the binary case by the implicit equation in $d_{\text{GV}}/n$:

$$1 - k/n = \mathcal{H}_2(d_{\text{GV}}/n), \tag{5.9}$$

where the binary entropy function $\mathcal{H}_2(\cdot)$ has been defined in Eq. (4.11). Only the increasing branch of this function is relevant, since the largest minimum distance is obviously a decreasing function of the code rate $R = k/n$: the more numerous the codewords, the smaller this distance. This remark implies $d_{GV} < n/2$. For instance, if we assume that $k = n/2$, $d_{GV}$ is very close to $0.11 \times n$. Notice that the left hand side of Eq. (5.9) measures the redundancy of the code and that $\mathcal{H}_2(\cdot)$ is an increasing function when its argument is smaller than 1/2, so the larger the redundancy, the larger $d_{GV}$, as expected. For a very redundant code, i.e., when $k/n$ approaches 0, $d_{GV}$ tends to $n/2$. Since it is $d_{GV}/n$ which is specified by Eq. (5.9), the Gilbert-Varshamov distance is proportional to the code length $n$. Taking it as a figure of merit shows that an error-correcting code is the better, the longer.

For a code of alphabet size $\alpha > 2$, the Gilbert-Varshamov bound becomes the solution of the implicit equation:

$$1 - k/n = \mathcal{H}_\alpha(d_{GV}/n) + (d_{GV}/n)\log_\alpha(\alpha - 1), \tag{5.10}$$

where

$$\mathcal{H}_\alpha(x) = -x\log_\alpha(x) - (1-x)\log_\alpha(1-x) = \frac{\mathcal{H}_2(x)}{\log_2(\alpha)}.$$

Notice the presence in Eq. (5.10) of the additive term $(d_{GV}/n)\log_\alpha(\alpha - 1)$ which cancels for $\alpha = 2$. For $k/n$ approaching 0, $d_{GV}/n$ approaches $(\alpha - 1)/\alpha$, a result easily deduced from Eq. (5.10).

### 5.4.5 *General Case: Feinstein's Lemma*

Up to now, we just gave the sketch of a proof of the fundamental theorem of channel coding for the binary symmetric channel and the binary erasure channel, which are both very simple examples. We now intend, for the sake of completeness, to state without proof a much more general result which is used for proving the fundamental theorem of channel coding. The simple case of the binary symmetric channel may be thought of as a particular instance of a general result named *Feinstein's lemma.* Assuming a stationary and ergodic source and a discrete stationary channel with finite memory, it states that a one-to-one correspondence between the typical sequences in the sense of the Shannon-McMillan theorem (see Sect. 4.3.5) and asymptotically disjoint sets of points in the space of the output sequences can be established provided the source entropy is less than the channel capacity, when the sequence length approaches infinity. The 'error spheres' in the case of the binary symmetric channel are mere examples of such asymptotically disjoint sets of points in the channel output space. (The proof of Feinstein's lemma is rather difficult and implies lengthy mathematical developments, so we omit it.) The statement of the fundamental theorem of channel coding then results from the lemma in a straightforward manner: observing the channel output enables uniquely identifying the channel input sequence if it belongs to the set of typical sequences. If it does not, an event of vanishingly small probability according to the Shannon-McMillan theorem, a specific distinguishable sequence is transmitted in order to warn the destination that an atypical sequence occurred.

## 5.5 Error-Correcting Codes

### 5.5.1 Defining an Error-Correcting Code

An $(n, k, d)$ binary error-correcting code, with $k < n$, has been defined in Sect. 5.4.1 as a subset of all $n$-bit long sequences or $n$-tuples), containing $2^k$ elements and such that a minimum Hamming distance of at least $d > 1$ exists between any two of them. The condition $k < n$ expresses that this code is necessarily redundant. Once $n$ and $k$ are chosen, $d$ cannot assume arbitrary values: the largest possible minimum distance of such a code, which is in general not explicitly known, exceeds the Gilbert-Varshamov bound $d_{GV}$ introduced in Sect. 5.4.4 but is probably only slightly larger than it. In geometrical terms, we may think of the $n$-dimensional Hamming space as the set of all $n$-bit sequences where a distance between them is defined by the Hamming metric as introduced in Sect. 5.4.1. The existence of a minimum distance $d$ between the codewords is intended to make them as different as to enable discriminating between them if the number of channel errors is low enough. We stated above that the optimum decoding rule is to 'choose the codeword the closest to the received word for the Hamming distance'. As a consequence, if less than $d/2$ channel errors occurred, the actually transmitted codeword can be identified with absolute certainty. This does not entail that a decoding error necessarily occurs in the event of more than $d/2$ channel errors, however, but that the probability of such an error conditioned on $t$ channel errors becomes strictly positive when $t$ becomes larger than $d/2$, while it is exactly 0 when $t < d/2$. This is especially true for very redundant codes, i.e., where the codewords constitute a very sparse minority among the $n$-tuples. For instance, if we let $k$ approach 0, the Gilbert-Varshamov bound of the minimum distance results in $d_{\mathrm{GV}}$ approaching $n/2$. If this distance is achieved, then two codewords can be distinguished with absolute certainty only if their distance is less than $n/4$. However, correct decoding is possible in many instances where their distance is larger than that. As an example, the capacity of a binary symmetric channel of symbol error probability $p_{\mathrm{su}}$, as given by Eq. (5.2), approaches 0 when $p_{\mathrm{su}}$ approaches 1/2. According to the fundamental theorem, this means that all symbols errors can be corrected if they occur with a probability up to about 1/2 provided the code rate $R = k/n$ vanishes. If no more than $n/4$ error patterns could be corrected in this case, the limiting value of the error probability would be $p_{\mathrm{su}} = 1/4$ and not $p_{\mathrm{su}} = 1/2$. It is indeed possible to design codes which correct errors occurring with a probability larger than 0.25 (of course, less than 0.5). Then only the rarity of the codewords among the $n$-tuples is relevant, and the minimum distance between them is comparatively unimportant. This rarity is indeed the key factor why error-correcting codes are efficient and it is why the codewords are distant from each others. We may define the 'dilution factor' of a code as the ratio of the number of codewords to the number of $n$-tuples, namely, $2^{-(n-k)} = 2^{-n(1-R)}$. The code is the better, the smaller this figure. As in the discussion of the Gilbert-Varshamov bound, we find that a code is the better, the larger its length $n$ and the smaller its rate $R = k/n$. Assuming its parameters $n$ and $k$ to be given, designing a good code aims at spreading the codewords as evenly as possible within the set of $n$-tuples.

During decades, the only criterion for designing an error-correcting code has been to maximize its minimum distance. The theory of algebraic codes entirely relies on this criterion. We questioned this criterion and suggested instead that codes should be designed in order to make their distance distribution mimic that which results in the average from random coding (Battail 1989). We proposed as a proximity criterion the Kullback-Leibler divergence, as defined in Sect. 4.2.6, between the normalized distance distribution of a code and that of random coding. We referred to codes designed according to this criterion as 'random-like'. Questioning the minimum distance criterion, become with time a kind of dogma, was deemed heretical by many. However, the invention of turbocodes and the rediscovery of Gallager's low-density parity-check codes, a few years later, blatantly showed that the codes with the best actual error correction performance were not designed so as to maximize their minimum distance. The proximity of their distance distribution with that of random coding was established (Battail 1993, 2000; Battail et al. 1993). The minimum distance of these best codes remains most often unknown.

From a geometrical point of view, a word of a random-like code has few closest neighbours in the Hamming space so the performance of these codes mainly relies on this scarcity. If the minimum distance of a random-like code is comparatively small an 'error floor' results, meaning that the improvement in decoding error probability due to the code varies much slower for large values of the signal-to-noise ratio than for smaller ones, because the performance is then dominated by the closest neighbours. Turbocodes exhibit this phenomenon but the floor decoding error probability is low enough for most applications.

### 5.5.2 Using Error-Correcting Codes: Decoding and Regeneration

We already met communication by means of an error-correcting code so the function of encoding has become familiar to us. Let us just recall that encoding, for an $(n, k)$ code $C$, with $n > k$, consists of establishing a one-to-one correspondence between the set of $k$-bit information messages and a set of $n$-bit words, or code, containing $2^k$ codewords. The code may be interpreted as a list of pairs (information message—codeword), whether such a list is actually used, or not, in order to perform the encoding. Encoding may thus be performed by merely reading a codeword at a specified address in a memory. The critical step of an encoded communication is met at the receiving end: dealing with a codeword received in the presence of symbol errors.

Processing a codeword received in the presence of errors in order to recover the transmitted one is referred to as *decoding*. It involves two steps. The first one is intended to determine the most likely transmitted codeword, or point in the Hamming space $S_n$ according to the geometrical description of an error-correcting code proposed in Sect. 5.4.1, given the received word or point. The success at this step is a chance event which implies a risk of wrong recovery. In communication engineering, the ultimate goal of this processing is the recovery of the encoded information message. Once the algorithm which performs the first step results in designating a codeword (right or wrong), the second step is thus intended to recover the information message which corresponds to it. It may be implemented by just reading the proper entry in the list of the codewords and is basically deterministic. The first step is the critical one in the whole decoding process and, at least when applied to a genomic error-correcting code, it will be referred to as *regeneration*. Contrary to the encoding process which is intrinsically simple, decoding is by far more difficult because its first step, regeneration, involves making hypotheses and comparing their results. Still more important than designing good codes, devising efficient decoding algorithms has been a major goal in the research about error-correcting codes.

If it is restricted to its first and crucial step, i.e., regeneration, the reception process no longer results in delivering an explicit information message, but in making the encoded sequence as a whole resilient to casual errors. Regeneration fails with a very small probability if the error-correcting code is well fitted to the channel. In the presence of symbol errors of constant probability per time unit, performing repeated regenerations at short enough time intervals then results in the almost sure conservation of this sequence during very long times. We examine at length in the second part how error-correcting codes can ensure by this means the conservation of genomes by means of repeated regenerations. Then the information message remains always hidden: only making the recovered genome identical to the original one is relevant.

### 5.5.3 Designing Error-Correcting Codes

As an important application of information theory to communication engineering, which moreover posed difficult problems to researchers, error-correcting codes gave rise to a plentiful literature. We shall give only a short account of it, restricted to the most successful code family which however remained unnoticed, at best marginal, during decades. We refer to this family as *random-like codes*. We entirely omit the algebraic codes which have long been the main research topic in the field (just using a simple Hamming code which belongs to this family in examples). It is not easy to explain the operation of algebraic codes which rely on advanced mathematics, and the codes which most closely approach the theoretical limit were designed by very different methods. Let us just say that the aim of researches about algebraic codes was to find codes having the largest possible minimum Hamming distance, at variance with the codes most closely approaching the channel capacity which can be interpreted as attempting to mimic random coding. We thus first examine how random coding could be implemented and look for non-random coding methods having a similar result.

Implementation of random coding in order to design a binary $(n, k)$ code would first consist of establishing once and for all a list of $2^k$ $n$-bit codewords. Each bit of a codeword is chosen at random, with probability 1/2, independently of the other bits of the word, and the other codewords are similarly obtained independently of the others. A one-to-one correspondence is established between the words of the list and the $k$-bit information messages, such a message being used as the address of a codeword in the list. Then, encoding a $k$-bit information message merely consists of reading a single entry of the codeword list at the appropriate address. Decoding, however, is highly impractical since a received sequence, consisting of a codeword affected by symbol errors, does not generally belong to the codeword list. There is no other decoding means than comparing *each* received sequence with *all* the $2^k$ $n$-bit codewords in order to determine which of them is the closest to it (according to the optimum rule stated in Sect. 5.4.1). For large values of $k$ (good performance demands numerous codewords hence $k$ should be large) $2^k$ is huge: decoding becomes so complex that it is practically impossible. If random coding is an outstanding theoretical tool, it cannot be used in practical instances.

However, the performance of an error-correcting code is determined by the set of distances between its words, not by the way it is generated. If a code is defined by a non-random rule such that the distance distribution of its words closely mimics that of random coding, its performance is close to that of random coding but its decoding can be made far simpler. This is the rationale of the search for such deterministic, *random-like codes*. (Both the concept and name of 'random-like codes' were introduced by me (Battail 1996) and are not of general use.) Random-like codes were sought without special consideration to their minimum distance, but a distance distribution identical to that of random coding implies a minimum distance which satisfies the Gilbert-Varshamov bound (5.9). The family of random-like codes mainly contains the turbocodes invented by Berrou and Glavieux in 1993 and the low-density parity-check codes, invented much earlier by (Gallager 1962) but rediscovered and recognized as achieving an outstanding performance only after the turbocodes were disclosed.

### 5.5.4 Recursive Convolutional Codes

Before dealing with turbocodes, we first consider their main components, namely, the recursive convolutional codes. We just describe how they are generated, from which we derive their main properties. We already met a recursive convolutional encoder since the example of an encoder given in Sect. 3.4.2 and illustrated by Fig. 3.4 is actually of this type. The main component of such an encoder is its rate-1 encoder which generates the redundancy sequence, or *check* sequence, to be appended to the information message so as to constitute an error-correcting code. The slightly more complicated example of Fig. 5.4 below represents a rate-1 encoder similar to that of Fig. 3.4 but having a memory $\mu = 3$ instead of $\mu = 2$. It will be helpful for understanding the properties of the sequences generated by recursive encoders.

We always assume that the initial content of the binary shift register memories is all-zero. For dealing with binary sequences, we introduce a convenient formalism,

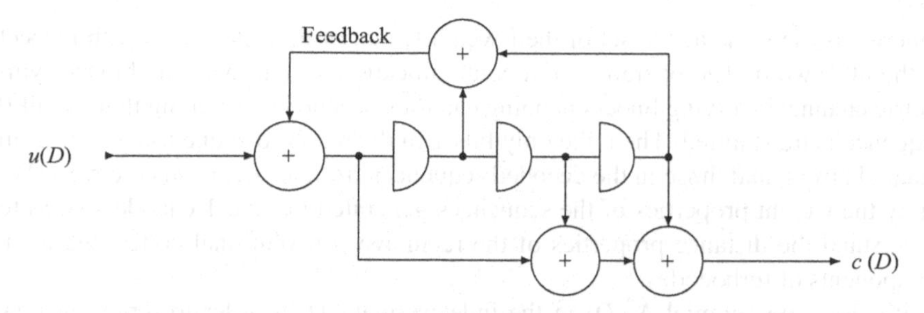


**Fig. 5.4** Example of a rate-1 encoder with memory $\mu = 3$. The input $u(D)$ and the output $c(D)$ represent the information message and the corresponding check sequence, respectively, in terms of the delay operator $D$. Each point may assume one of two states, denoted by 0 and 1. The elements drawn as half-circles are delay operators (of an information bit duration); they form together a shift register, and + denotes addition modulo 2. This encoder is referred to as 'recursive' because it involves a feedback

referred to as the '$D$-transform', according to which a sequence of finite length is represented by a polynomial where the indeterminate, say $D$, may be interpreted as a delay of one bit duration with respect to some origin of time; $D$ is thus referred to as 'delay operator'. The coefficients of the polynomial represent the symbols of the sequence; they are elements of the binary field, denoted by 0 and 1 and endowed with addition modulo 2. As in the ordinary algebraic notation, only the terms with a non-zero coefficient are explicitly written. For instance, the binary sequence 1101 is represented by $1 + D + D^3$, 0011 is represented by $D^2 + D^3$, etc. Notice that, due to the fact that the bit 0 is not explicitly written, these polynomials represent the written sequences up to an arbitrary number of bits 0 appended to them. This dissymmetry between 0 and 1 results from the usual assumption that the registers are initially in the all-0 state, so the occurrence of bits 1 means a departure from the initial state. The representation of sequences by polynomials extends immediately to the case of an infinite number of terms, polynomials then becoming 'formal series'. We will meet sequences with infinitely many symbols (although, of course, they need to be truncated in any practical situation) and formal series will be useful for denoting them. As an example, the formal series $1+D+D^2+\dots$ represents the all-1 sequence (starting from the origin of time). Any polynomial in $D$ can be interpreted as the infinite series which results from appending infinitely many bits 0 to the finite sequence they represent. For instance 1, interpreted as a formal series, represents a single bit 1 followed by infinitely many bits 0.

The number of bits 1 in a sequence is referred to as its *weight*. The encoding performed by a device like that of Fig. 5.4 is referred to as *linear* in the mathematical meaning of the word: it involves only additions and multiplications in the binary field, i.e., additions modulo 2 and multiplications by 0 (no connection) or 1 (connection). Restricting the considered codes to linear ones does not significantly restrict generality but provides a great simplification. The all-0 word belongs to any linear code and it is generated in response to the all-0 information message. It is easily shown that the set of distances between all the sequences that a linear encoder can generate is the same as the set of their weights, i.e., of their distances with respect to the all-0 word. The operation of a communication system over the binary symmetric channel involving linear encoding can thus be studied assuming that the all-0 sequence is transmitted. Then, the only bits 1 in the received sequences result from channel errors, and those in the decoded sequences from an erroneous recovery. We study the weight properties of the sequences generated by rate-1 encoders so as to understand the distance properties of the recursive convolutional codes, the main components of turbocodes.

We use a polynomial $\mathcal{N}(D)$ in the indeterminate $D$ in order to represent how the output is computed as the sum modulo 2 of bits contained in the shift register. The coefficients of $\mathcal{N}(D)$ are 1 or 0, depending on the output of the memory cell corresponding to the degree of the indeterminate being connected or not to the modulo 2 adder which computes the encoder output. For instance, $\mathcal{N}(D) = 1 + D^2 + D^3$ for the encoder depicted in Fig. 5.4. Similarly, a different polynomial $\mathcal{Q}(D)$ specifies the recursion bit, where the coefficients are 1 or 0, depending on the output of the corresponding memory cell being used or not to compute the recursion bit fedback to the register input ($\mathcal{Q}(D) = 1 + D + D^3$ for the encoder depicted in Fig. 5.4). We moreover assume that the degree of both polynomials $\mathcal{N}(D)$ and $\mathcal{Q}(D)$ equals the register memory $\mu$ and that both polynomials are irreducible, i.e., cannot be expressed as a product of polynomials. Then the sequence generated by the encoder made of the shift register with its feedback determined by $\mathcal{Q}(D)$ and its output computed according to $\mathcal{N}(D)$, provided the initial state, i.e., the content of all the memory cells of the register, is all-0 at the origin of time, is represented by:

$$c(D) = u(D)\frac{\mathcal{N}(D)}{\mathcal{Q}(D)}, \tag{5.11}$$

where $u(D)$ represents the input (information sequence) and $c(D)$ the output (check sequence). For instance, if $u(D) = 1$, i.e., if the input sequence consists of a single bit 1 at the origin of time followed by infinitely many bits 0, the encoded sequence is represented by

$$c(D) = \frac{\mathcal{N}(D)}{\mathcal{Q}(D)}.$$

Since $\mathcal{Q}(D)$ does not divide $\mathcal{N}(D)$, this fraction can be expanded into a formal series with infinitely many non-zero terms. The generated sequence has thus an infinite weight. For instance, we have in the above example:

$$c(D) = \frac{1 + D^2 + D^3}{1 + D + D^3}.$$

The reciprocal of the polynomial $\mathcal{Q}(D)$ is the formal series

$$\frac{1}{1 + D + D^3} = 1 + D + D^2 + D^4 + D^7 + D^8 + D^9 + D^{11} + D^{14} + \cdots$$

where the right hand side is periodic (the period in this example is the largest possible with a shift register of memory $\mu = 3$, namely $2^3 - 1 = 7$), and its multiplication by $1 + D^2 + D^3$ results in

$$c(D) = 1+D+D^4+D^5+D^6+D^8+D^{11}+D^{12}+D^{13}+D^{15}+D^{18}+D^{19}+D^{20}+\cdots$$

which is also 7-periodic beyond the first two terms. Then the periodically repeated motif has weight 4.

For a shift register of memory $\mu$, the length of the periodically repeated motif is at most $2^\mu - 1$, in which case its weight is $2^{\mu-1}$. It is obtained if the polynomial $\mathcal{Q}(D)$ which describes the feedback is 'primitive'[1]; there always exists at least a primitive polynomial of any degree $\mu$. In this case, the content of the register successively represents in binary numeration all possible $\mu$-bit nonzero integers from 1 to $2^\mu - 1$ (in an order which depends on $\mathcal{Q}(D)$) and we get a 'maximum-length' sequence, of the longest possible period that can be generated by a register of length $\mu$, namely $P = 2^\mu - 1$. Then $\mathcal{Q}(D)$ is a factor in the decomposition of $1 + D^{2^\mu - 1}$ as a product of irreducible polynomials. In the example of Fig. 5.4, one easily checks that $1 + D^7 = (1 + D + D^3)(1 + D^2 + D^3)(1 + D)$ (do not forget that the coefficients are computed modulo 2), where both $1 + D + D^3$ and $1 + D^2 + D^3$ are primitive. It turns out that a sequence thus generated by a shift register endowed with a proper feedback, when $\mu$ is large, mimics in some sense a binary sequence generated at random and therefore is referred to as *pseudo-random*. Such sequences are currently used instead of truly random sequences in simulations intended for instance to assess the performance of communication systems. Long pseudo-random sequences can moreover be obtained by combining several sequences of shorter period generated by small-memory shift registers.

Due to its feedback, a recursive rate-1 encoder delivers infinite-weight sequences in response to certain finite-weight ones, especially of weight 1, which would be impossible without a feedback. A sequence generated by a rate-1 encoder is however not directly useful for correcting errors. There is a one-to-one correspondence between the input and output sequences, and any change in the output sequence would result in a sequence which can be generated by another input sequence. Remember that an error-correcting code is necessarily redundant, which is not the case for the set of sequences generated by a rate-1 encoder. It is why the full encoder of Fig. 3.4 appends the input sequence to the check sequence generated by the rate-1 encoder, resulting in a rate-(1/2) code which can actually correct errors. An encoder like this one, where the information message is a part of the output, is referred to as *systematic*. Then, the full encoded sequence, made of the serial combination of the information message and the check sequence delivered by the rate-1 encoder, has as weight the sum of their weights. When the check sequence has an infinite weight, there is an infinite distance between the whole encoded sequence and the all-0 sequence, hence no decoding error is possible. The check sequences generated by the rate-1 encoder have most often an infinite weight (actually limited in practice by its truncation to a finite arbitrary value) since only a fraction $2^{-\mu}$ of the information messages result in finite weight encoded sequences (Battail et al. 1993). For

[1] A polynomial of degree $\mu$ is said to be primitive if taking the successive powers of one of its roots generates all the $2^\mu - 1$ non-zero elements of the $\mu$-th extension of the binary field.

randomly chosen information messages, we may interpret the fraction $2^{-\mu}$ as the probability that a finite weight sequence is generated by the encoder.

The solution to the problem of errorless communication seems then obviously to use a recursive systematic convolutional encoder with a large memory $\mu$ so as to almost always generate infinite weight sequences. Unfortunately, the complexity of decoding varies as $2^{\mu}$, which severely limits the possible values of $\mu$ hence the practical usefulness of such encoders. Excellent distance properties can however be obtained if several short encoders are combined according to the *turbocode* scheme, to be described in the next section. Performance close to the theoretical limit can be obtained by this means with only two combined registers. It turns out moreover that turbocodes can be almost optimally decoded with fairly low complexity, as will be shown in Sect. 5.5.9.

The requirement that the initial state of the shift register be zero implies that means for controlling the initial state are employed. For instance, a termination made of $\mu$ properly chosen bits may be appended to an $N$-bit information message so as to make the register content return to the zero state. Another solution to the register initialization problem consists of choosing a nonzero initial state such that the same state is reached after an $N$-bit message has been fed to the register, resulting in a 'tail-biting' convolutional code. The initial state needs then to be pre-computed in terms of the message. This is a smart solution to the problem of initializing both component encoders of a turbocode (see next section).

### 5.5.5 *Turbocodes*

The basic idea of turbocodes is to encode twice the information message, firstly as it is and secondly after it has been scrambled by an interleaver. Both encodings are performed by systematic recursive convolutional encoders like that of Fig. 3.4; their rate-1 encoders may be identical or different and both have a same memory $\mu$ of moderate value, e.g., $\mu = 3$ as in Fig. 5.4. Due to its feedback, each of the rate-1 encoders generates sequences of weight tending to infinity as the message length increases, except for small-weight sequences generated by a fraction $2^{-\mu}$ of all the input sequences. Thanks to the interleaver, the fraction of input sequences which eventually result in an overall transmitted small-weight sequence, interpreted as a probability, becomes $2^{-2\mu}$ and thus becomes negligible for values of $\mu$ remaining as small as to keep the decoding complexity reasonably low.

The basic scheme for generating a turbocode is represented in Fig. 5.5. It combines two[2] convolutional systematic recursive encoders with an interleaver. The rate-1 encoders have been defined in the previous section. An *interleaver* of length $N$ is a device having sequences of length $N$ as input and output, such that the output contains the same symbols as the input but in a different order. If its input sequence

[2] Or more, but then some specific difficulties are met; two-component codes suffice for obtaining results close enough to the theoretical limit for most practical purposes.

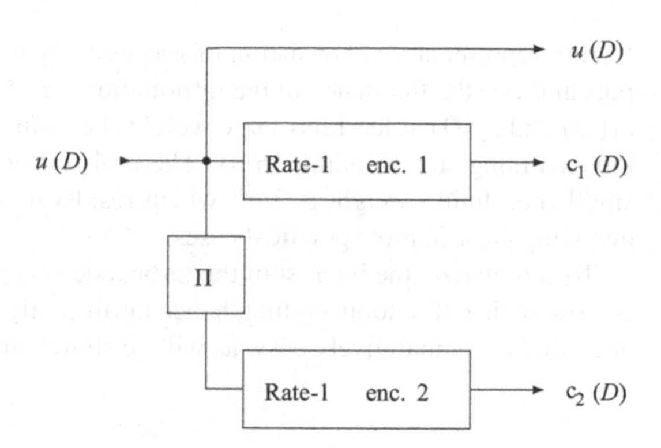


**Fig. 5.5** Encoder of a rate-1/3 turbocode. The blocks labelled 'Rate-1 enc.' designate rate-1 encoders as represented in Fig. 5.4. The block labelled Π represents the interleaver which implements the permutation Π

is $u_1, u_2, \ldots, u_N$, where $u_i$ is an element of some alphabet, for instance the binary one, its output is $u_{j_1}, u_{j_2}, \ldots, u_{j_N}$, meaning that the $j_i$-th element of the input sequence became the $i$-th in the output sequence. An interleaver $\Pi$ is completely defined by a set of $N$ different indices $\{j_i, 1 \leq j_i \leq N, i = 1, 2, \ldots, N\}$. For example, an interleaver of length $N = 7$ defined by the set of indices $\{3, 6, 1, 2, 7, 5, 4\}$ transforms the word 0101100 into 0001011 and the word 1101011 into 0111110. The encoders currently used in turbocodes have a small memory (e.g., $\mu = 3$) but, unlike the above example, the interleaver length $N$ is long (say, a few thousands of bits), so there are as many possible interleavers as permutations of $N$ objects, namely, $N! = 1 \times 2 \times \ldots \times (N-1) \times N$, a huge number for the usual values of $N$. When operating on some sequence $u(D)$ (represented by a polynomial of degree $N-1$ in the indeterminate or delay operator $D$ as defined in Sect. 5.5.4), the interleaver $\Pi$ transforms it into $\Pi[u(D)]$. The device which recovers any $u(D)$ from its interleaved version $\Pi[u(D)]$, often referred to as the de-interleaver, is another interleaver of length $N$, to be denoted by $\Pi^{-1}$, such that $\Pi^{-1}\{\Pi[u(D)]\} = u(D)$.

The encoder which generates a turbocode operates as follows. The input sequence of length $N$, represented by the polynomial $u(D)$ of degree $N-1$, is also one of its output sequences since it is systematic. At the same time, each of two recursive rate-1 encoders like that of Fig. 5.4 computes check bits. These encoders may be identical to each other, or not. One rate-1 encoder is fed directly by the input sequence and the check sequence it generates is denoted by $c_1(D)$, while the other one is fed by the output of the interleaver, the input of which is the information sequence $u(D)$. The check sequence it generates is denoted by $c_2(D)$. This basic scheme has inherently a rate of 1/3, but several tricks can be used so as to obtain higher rates (the most usual of them, referred to as 'puncturing', consists of suppressing a fraction of the generated check bits according to a given periodic pattern). Considering the case of a rate-1/3 turbocode will however suffice to understand the turbocode principle so we do not consider other rates. We denote the output of this rate-1/3 encoder by $[u(D), c_1(D), c_2(D)]$. It is a sequence of length 3 $N$, the weight of which is the sum of the weights of $u(D)$, $c_1(D)$ and $c_2(D)$. The encoder output in Fig. 5.5 is represented in 'parallel' but in practice the encoder would generally involve a conversion into a 'serial' representation, consisting of successively transmitting the three binary outputs at thrice the frequency of the information bits. Most often, one of the sequences $c_1(D)$ and $c_2(D)$ at least has large weight (i.e., infinite would no truncation occur), thus avoiding any decoding errors. The probability that $[u(D), c_1(D), c_2(D)]$ has a small (i.e., finite) weight is $2^{-2\mu}$ which results in a small enough probability of a decoding error in most practical cases.

To summarize, the interest of the turbocode scheme is that its weight distribution is close to that of random coding, hence intrinsically good, while its almost optimum decoding is comparatively easy, as will be shown in Sect. 5.5.9.

### *5.5.6 Low-Density Parity-Check Codes*

The low-density parity-check (LDPC) binary codes are defined by a number of parity-check equations, each involving only a small number of information and parity-check bits. At least two information bits and one parity-check bit are randomly associated in each parity-check equation. A parity-check equation expresses that the sum modulo 2 of certain bits equals 0 (equivalently, that the number of '1's in it is even). Parity-check equations are standard means for expressing constraints in binary coding.

For instance, an LDPC code is defined by a set of parity-check equations involving each two information bits and a check bit (the minimum possible values) as follows:

$$\begin{aligned} i_a \oplus i_b \oplus c_u &= 0 \ (C_u) \\ i_a \oplus i_c \oplus c_v &= 0 \ (C_v) \\ \cdots\cdots\cdots & \ \cdots\cdots \end{aligned} \tag{5.12}$$

where $i_a$, $i_b$, $i_c$, ... denote three arbitrary information bits, $c_u$, $c_v$, ... denote check bits associated with information bits according to the parity checks labelled $(C_u)$, $(C_v)$, ..., and $\oplus$ denotes addition modulo 2. The parity-check equations may be quite numerous, and they are so for a long code. We assume that each information bit is checked by at least two of these equations.

One may interpret LDPC codes as actually implementing random coding, but by establishing at random parity-check equations instead of a list of codewords. That each of them involves only a few information bits makes their separate decoding easy, and the reliability assessment about the information bits acquired by the decoding of a parity-check equation helps decoding the other ones, thanks to the decoding process using soft decisions to be expounded in Sect. 5.5.8 and its iteration. It turns out that combining in the parity-check equations a small number of bits, as small as to make decoding tractable, still results in an overall distance distribution of the code which mimics that of random coding properly said (Battail 2000).

The LDPC codes are extremely flexible since further constraints can be added at will. It will be clear moreover in Sect. 5.5.8 that their decoding does not demand that all constraints are accounted for. The key to their efficient decoding is symbol-by-symbol decoding with an entirely analog processing aimed at reassessing the symbol probabilities.

### 5.5.7 Decoding Random-Like Codes: Principles

We deal here only with the decoding of turbo- and LDPC codes. We already noticed that these codes may be interpreted as endowing the set of sequences generated by the encoder with a distance distribution close to that which results in the average from random coding. A further reason of their interest, which is not the least, is that they lend themselves to almost optimal decoding, which moreover is comparatively simple.

Decoding relies on the same general principles for both code families. We first redefine the very function of decoding according to the following two points (Battail 1987b). Both contradict the principles which were earlier in force: decoding was intended to determine codewords, not separate symbols, which made 'hard decisions' mandatory since a codeword is a sequence of alphabet symbols. When researches about coding began, it was not realized that hard decisions imply a prohibitive information loss, and that word-by-word decoding of long codes is inherently very complex. Algebra of finite fields was then the main framework, not probabilities. In sharp contrast, the principles which underlie the decoding of random-like codes are:

1. Decoding deals with a single symbol at a time: it is intended to find the most likely transmitted symbol at a particular location in a word, not the most likely transmitted codeword. This is referred to as *symbol-by-symbol* decoding. Notice that doing so does not comply with the optimum word-by-word decoding rule stated in Sect. 5.4.1.
2. Such decoding does not merely consist of determining the most likely transmitted symbol, but of assessing for each received symbol the probabilities that each of the alphabet symbols has been transmitted, in terms of all the received symbols (the one to be decoded as well as the others) and taking the code constraints into account. It is thus intended to take *soft decisions* according to the vocabulary introduced in Sect. 3.3: it deals with *analog* data, which in the binary case are the log-likelihood ratios defined by Eq. (3.8), not with the discrete symbols of the alphabet.

The interest of doing so is that provisional 'decisions' (actually, probability assessments) about some bits can be made, accounting for only some of the coding constraints. Since these provisional decisions do not incur in principle any information loss, further taking into account other constraints becomes possible. Then decoding a large and complex code can be split into a number of lossless decodings of elementary component codes. We noticed in Sect. 3.3 that hard decisions should be avoided unless they are absolutely necessary because they entail an irreversible loss of information. The decoding process of turbo- or LDPC codes involves soft decisions throughout. This kind of decoding is usually referred to in the engineering literature as 'soft-input, soft-output' (abbreviated as SISO) decoding.

We begin with expounding how to decode a binary LDPC code. The constraint expressed by one of its parity-check equation enables first reassessing the log-likelihood ratio of an information symbol it contains in terms of the *a priori* log-likelihood ratios of all its other symbols. Considering another parity-check equation containing the same information symbol, its reassessed log-likelihood ratio improves the reassessment of the log-likelihood ratio of the other information symbols it contains. Repeating this process enables reassessing the log-likelihoods of all the information symbols according to a kind of oil stain effect. Not only this process can be repeated until all the parity-check equations have been taken into account, but it can be *iterated*: the log-likelihood ratios improved in a previous decoding step are used to still improve further ones. For this process to work, however, care must be taken for avoiding 'self reinforcement' of the log-likelihood ratios. This will be better understood after how log-likelihood ratios are reassessed is more formally described.

### 5.5.8 Decoding an LDPC Code

We do not describe here the decoding process of LDPC codes as originally expounded by (Gallager 1963), but we rather use the concepts and vocabulary of replication decoding (Battail and Decouvelaere 1976; Battail et al. 1979), which may be interpreted as a variant of Massey's threshold decoding (Massey 1963) such that the threshold is systematically set to 0, and of iterative decoding. We believe that these concepts are well fitted to an intuitive understanding.

Replication decoding turns the redundancy due to encoding constraints into explicit repetition, thereby transforming decoding into 'diversity reception', an intuitive method for improving reception known by radio engineers well before error-correcting codes were invented and thus validated by a many-year experience. It is based on the fact that two or more unreliable copies of a same signal can be combined into a single one which is more reliable than each of them separately. Let a bit be transmitted over two independent binary symmetric channels and let $\ell_1$ and $\ell_2$ denote the log-likelihood ratios associated with their respective outputs, referred to as *a priori*. It can easily be shown (by a direct computation or by applying Kullback's principle; see Sect. 4.2.6) that a decision jointly based on both channel outputs, i.e., the most likely assumption as regards the transmitted bit given $\ell_1$ and $\ell_2$, has as *a posteriori* log-likelihood ratio the sum $\hat{\ell} = \ell_1 + \ell_2$. This result extends immediately to an arbitrary number $n$ of replicas:

$$\hat{\ell} = \sum_{i=1}^{n} \ell_i. \tag{5.13}$$

The additivity of the log-likelihood ratios of independent replicas can be thought of as the main reason why they are useful. It is easily shown that the reliability $|\hat{\ell}|$ is larger in the average than that of the individual replicas, meaning that the reliability of the binary decision associated with the sign of $\hat{\ell}$ is improved. We moreover recall that, in the presence of additive white Gaussian noise, the log-likelihood ratio of any received bit is immediately available as proportional to the output of the receiver matched filter (see Sect. 3.3). Decoding then can directly process this analog output.

Would the log-likelihood ratios of the replicas not be available, but only their binary values, replacing the binary symbol 0 by +1 and 1 by $(-1)$ would define 'false' log-likelihood ratios $\{\ell'_i\}$ (the magnitude of which no longer measures a reliability) which may be used instead of $\{\ell_i\}$ in the decision rule (5.13), resulting in a hard decision equivalent to majority voting: the sign $\pm$ of the sum meaning the binary symbol 0 or 1; in the case where this sum is 0, which can occur only if the number of replicas is even, the two binary symbols are equally probable and the result can be chosen at random. Assigning values $\pm 1$ to given binary symbols is moreover a means to measure the relative reliability of intermediate decisions. Then the module of a log-likelihood ratio no longer provides an absolute estimate of the reliability of a decision, but a decision is still the more reliable, the larger is its log-likelihood ratio.

Let the LDPC code to be decoded be that described in Sect. 5.5.6, obeying the set of parity-check equations Eq. (5.12). Let us consider the first of them, labelled $(C_u)$. Solving it with respect to $i_a$ results in:

$$i_a = i_b \oplus c_u, \tag{5.14}$$

which shows that $i_b \oplus c_u$ is a replica of $i_a$. Two replicas of $i_a$ are thus available, and they are independent as involving different bits transmitted over a channel assumed memoryless, hence where errors separately affect each symbol. We refer to $i_a$ as the *trivial replica* and to any combination of other bits equal to $i_a$ in the absence of error, like $i_b \oplus c_u$ above, as a *compound* replica.

Applying the decision rule (5.13) to these two replicas needs computing the log-likelihood ratio of the sum modulo 2 of two received bits, say $b_1$ and $b_2$, of respective log-likelihood ratios $\ell(b_1)$ and $\ell(b_2)$. Let $p_{b_1} \triangleq \Pr(b_1 = 1)$ and $p_{b_2} \triangleq \Pr(b_2 = 1)$. The probability $p_{b_1 \oplus b_2} \triangleq \Pr(b_1 \oplus b_2 = 1)$ equals $p_{b_1}(1 - p_{b_2}) + (1 - p_{b_1})p_{b_2}$, which entails that

$$1 - 2p_{b_1 \oplus b_2} = (1 - 2p_{b_1})(1 - 2p_{b_2}).$$

In terms of the log-likelihood ratios, this equality becomes

$$\tanh\left[\ell(b_1 \oplus b_2)/2\right] = \tanh\left[\ell(b_1)/2\right] \tanh\left[\ell(b_2)/2\right],$$

where

$$\tanh(x) \triangleq \frac{\exp(x) - \exp(-x)}{\exp(x) + \exp(-x)}.$$

For brevity's sake, we introduce the function $\mathrm{t}(\cdot) \triangleq \tanh(\cdot/2)$, which enables more compactly rewriting the above implicit equality which expresses $\ell(b_1 \oplus b_2)$ as:

$$\mathrm{t}[\ell(b_1 \oplus b_2)] = \mathrm{t}[\ell(b_1)]\mathrm{t}[\ell(b_2)].$$

This equality extends to the sum modulo 2 of an arbitrary number $m$ of bits:

$$\mathrm{t}[\ell(b_1 \oplus b_2 \oplus \ldots)] = \prod_{i=1}^{m} \mathrm{t}[\ell(b_i)].$$

We also introduce the inverse function $\mathrm{t}^{-1}(\cdot)$ of $\mathrm{t}(\cdot)$, i.e., the function such that $\mathrm{t}^{-1}[\mathrm{t}(x)] = x$, which explicitly reads

$$\mathrm{t}^{-1}(\cdot) = \ln\left(\frac{1+\cdot}{1-\cdot}\right). \tag{5.15}$$

Due to the additivity of the log-likelihood ratios of independent replicas, and using Eq. (5.15) to express the log-likelihood ratio of the sum modulo 2 (5.14), we may write the *a posteriori* log-likelihood ratio of $i_a$ with respect to the first parity-check equation $C_\alpha$ as the sum:

$$\hat{\ell}_u(i_a) = \ell(i_a) + \mathrm{t}^{-1}\{\mathrm{t}[\ell(i_b)]\mathrm{t}[\ell(c_u)]\} \tag{5.16}$$

where we use a caret for denoting an *a posteriori* log-likelihood ratio; the subscript tells what parity check equations have been taken into account: a single one, namely $(C_u)$, in the above expression. This equality shows that $\hat{\ell}_u(i_a)$ consists of the sum of the initial log-likelihood ratio $\ell(i_a)$ of $i_a$, to be referred to as 'intrinsic', and an 'extrinsic' term which only depends on bits other than $i_a$.

Since the second parity-check equation $(C_v)$ also involves the information bit $i_a$, we may use it for further improving the estimate of the log-likelihood ratio of $i_a$. In order to account for $(C_v)$, we may replace the *a priori* log-likelihood ratio $\ell(i_a)$ with $\hat{\ell}_u(i_a)$ which has already been improved by accounting for the parity check bit $c_u$. We thus obtain:

$$\hat{\ell}_{u,v}(i_a) = \ell(i_a) + \mathrm{t}^{-1}\{[\mathrm{t}[\ell(i_b)]\mathrm{t}[\ell(c_u)]\} + \mathrm{t}^{-1}\{\mathrm{t}[\ell(i_c)]\mathrm{t}[\ell(c_v)]\}. \tag{5.17}$$

Again, this *a posteriori* log-likelihood ratio is the sum of the intrinsic *a priori* log-likelihood ratio $\ell(i_a)$ and an extrinsic term which involves a larger number of received bits.

The process of computing the *a posteriori* log-likelihood ratio can be extended to all parity-check equations which involve $i_a$ and, of course, it can be used for similarly reassessing the log-likelihood ratios of the other information bits $i_b, i_c, \ldots$

Moreover, every time the log-likelihood ratio of an information bit is reassessed, that it can improve the log-likelihood ratio estimates of other information bits enables in turn further improving the estimate of its own log-likelihood ratio. Iterated decoding thus becomes possible: the *a posteriori* log-likelihood ratio obtained using some of the parity-chek equations (5.12) is substituted for the corresponding *a priori* one in another parity-check equation before a new *a posteriori* log-likelihood ratio is computed. This process may be repeated several times but its benefit decreases with the number of iteration steps. When doing so, care should be exerted for avoiding that the *a priori* (intrinsic) log-likelihood ratio of an information bit appears more than once in the expression of its *a posteriori* log-likelihood ratio. For lack of this precaution, the condition that the computation of the *a posteriori* log-likelihood ratios should involve independent replicas would be violated and would result in meaningless values. How such an iteration is implemented will be expounded in the next section.

### 5.5.9 Decoding a Turbocode

The name 'turbocode' coined by Berrou and Glavieux actually refers to the iterated process used for decoding it, which resembles the way a turbo-compressed engine uses its exhaust gas.

As for an LDPC code, the aim of decoding a turbocode is to reassess the log-likelihood ratios of the information bits, given the *a priori* log-likelihood ratios of the incoming bits and taking into account the constraints of the code. The reassessed probabilities will be referred to as *a posteriori* probabilities. Such 'soft-input soft-output decoding' (SISO) has been introduced above in Sect. 5.5.7. Decoding an LDPC code as described in Sect. 5.5.8 was already of this kind. Then, we may think of the decoding process as consisting of computing the *a posteriori* log-likelihood ratio of each information symbol in terms of the *a priori* log-likelihood ratios of all the received symbols. By *a priori* log-likelihood ratio, we now mean the log-likelihood ratio known at some present stage of the decoding process, prior to further processing, while an *a posteriori* log-likelihood ratio means that it resulted from taking into account some constraints yet ignored, or somehow improving the way constraints were accounted for.

The dependence between the encoded bits created by the rate-1 encoders of Fig. 5.5 is not as simple as that expressed by a parity-check equation since it involves the entire past of the information message. Let us just mention that algorithms exist which, similarly to Eq. (5.16), can exploit the encoding constraints in order to reassess the log-likelihood ratios of the information bits: the soft-output Viterbi algorithm (Battail 1987a, Hagenauer and Hoeher 1989) and the Bahl, Cocke, Jelinek and Raviv (BCJR) algorithm (Bahl et al. 1974). These algorithms enable writing the *a posteriori* log-likelihood ratio of any symbol, say $\hat{\ell}$, similarly to Eq. (5.16), as the sum of two terms, referred to as 'intrinsic' and 'extrinsic', respectively. The first one, $\ell_{\text{in}}$, is the *a priori* log-likelihood ratio of the symbol itself (which was referred to as the trivial replica in Sect. 5.5.8). The second one, $\ell_{\text{ex}}$, involves other symbols through the constraints newly taken into account and can be interpreted as the log-likelihood ratio of a 'compound' replica. The trivial and compound replicas are independent as written in terms of disjoint sets of symbols. According to Eq. (5.13) but with a simplified notation, the corresponding *a posteriori* log-likelihood ratio of an information bit is thus the sum of the intrinsic and extrinsic log-likelihood ratios:

$$\hat{\ell} = \ell_{\text{in}} + \ell_{\text{ex}}. \tag{5.18}$$

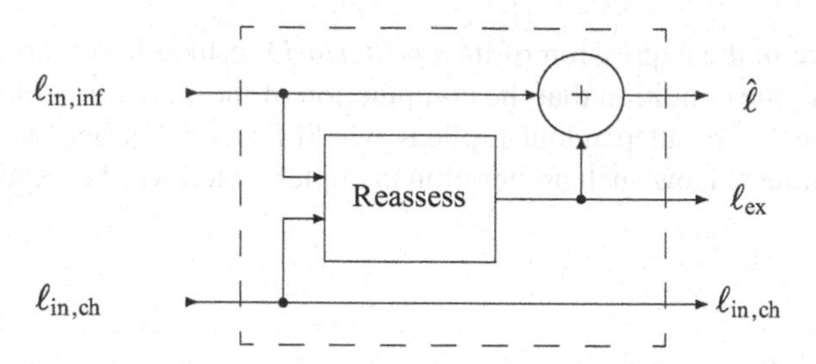


**Fig. 5.6** Elementary SISO decoder. Its two inputs are the sequence of *a priori* log-likelihood ratios of the information bits and that of the check bits, denoted by $\ell_{\text{in,inf}}$ and $\ell_{\text{in,ch}}$, respectively. Its three outputs are the *a posteriori* log-likelihood ratio of the information bits, $\hat{\ell}$, its extrinsic log-likelihood ratio $\ell_{\text{ex}}$ and the *a priori* log-likelihood ratio of the check bit which passes through the device without change. All these quantities are analog and + denotes ordinary addition. The *box* labelled 'Reassess' computes the extrinsic log-likelihood ratio $\ell_{\text{ex}}$

We consider in the sequel elementary SISO decoders which implement the decoding rule (5.18) and are organized as in Fig. 5.6. For interpreting this figure, it should be kept in mind that the decoder outputs at a given instant actually depend on the past input *sequences* of log-likelihood ratios.

Let us now assume that we have an optimum (or nearly optimum) decoding rule for each of the two systematic codes (to be referred to as 'component codes') consisting of the two check sequences generated by the rate-1 encoders of the turbo-encoder each associated with the corresponding information sequence. Both information sequences contain the same information bits and differ only as regards their order. According to Eq. (5.18), we may write the *a posteriori* log-likelihood ratio of any symbol which results from decoding the first code as:

$$\hat{\ell}_1 = \ell_{\text{in}} + \ell_{\text{ex},1}, \tag{5.19}$$

where $\ell_{\text{ex},1}$ only takes account of the constraints of the first code. The same information symbol is present in the input sequence to the second encoder, at a location determined by the interleaver operation. We may thus consider the *a posteriori* log-likelihood ratio from the first decoder (fed by the check sequence generated by the first rate-1 encoder) as the *a priori* log-likelihood ratio of this symbol for the decoder which operates on the second component code, and write the decoding rule of this symbol with respect to the second code:

$$\hat{\ell}_2 = \hat{\ell}_1 + \ell_{\text{ex},2} = \ell_{\text{in}} + \ell_{\text{ex},1} + \ell_{\text{ex},2}, \tag{5.20}$$

where the extrinsic log-likelihood ratio $\ell_{\text{ex},2}$ is computed in terms of *a priori* log-likelihood ratios consisting of the *a posteriori* log-likelihood ratios delivered by the first decoder. The second equality results from Eq. (5.19).

A very important fact is that the decoding process can now be *iterated*. Indeed, once all the symbols have been decoded by the second decoder, we may think of the *a posteriori* log-likelihood ratios from the second decoder as being improved estimates which can be used as *a priori* log-likelihood ratios in the first one. However,

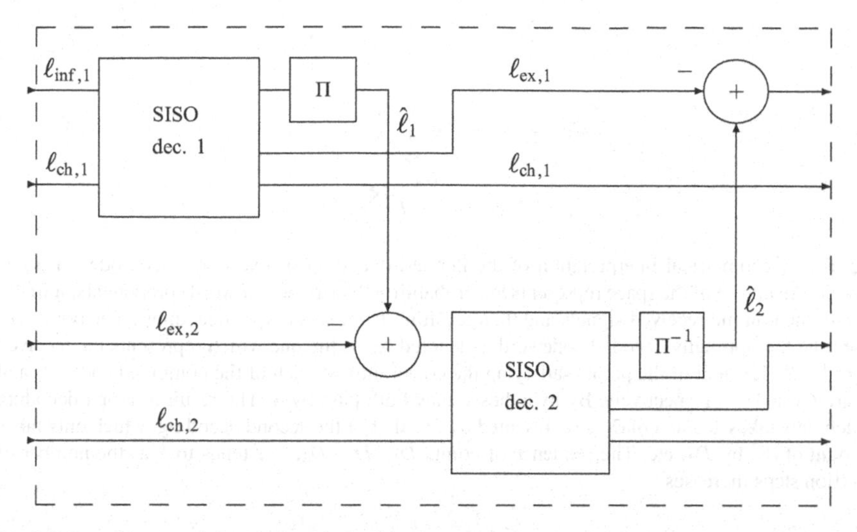


**Fig. 5.7** Decoding module (*dashed box*) for the turbocode of rate 1/3 generated by the encoder of Fig. 5.5. Iteration results from cascading a number of such modules. Contrary to Fig. 5.5, all the quantities considered are analog: the log-likelihood ratios $\ell_{inf,i}$ and $\ell_{ch,}$ of the information and check bits, respectively, and extrinsic log-likelihood ratios as defined in the text, $\ell_{ex,i}$, with $i = 1$ or 2. The boxes labelled 'SISO dec.' are identical to the elementary SISO decoder of Fig. 5.6. The extrinsic log-likelihood ratio input is 0 in the first module, since then no extrinsic log-likelihood ratio has yet been generated

the term $\ell_{ex,1}$ in the *a posteriori* log-likelihood ratio as expressed by Eq. (5.20) has already been calculated when the the first decoder was used, so it should be *subtracted* from the log-likelihood ratios fed to it. Provided the extrinsic log-likelihood ratio originating from a decoder is systematically subtracted[3] from the *a posteriori* log-likelihood ratio from the other one used as input to this decoder, as shown in Fig. 5.6, the decoding process can in principle involve an arbitrary large number of iteration steps. In practice it almost always converges, although expressing theoretical conditions of its convergence is very difficult. The number of iteration steps has actually to be limited, and criteria for stopping the iteration as soon as a good enough result has been obtained can be used. The complex task of decoding the turbocode as a whole has been reduced to a succession of alternate decodings of the component codes, each of which is easily performed.

As regards the implementation of the iterative decoding process, we may schematically think of cascading several decoding modules, each performing an iteration step. One of these modules is represented in Fig. 5.7. Its inputs can be connected to the outputs of the previous one, and its outputs to the inputs of the next one. Besides SISO decoders for each of the component codes, this elementary decoder involves an interleaver and a de-interleaver so as to ensure that both SISO decoders operate on

[3] Failing to do so would increase the magnitude of the computed *a posteriori* real value without improving its reliability; remember that the magnitude of a log-likelihood ratio is intended to measure the reliability of the corresponding bit.

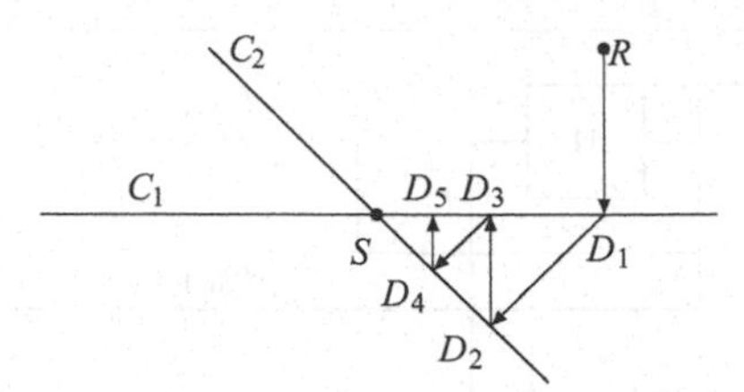


**Fig. 5.8** A geometrical interpretation of the iterated decoding of two combined codes of a turbocode. Any point of the space represents the probability distribution of word components, initially conditioned on the received signals and then resulting from soft-output decodings. The point representing the optimally decoded codeword is denoted by $S$, the one which represents the received word by $R$. The loci of the points satisfying the constraints of each of the combined codes $\mathcal{C}_1$ and $\mathcal{C}_2$ are $C_1$ and $C_2$, respectively. By hypothesis, they both pass by $S$. The result of a first decoding which only takes account of $\mathcal{C}_1$ is represented by $D_1$, that of the second decoding, which only takes account of $\mathcal{C}_2$, by $D_2$, etc. The sequence of points $D_1$, $D_2$, $D_3$, ... tends to $S$ as the number of iteration steps increases

the same information bit. The *a posteriori* log-likelihood ratios of successive bits are correlated due to the dependency created by encoding and, in particular, the errors of each SISO decoder appear as bursts of decoded symbol errors. The interleavers used in the decoding device spread out the *a posteriori* log-likelihood ratios (and their possible errors occuring in a burst) before the sequence is fed to the other one, thus ensuring that the successive input *a priori* log-likelihood ratios are uncorrelated, a condition for SISO decoding being valid. This is also an important role of the interleaver, besides that of shaping the overall weight distribution of the turbocode which we already mentioned.

The iteration process can be given a very simple picture in order to illustrate why it improves decoding (Fig. 5.8). We assume that we can define a space where a probability distribution is represented by a point and where the closeness of two distributions can be interpreted as the distance between the corresponding points (this can be done using the concept of cross-entropy of two distributions introduced in Sect. 4.2.6 above). Then the coding constraint of a code can be represented by a subspace, say a line, and a probability distribution which satisfies the constraints of both codes is represented by the point where the corresponding lines intersect. The decoding firstly performed may be thought of as determining the point of the line which represents the constraint of the first code the closest to the one which represents the received point. To take account of the constraint due to the second encoding, we must start from this point and determine the point on the second line the closest to it. Since the second line only represents the constraint due to the second code, we must again determine the point on the first line the closest to the lastly obtained one, etc. After a sufficient number of iteration steps, a point close enough to the intersection of the two lines is reached and almost optimum decoding of the two combined codes has been performed. At variance with the figure, the two loci associated with the codes have more than a single intersecting point. There are indeed as many intersecting

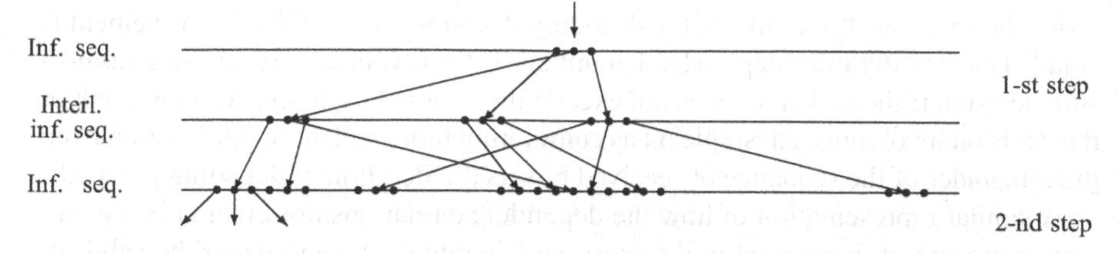


**Fig. 5.9** Diffusion of the dependence in the iterated decoding process. The lines labelled 'Inf. seq.' and 'Interl. inf. seq.' refer to the original information sequence and the one which results from its interleaving, respectively, at the 1-st iteration step and the 1-st half of the second one, from *top* to *bottom*. The points locate particular bits in the sequence. See the text for comments

points as codewords, so the convergence illustrated by Fig. 5.8 is merely a local property.

We can also illustrate how the iterated decoding proceeds in the information sequence with the help of Fig. 5.9. Remember that we interpreted the encoding as a kind of indirect repetition, i.e., where an information bit is repeated as combined modulo 2 with other information and check bits. Let us assume that this combination only concerns the two neighbours of each information bit (the combination also involves check bits which are not represented in the figure). The top horizontal line represents the information sequence at the beginning of the first iteration step. We consider the particular bit indicated by the vertical arrow at the top of the figure. As assumed, the encoding by the first component code made it dependent on its two neighbours. This means that, at the first half of the first iteration step (i.e., decoding in terms of the 1-st component code), the log-likelihood ratio of the bit initially considered as well as those of its neighbours have been recomputed in terms of their own *a priori* log-likelihood ratios and that of the corresponding checks bits. The *a posteriori* log-likelihood ratios thus obtained are used as *a priori* log-likelihood ratios for the second half of the first iteration step, which consists of decoding the interleaved information sequence in terms of the second component code. The arrows indicate the location of the interleaved bits. Due to the encoding by the second code, the neighbours of each of these bits in the interleaved information sequence have been made dependent and their *a posteriori* log-likelihood ratios have been recomputed according to the second component code. At the first half of the second iteration step, the information bits are again ordered according to the original sequence, thanks to the de-interleaver operation. The bits initially considered have recovered their original place, but the neighbours of the corresponding bits in the interleaved sequence are located somewhere, generally far apart in the de-interleaved sequence. Iteration consists of repeating this process. Clearly, there is a diffusion of the dependence relationships through the entire information sequence, which eventually makes any of the *a posteriori* log-likelihood ratio of an information bit depend on increasingly many other ones in an increasingly more complex fashion. We have a kind of 'diffusion of the dependence' as the decoding iteration proceeds according to a kind of oil stain effect. Remember that, in the average, the decoding process results in an increase of the magnitude of the *a posteriori* log-likelihood ratios, hence of the reliability of the decoding decisions. Even if the improvement is small at the first iteration steps, a large number of steps eventually results in an almost sure decision if the code rate does not exceed the channel capacity. We may think of the decision involving each single bit as cumulating more and more information from the remainder of the sequence of received bits as the iteration of decoding proceeds.

A similar representation of how the dependence relationships created by the encoding are successively used in the process of iterated decoding would be valid for low-density parity-check codes, except that there is a single sequence of information bits. Randomness is not provided by an interleaver, but by the random association of the information bits in parity checks equations.

### *5.5.10 Variants and Comments*

In the device represented in Fig. 5.7, we assumed that the *a priori* log-likelihood ratios of the check bits pass through the elementary SISO decoders without change. As represented in Fig. 5.6, these decoders only update the *a posteriori* log-likelihood ratios of the information bits. However, the formulas which express the *a posteriori* log-likelihood ratio of any received bit in terms of the *a priori* log-likelihood ratios of all the received bits enable computing the *a posteriori* log-likelihood ratios of the check bits, too. The choice of not doing so has the advantage of keeping samples of the channel output throughout the iterated decoding process, to the benefit of its stability, but schemes where both the log-likelihood ratios of the information and check bits are updated can be contemplated.

The iterated decoding of turbocodes can actually be implemented quite differently from the scheme of Fig. 5.7 since the same device may be used several times at a speed greater than the input bit rate. Then the iteration can be implemented using a single elementary decoder connected in a feedback loop.

We already stated that the information bits and the check bits, which appear as separate outputs in the encoder of Fig. 5.5, are generally transmitted 'serially', i.e., in alternation according to a regular time pattern. Inspired by turbocodes, we suggest the rather general scheme of Fig. 5.10 for representing a random-like code. Three blocks are serially connected to each other: an $n$-replicator, an interleaver and a rate-1 encoder. Of course, each of these blocks results from a serial rearrangement of the elementary components (memory cells, connections, ...) of the corresponding blocks of Fig. 5.5. Then, each of the blocks of Fig. 5.10 performs one of the three functions which can be expected from a random-like encoder since they provide *redundancy, randomness* and *mutual dependence*, respectively. We may thus think of Fig. 5.10 as describing a kind of paradigmatic random-like encoder. Moreover, it turns out that some improvements can be obtained from variants of this scheme, especially if the number of copies generated by the replicator is made irregular, some symbols being repeated more than others while keeping the overall rate constant. The ease of decoding demands however that the overall code can be decomposed into simple

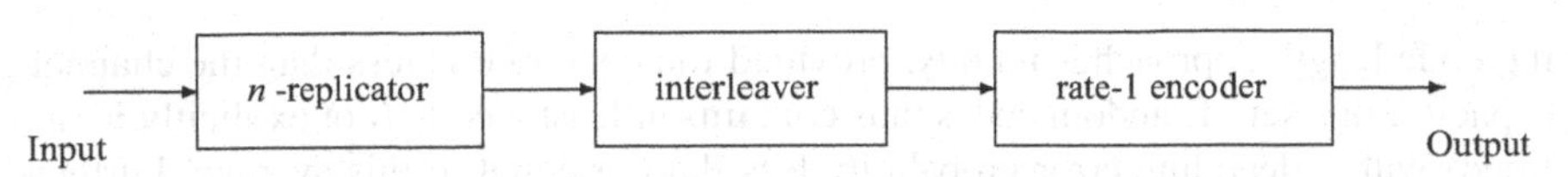


**Fig. 5.10** A fully serial schematic representation of an encoder of rate $1/n$ generating a random-like code. The box named $n$-replicator designates a device which repeats $n$ times its input symbol. The interleaver scrambles the sequence it receives and the rate-1 encoder creates dependency between a number of successive symbols. Among the three devices, only the $n$-replicator generates redundancy

component codes, so as to enable splitting the whole task of decoding into simple alternate iterated decodings.

### *5.5.11 Error-Correcting Codes Defined by Non-Mathematical Constraints: Soft Codes*

The flexibility of LDPC codes, at least as regards their structure and decoding means, makes them likely candidates to the yet unidentified genomic error-correcting codes (which are shown to be needed in Sect. 8.1), except that mathematically defined constraints seem to be quite foreign to the living world. An error-correcting code needs to be redundant and, in any practical system, this result is obtained by an encoding rule which specifies *constraints*. In engineering, these constraints are *mathematical*, thus precisely defined and easily implemented by electronic devices. Any set of constraints, however, similarly defines a subset among all the sequences of a given length, hence a potential error-correcting code. Constraints of mathematical character are obviously very convenient for human engineers, but a set of words is endowed with error correction ability by any kind of constraints. In error-correcting codes of biological origin, these constraints concern the *support* of symbolic sequences, e.g., the DNA molecule in the case of genomes, as well as the sequence of symbols itself. The constraints on DNA molecules are of physical-chemical character and those concerning the sequence of symbols may be of linguistic character, not necessarily limited to mathematically defined equalities as the parity-check equations. Most of them consist of exclusion rules which forbid certain symbol sequences. Restricting the possible sequences to some subset thus provides redundancy. Error-correcting codes defined by such non-mathematical constraints will be referred to as *soft codes*. How they possibly look like will be examined in Sect. 8.4 of the second part.

The efficiency of soft codes should first be established. Have they any error-correction ability? Rather surprisingly, the answer is *yes*. Soft codes may even be expected to be good in this respect. Constraints which define soft codes are presumably foreign to any intended error correction. However, it turns out that the most successful means of error correction is not explicitly designed for this purpose: we saw in Sect. 5.4.2 that Shannon used *random coding* for proving the fundamental theorem of channel coding (Shannon 1948). He could not design a code reaching the limit of what is possible, i.e., the channel capacity, but he was able to show that the *average* probability of decoding error of a set of randomly chosen codes vanishes as the code length approaches infinity, provided the code rate is less than the channel capacity; this set of random codes thus contains at least a code (not explicitly identified) with a decoding error probability less than or equal to this average. Further studies revealed moreover that a code chosen at random has with high probability a good error-correcting ability. Codes defined by arbitrary constraints are thus likely to efficiently correct errors. Communication engineers believed during many years that the folk theorem 'all codes are good, except those we can think of' was true. This statement was even given a formal expression in (Coffey and Goodman 1990). Although the exception was denied in 1993 when the turbocodes were invented, it remains generally true. Thus, paradoxically, being defined by outer constraints rather than designed for communication purpose, the genomic soft codes assumed to exist in Sect. 8.1.3 are likely to be good. That the error-correction ability is a by-product of devices having another function in the life processes may be thought of as an example of 'tinkering', illustrating a typical approach of Nature (see Sect. 9.4).

Another reason why soft codes should efficiently correct errors can be drawn from the Shannon-McMillan theorem (see Sect. 4.3.5). Indeed, this theorem shows that the set of typical sequences generated by a stationary and ergodic source is similar to the codewords of a code of rate $H$, where $H$ is the source entropy.

## References


Bahl, L. R., Cocke, J., Jelinek, F., & J. Raviv, J. (1974). Optimal decoding of linear codes for minimizing symbol error rate. *IEEE Transaction on Information Theory*, IT20(2) 284–287.

Battail, G. (1987a). Pondération des symboles décodés par l'algorithme de Viterbi. *Annales Télécommunications*, 42(1–2), 31–38.

Battail, G. (1987b). Le décodage pondéré en tant que procédé de réévaluation d'une distribution de probabilité. *Annales Télécommunications, 42*(9-10), 499–509.

Battail, G. (1989). Construction explicite de bons codes longs. *Annales Télécommunic., 44*(7–8), 392–404.

Battail, G. (1993). Pseudo-random recursive convolutional coding for near-capacity performance. 2nd International Symposium on Communication Theory and Applications, Ambleside (UK), 12-16 July 1993. (Communications theory and applications II, B. Honary, M. Darnell, P. Farrell, Eds., HW Communications Ltd., pp. 54–65).

Battail, G. (1996). On random-like codes. *Proceeding 4-th Canadian workshop on information theory*, Lac Delage, Québec, 28–31 May 1995. (Information Theory and Applications II, J.-Y. Chouinard, P. Fortier and T. A. Gulliver, Eds., Lecture Notes in Computer Science No. 1133, pp. 76–94, Springer).

Battail, G. (2000). On Gallager's low-density parity-check codes. International symposium on information theory. *Proceeding ISIT 2000*, p. 202, Sorrento, Italy, 25–30 June 2000.

Battail, G., Berrou, C., & Glavieux, A. (1993). Pseudo-random recursive convolutional coding for near-capacity performance. *Proceeding GLOBECOM'93, Communication Theory Mini-Conference*, Vol. 4, pp. 23–27, Houston, U.S.A.

Battail, G., Decouvelaere, M. (1976), "Décodage par répliques", *Ann. Télécommunic.*, Vol. 31, No. 11-12, pp. 387–404.

Battail, G., Decouvelaere, M., & Godlewski, P. (1979). Replication decoding. *IEEE Transaction on Information Theory, IT-25*(3), 332–345.

Coffey, J. T., & Goodman, R. M. (1990). Any code of which we cannot think is good. *IEEE Transaction on Information Theory, IT-36*(6), 1453–1461.

Gallager, R. G. (1962). *Low-density parity-check codes. IRE Trans. on Inf. Th.*, Vol. IT-8, pp. 21–28.

Gallager, R. G. (1963). *Low-density parity-check codes*. Cambridge: MIT Press.

Gallager, R. G. (1965). A simple derivation of the coding theorem and some applications. *IEEE Transactions on Information Theory, IT-13*(1), 3–18.

Hagenauer, J., & Hoeher, P. (1989). A Viterbi algorithm with soft-decision outputs and its applications. *Proceeding GLOBECOM'89*, pp. 47.1.1–47.1.7 (Nov.). Dallas, U.S.A.

Khinchin, A. I. (1957). *Mathematical foundations of information theory*. Ney York: Dover.

Kolmogorov, A. N. (1956). On the Shannon theory of information transmission in the case of continuous signals, in (Slepian 1974, pp. 238–244).

Massey, J. L. (1963). *Threshold decoding*. Cambridge: MIT Press.

Shannon, C. E. (1948). A mathematical theory of communication. *The Bell System Technical Journal, 27,* 379–457, 623–656. (Reprinted in Shannon and Weaver 1949, Sloane and Wyner 1993, pp. 5–83 and in Slepain 1947, pp. 5–29).

Shannon, C. E. (1949). Communication in the presence of noise. *Proceeding IRE*, pp. 10–21. (Reprinted in Sloane and Wyner 1993, pp. 160–172 and in Slepian 1974, pp. 30–41).

Shannon, C. E., & Weaver, W. (1949). *The mathematical theory of communication*. Urbana: University of Illinois Press.

Slepian, D. (Ed.). (1974). *Key papers in the development of information theory*. Piscataway: IEEE Press.

Sloane, N. J. A., & Wyner, A. D. (Eds.). (1993). *Claude Elwood Shannon, collected papers*. Piscataway: IEEE Press.

# Chapter 6
# Information as a Fundamental Entity

**Abstract** Chapter 6 considers information as a fundamental scientific entity. It begins with briefly expounding the variant of information theory referred to as algorithmic, which relies on computer science rather than communication engineering and gives more insight into the information concept. The main specific property of information is that it can be annihilated, especially when the medium which bears it is destroyed, but also shared, which entails that it can proliferate. Moreover, it appears as an emergent entity in a population of clones. The relationship of information and physical entropy is examined. Physical entropy can be interpreted as measuring the amount of information which is lost because the actual configuration of a physical system at the atomic or molecular scale is unknowable. We interpret Boltzmann's constant as a signal-to-noise ratio. We show that our definition of information entails the non-existence of the omniscient Laplace's demon, and confirm that information is non-physical. As an abstract entity which necessarily resides in the physical world, information actually appears as a *bridge* between the *abstract* and the *concrete*.

This chapter is devoted to investigate the status of information as a fundamental entity and its relationship with the physical world. Section 6.1 contains a short discussion of the algorithmic information theory, which does not rely on probabilities and provides a renewed view on the information concept. It turns out, however, that the algorithmic complexity which it uses for measuring the information quantity borne by a single sequence is generally not computable. The problems often become tractable if probabilities are re-introduced, which moreover provides a link with Shannon's information theory and underlines the unicity of the information concept.

Section 6.2 examines the consequences of the *sharing property* which is specific to information: an information can be copied, so repeatedly copying some ancestral information results in a population of clones. We examine the information borne by such a population and show that new information is created only when errors make differences within its members.

In order to understand the connection of information with the physical world, we briefly examine in Sect. 6.3 the relationship of information with the physical entropy. Schrödinger and Brillouin proposed that the information quantity should be defined as the negative of the entropy of physics; they named it *negentropy*. This proposal results however in a contradiction with the sharing property of information, which leads us to reject it. We propose instead to take information as a fundamental,

G. Battail, *Information and Life*, DOI 10.1007/978-94-007-7040-9_6,

non-physical entity, the negative of its quantitative measure being then likened to the physical entropy.

Finally, and as a conclusion of the first part, Sect. 6.4 presents information as a *bridge* between the abstract and the concrete. The second part of the book in its entirety will illustrate this statement.

## 6.1 Algorithmic Information Theory

Chapter 4 above was devoted to basic concepts of Shannon's information theory, based on probabilities and deriving from the analysis of communication engineering. We now consider an alternative way of introducing information which derives from computer science. It is usually attributed, independently, to the great mathematician Andreï N. Kolmogorov and to Gregory Chaitin (who then was 15). A recent document containing an ample bibliography (Gàcs and Vitànyi 2011) actually states that a third man, Raymond J. Solomonoff, who died in 2009, published in 1964 the very basic ideas of the theory, earlier than Kolmogorov (who stated that he arrived at the same conclusions before he was aware of Solomonoff's works). Chaitin expounded the algorithmic information theory in (Chaitin 1988); a recent, very readable and fascinating book on the subject, by the same author, is (Chaitin 2005). Far from opposing algorithmic and Shannon's information theory, we show that these seeming different concepts actually pertain to a single fundamental entity. The algorithmic information theory will actually provide further insight on the very concept of information and on its relationship with semantics. Despite its interest, it cannot replace everywhere Shannon's information theory which remains needed for dealing with literal communication and other applications which involve probabilities.

Let us first state the idea which founds the algorithmic information theory:

> the *information message* associated with some sequence is the *shortest binary input* to a universal computer which instructs it in *how to generate* the given sequence. The length $k$ of this message is referred to as the *algorithmic complexity* of the sequence and measures the information quantity it bears.

A computer is referred to as universal if it can compute any mathematical function, as does any modern computer. The first universal computer has been introduced by Alan Turing and it is referred to as the 'Turing machine'. The definition of the algorithmic complexity just given seems to depend on a particular computer and on its programming language, but it is actually almost independent of them (at least for long enough sequences). We may think of the computer as 'encoding' the input information message into the sequence it generates, since this case is not fundamentally different from that considered at the beginning of Sect. 3.4.2. We now illustrate this definition by some examples, assuming that the instructions to the computer are written in plain English and that its output is a printed binary sequence.

**Example a).** The binary sequence:

10101010101010101010101010101010

is made of the periodic repetition of '10', 16 times. Clearly, a computer input for generating this sequence may be as short as: "Print 16 times '10'". Another equivalent one would be "Print the first 32 digits of the binary development of 2/3". Would the sequence result from repeating '10' infinitely many times, then the first input would become "Print indefinitely '10'" and the second one would become *shorter*, namely, "Print the binary development of 2/3".

**Example b).** The sequence:

01101010000010011110011001100 11

is not periodic and looks random. However, it consists of the first 32 binary digits of the development of the irrational number[1] $\sqrt{2}-1$ so the computer input "Print the first 32 digits of the binary development of $\sqrt{2}-1$" suffices for generating it. Indefinitely continued, the sequence is generated by the shorter input "Print the binary development of $\sqrt{2}-1$".

**Example c).** The sequence:

10101110110001111100110100100001

looks random, too, but is not actually so as generated by the following deterministic process. Provided its initial content is non-zero, a binary shift register of length $\mu$ (as depicted in Fig. 5.4 and defined in its caption), endowed with a feedback and having as input the all-0 sequence, generates a periodic, infinitely repeated motif of length at most $2^{\mu}-1$. The first 31 bits of the above sequence is such a motif of maximum length, often referred to as 'pseudo-random', that a 5-bit shift register generates. The last bit, 1, is the first one of the repeated motif. The computer input for generating this sequence, periodically repeated or not, is short.

**Example d).** The sequence:

01001010101111000011101100101100

was obtained by drawing the successive digits '0' and '1' at random with the same probability 1/2, independently of each others. It is likely that no computer input for generating it exists which is much shorter than

"Print '01001010101111000011101100101100'"

Furthermore, one could not describe a random continuation of this sequence without lengthening the input. Writing the following outcomes is the only general means for doing so.

---

[1] Chosen as example of an irrational number smaller than 1.

The sequences of examples a), b) and c) have thus a description almost independent of their length, at variance with the sequence of example d), for which one may not expect to give a description shorter than itself.

**Example e).** The sequence

$$01101011111111011011111100101110$$

results from drawing bits at random, too, but the bit '1' was drawn with a probability of 3/4. One may give a description of it shorter than its length $n$, for instance by counting the number of '1's, say $m$, and by indicating the rank of this sequence in the list of all binary sequences of length $n$ comprising $m$ '1's, ranging in lexicographic order. Describing the original sequence then demands a number of bits equal to $\log_2(n) + \log_2\binom{n}{m}$, close to $\log_2(n)+n\mathcal{H}_2(m/n)$, where $\mathcal{H}_2(x)$, defined according to Eq. (4.11) as $\mathcal{H}_2(x) = -x\log_2(x)-(1-x)\log_2(1-x)$, is the binary entropy function. Applying Stirling formula shows indeed that $n\mathcal{H}_2(m/n)$ is an approximation of $\log_2\binom{n}{m}$ for $n$ and $m$ large enough. In this case, describing the sequence does not demand as many bits as the sequence itself, namely $n$, but except for the additive term $\log_2(n)$ this length equals $n$ multiplied by the factor $\mathcal{H}_2(m/n)$ which is less than 1, and the smaller, the frequencies of '0' and '1' are more unequal. In this case, the length of the computer input increases when that of the sequence increases, as in Example d) but not as fast.

**Example f).** Let us now consider the algorithmic complexity of the sequence generated by the encoder of Fig. 5.4. We may first write a program telling what operations the computer must perform on the bits of the input sequence so as to generate the encoded (output) sequence, then enter both this program and the encoder input sequence, i.e., the information message. The input to the computer can thus be split into two parts: a *program* and a *data string*, such that the former instructs the computer in how to process the latter. In the Examples a) to e), the same distinction can be made between a program and a data string. In any case, the program length is constant (and short for such simple examples). The length of the data string is very short in Examples a), b) and c), as long as the sequence to be generated in Example d), approximately proportional to its length but shorter in Examples e) and f).

The algorithmic complexity in Example f) is $k + \nu$, i.e., the length $k$ of the data string plus a constant $\nu$ which denotes the length of the program, understood as the message which specifies how the data should be processed. The length $k$ of the data string has been used in Sect. 3.4.2 for measuring the information quantity borne by the encoded sequence. We may think of the added constant $\nu$ as measuring the information which has been used for building the encoder and decoder. In the interpretation of information in a biological context given in (Battail 2009, 2011) and expounded above in Sect. 4.3.5, it corresponds to what is referred to as 'structural information', while the information quantity $k$ is that of the information which can be actually communicated over the channel, referred to as 'symbolic' in the same papers and above. Therefore, one may think of the algorithmic information theory as providing here the same information measure as the conventional (Shannonian) one, except that a constant is added to it, which corresponds to the structural information associated with the communication hardware. An arbitrary large number $k$ of information bits, hence $k$ shannons, can be communicated over the channel, while the information quantity $\nu$ is actually 'frozen' in the transmitting and receiving devices.

The remark above has moreover a broader validity than the Example f) since we already noticed in Sect. 4.3.5 that, thanks to the Shannon-McMillan theorem, channel coding can be considered as adequately (but approximately) representing the more general case of a stationary and ergodic source. The complexity measure of algorithmic information theory can actually be identified with Shannon's entropy (in cases where they are both relevant) plus an additive term which can be ignored for large enough sequences and can moreover be likened to structural information.

The algorithmic complexity appears as a valid measure of information quantity. In general, unfortunately, it is not computable. Of course, this feature severely limits its practical usefulness and mainly confines it to abstract reasonings.

**Some further comments** Further insight as regards the relationship of information and semantics will be provided by reflections about the algorithmic information theory, in addition to those prompted by source coding, discussed in Sect. 4.3.5. We may interpret indeed the shortest input to the computer which results in a given sequence being generated as performing the universal source coding of this sequence. We may interpret a program as having a *meaning* for the computer, that of specifying its output (for this peculiar computer with its peculiar programming language). The relationship between the input to a computer and its output can then be interpreted as a semantic rule, similarly to the relationship between the source symbols and the codewords in Huffman coding (see Sect. 4.3.4). The computer appears as a means for generating a symbolic sequence in response to another symbolic sequence. The algorithmic information theory thus provides an easy bridge between information and semantics.

Besides establishing relations between its input and output, which are symbolic sequences, a computer is a material object which can, and does in many instances, interact with the physical world by the agency of its output, according to rules which can be interpreted as semantic. Everything which was written about the relation of information and semantics in Sect. 4.3.5 remains true in the framework of the algorithmic information theory, which confirms the unicity of the information concept.

The length of the shortest program which instructs the computer in how to generate the given sequence, i.e., the algorithmic complexity of this sequence, seems unrelated to the probabilistic information measure of Shannon's information theory. We must remember, however, that this shortest program is most often unknown, unlike Example f) above. As stated above even its length is uncomputable. We are thus led to deal with programs as *random* sequences and, more precisely, as binary sequences where each bit is chosen at random with probability 1/2, independently of the others. Hence, a $k$-bit sequence has the probability $2^{-k}$ of being chosen as a tentative program. Once the computer recognizes that a sequence is a meaningful program, it is immediately executed: the computer prints the output sequence specified by this program, halts and resets itself. Once reset, it can execute another program.

A meaningful program thus cannot be a prefix to another meaningful program, which entails that the programs which specify the computer outputs constitute an irreducible set of binary sequences. In any way, they must be separated from each others so programs must be decipherable, just like the words of a code. Their lengths $\{k_i\}$ thus satisfy the Kraft inequality (4.32) extended to an infinite number of terms, namely:

$$\sum_{i=1}^{\infty} 2^{-k_i} \leq 1. \tag{6.1}$$

We may thus contemplate the following thought experiment: successively try random binary sequences subjected to the prefix condition as inputs to the computer and keep those which actually result in the computer printing an output sequence. It would result in a list of couples (binary program—generated binary sequence). The prefix condition then guarantees that the first program found which entails the printing of a particular binary sequence is actually the shortest one which has this result, so only the first found program which entails the printing of a sequence is its information message and needs to be kept in the list. Since the generated sequence itself can be a part of the program (as in Example d) above), this list would contain programs which can generate all the possible binary sequences. It is thus infinitely long, so it can be known only, at best, up to some finite rank. Moreover, it turns out that when fed by certain programs the machine will never halt, and a theorem[2] of theoretical computer science (similar to Gödel's in numbers theory) tells that it is impossible to predict whether a given program will behave so. This is obviously a long, complicated (and tedious) task, which implies in principle an infinitely long time interval. It is why we refer to it as a thought experiment.

The *a priori* probability of the $i$-th input sequence such that the computer halts, of length denoted by $k_i$, is $p_i = 2^{-k_i}$. It is the probability that the $i$-th output sequence in the list is generated by the computer. It is thus possible to associate a probability with each output sequence thus creating a link between Shannon's and algorithmic information theories.

Among the random binary inputs, certain result in a computer output. The programs which belong to this category obey the prefix condition and their lengths $\{k_i\}$, where $i$ is the rank of a program, satisfy the extended Kraft inequality (6.1). The other random inputs do not result in an output, however long they are. It thus makes sense to consider the probability that a random program results in an output, as did Chaitin (Chaitin 2005). As the lengths $\{k_i\}$ verify the extended Kraft inequality, it is less than 1, and even strictly less since some input sequences do not result in an output. Chaitin refers to it as the *halting probability* and denotes it by $\Omega$. Thus:

$$\Omega = \sum 2^{-k_i}, \tag{6.2}$$

[2] Chaitin credits Alan Turing for this theorem (Chaitin 2005).

where the sum should be performed for the lengths of all programs such that the computer halts. This number has a lot of properties which are discussed in (Chaitin 2005).

## 6.2 Emergent Information in Populations

Probably the most important property of information, which deeply differentiates it from physical entities, is its *sharing property*: an information can be copied, i.e., transferred onto a support different from the original one, which however does not lose it. Repeatedly copying some ancestral information thus results in a population of clones: information shares with life the ability to proliferate. We intend here to develop this remark, which leads us to consider information in populations.

Let us first consider a set of $N$ objects, regardless of their order[3]. It is assumed to contain $n_1$ identical objects of a 1-st type, $n_2$ identical objects of a 2-nd type, ..., and $n_m$ identical objects of an $m$-th type, with of course $m \leq N$ and $\sum_{i=1}^{m} n_i = N$. The following entropy-like formula enables measuring in shannons the information it bears:

$$H_{\text{set}} = -\sum_{i=1}^{m} (n_i/N)\log_2(n_i/N), \tag{6.3}$$

since the proportion $n_i/N$ of each type of element is an estimate of the probability of an element of this type. Thus, $H_{\text{set}}$ is the only available measure of the information borne by this set of objects.

Let us now consider an arbitrary sequence borne by some support, to be referred to as ancestral. It can be copied on another support, and its copies themselves can be copied. Let us refer to a set of $N$ copies of the given sequence as a *population*. This set is not ordered; in other words, it is not a sequence of sequences. We may use Eq. (6.3) as a measure of the information it contains due to the diversity of its elements. If the copying process is absolutely faithful and if the copies do not incur any modification, all its $N$ elements are identical, so $m = 1$, $n_1 = N$ and Eq. (6.3) entails $H_{\text{set}} = 0$. Then the population of clones, considered as a set of objects, contains no information. If the copied sequence is the representative of some information measured by $H$, the population of clones as a whole does not contain more information than each of its elements, namely, $H$. However, if at least one of the copies is not faithful to the original, an increase in the information borne by the population as a whole (besides that borne by each of its elements) results since its elements are no longer identical, and it is measured by $H_{\text{set}}$ according to Eq. (6.3). For instance if a single copy differs from the original, $H_{\text{set}} = (1/N)\log_2(N) - (1 - 1/N)\log_2(1 - 1/N) = \mathcal{H}_2(1/N)$, where the binary entropy function $\mathcal{H}_2(\cdot)$ has been defined by Eq. (4.11). As expected, this information quantity is small if $N$ is large.

[3] At variance with a sequence where the order of the symbols is highly relevant.

We may thus consider the population as bearing, regardless of that borne by its individual members, an information due to the *errors* which possibly make its elements distinct from each other. This creation of information is an *emergence phenomenon* at the population level. Copying one of the elements of a population become distinct from the initial clones originates in a distinct population, similar to that initially generated by the ancestral sequence, which is then similarly endowed with the information brought by errors. A simple example of such populations and of their evolution is given in Sect. 8.3.

That errors have a constructive role may look strange and even contrary to the common sense. It should be remembered, however, that the initial sequence is arbitrary, so an erroneous copy of it is just as arbitrary as the original. We noticed in Sect. 4.2.1 that there is no objective difference between 'useful information' and 'perturbing noise'. The difference only lies in what interests the destination as an actor in Shannon's paradigm. The occurrence of an error is a random event among others and the word 'error' should not be given a negative connotation.

From another point of view, a population of $N$ copies of a *same* ancestral sequence, where some of them are possibly affected by errors, constitutes in itself a kind of rudimentary error-correcting code. Would no errors have occurred, $N$ copies of each symbol of the sequence would be available. In the case of binary sequences, assuming that a symbol error occurs with probability $p_{su}$, $(1 - p_{su})N$ of these copies are in the average identical to the original one (provided $p_{su} < 1/2$, but the labelling of the symbols by 0 and 1 is arbitrary, so it suffices to swap them to fulfill this condition), comparing the number of identical symbols with a threshold equal to $\lceil N/2 \rceil$, where $\lceil x \rceil$ means the smallest integer strictly larger than $x$, provides the best estimate of the original symbol. The optimum decision about it thus results from majority voting. Replication decoding, as discussed in Sect. 5.5.8, is fully relevant here and majority voting could be replaced by the addition of the log-likelihood ratios of the symbols if they are available, according to Eq. (5.13). For a sequence with an alphabet of size $\alpha$, the threshold above which the number of identical sequences determines the most likely one becomes $\lceil N/\alpha \rceil$ since an error then results in the average in $(\alpha - 1)$ different symbols. Comparing the number of identical symbols in each position of the repeated sequence to the indicated thresholds results in the best estimate of the symbol at this position, and repeating this process for all symbols results in the best estimate of the original sequence as a whole. Of course, for any alphabet size, the probability that the decision about the original symbol is the more reliable, the larger the number of identical symbols which exceed the threshold. This rough error-correcting means is not foreign to biology, e.g., when decisions are taken by a 'parliament of cells'.

To summarize, a population of copies subjected to random errors bears some information in itself, besides that borne by the ancestral sequence, but also enables determining the most likely estimate of this ancestral sequence.

## 6.3 Physical Entropy and Information

### 6.3.1 Thermodynamics and Physical Entropy

The first law of thermodynamics (or Meyer's principle) states that the total energy of an isolated system is conserved, regardless of the form it may take (thermal, chemical, mechanical, electrical, ...). These forms can be converted into each other and therefore can be expressed using the same unit. It is why they are dealt with as distinct forms of a same entity. There is however a qualitative difference between them which may be expressed as a degradation of energy. Mechanical and electrical forms of energy are at the highest grade and heat at the lowest because in spontaneous physical processes the highest forms of energy are invariably degraded into heath (e.g., a car which brakes converts its kinetic energy into heat, or an electrical current produces heat when it flows through a resistor). The conversion of heat into mechanical or electrical energy, on the contrary, does not spontaneously occur but always needs manufactured devices which moreover convert only a fraction of the available heat into a higher form of energy. As regards heat itself, when two objects at different temperatures exchange heat, it spontaneously flows from the highest-temperature object to the other one, never the other way round, thus tending to equalize temperatures.

Thermodynamics is the branch of physics which initially aimed at understanding how the steam engine could use heat in order to produce mechanical work. Just like information theory, it originated in a reflection about engineering achievements. In both cases, thus, engineering came first and prompted scientific investigation, not the other way round. Sadi Carnot first understood in 1824 (Carnot 1824) that two heat sources at different temperatures are necessary for operating any thermal engine, and that its energetic efficiency is proportional to the difference between these temperatures. Rudolf Clausius formally defined the entropy and stated the second law circa 1850. He and many other researchers, among whom Maxwell, progressively elaborated the science of energy exchanges, thermodynamics, which was further applied to all domains of macroscopic physics: electricity, electromagnetism, chemistry, ...

The second law of thermodynamics is formulated in terms of a quantity referred to as *entropy*[4]. According to Clausius' definition, a system incurs the infinitesimal increase of entropy $\mathrm{d}S_{\mathrm{th}}$ when it exchanges (gives up or receives, depending on its sign $\pm$) an infinitesimal heat quantity $\mathrm{d}q$ with a heat source at the absolute temperature $T_{\mathrm{abs}}$:

$$\mathrm{d}S_{\mathrm{th}} \triangleq \frac{\mathrm{d}q}{T_{\mathrm{abs}}}. \tag{6.4}$$

Intuitively, heat spontaneously leaves the regions where the temperature is high for those where it is lower, so it is the easier to receive heat from a source, the higher its

[4] Not to be confused with the entropy as defined by information theory, e.g., in Sect. 4.2.2 above; the relationship between the physical and informational entropies is examined farther in this section.

temperature. Thus, $dS_{th}$ measures in a sense the 'cost' of exchanging the infinitesimal heat quantity $dq$ between the system and the source.

The entropy results from integrating the infinitesimal increases of entropy $dS_{th}$ so as to encompass all heat exchanges incurred by the system until it reaches its present state, which can be formally written as

$$S_{th} = \int \frac{dq}{T_{abs}}. \tag{6.5}$$

The meaning of the integral in Eq. (6.5) is rather difficult to make precise and the conditions of validity of this definition are rather restrictive.

The second law of thermodynamics states that the entropy of an isolated system can but increase. It quantitatively measures the degradation of energy, meaning that the energy has been transferred from the macroscopic to the microscopic scale. For the first time, a physical law predicted an irreversible evolution: the *arrow of time* entered physics. It turns out that the entropy measures how uniform is a system. The second law of thermodynamics thus tells that differences tend to vanish, a phenomenon observed in the trend of objects to equalize their temperatures as well as in the diffusion of a drop of ink in a glass of water, or that of gas molecules tending to occupy the total volume of a given enclosure. In other words, the increase of physical entropy stated by the second law means that the disorder of any system tends to increase. For instance, when friction converts the mechanical movement of some object into heat, this means that the movement of this object is converted into the disorderly movements of particles. A macroscopic observable displacement is then exchanged with very many microscopic uncoordinated movements which merely manifest themselves, statistically, as an increase of temperature.

A consequence of the second law of thermodynamics is that an isolated physical system tends to thermal equilibrium. In a system where a monatomic gas and a solid are simultaneously present in some isolated enclosure, they eventually equalize their temperatures. The gas atoms travel long distances at high speed. The absolute temperature of the gas measures the average kinetic energy of these atoms. In the solid, molecules no longer move this way but incur vibratory movements at the atomic scale, of average kinetic energy equal to that of the gas at thermal equilibrium. Moreover the enclosure 'contains' an electromagnetic field at the same temperature, now measuring the average energy of its photons. This field actually exists everywhere, even in the absence of any matter.

Restricting its validity to an isolated system, which is actually a limiting case, seems to severely restrict the validity domain of the second law. However, relaxing this condition broadens to some extent its validity domain. Besides its formal statement, it can moreover be loosely interpreted and understood as stating that physical systems generally tend towards a state of maximal entropy, i.e., tend to mixing and uniformity. The daily experience itself clearly shows that the trend towards disorder is a universal reality of the inanimate world, even for physical systems which cannot be considered as isolated. This trend can be counteracted only by a deliberate control. A fictitious agent having the ability of perceiving individual molecules can perform this task; it has been dubbed Maxwell's demon. Section 10.2 below is devoted to it.

Ludwig Boltzmann interpreted around 1877 the entropy of some system according to the famous formula[5] already written in Sect. 4.2.7, namely,

$$S_{th} = k_B \ln(W), \tag{6.6}$$

where $W$ denotes the number of microscopic states (or 'complexions') which the system can assume but which are indistinguishable at the macroscopic scale and $k_B$, referred to as Boltzmann constant equals $1.38 \times 10^{-23}$ JK$^{-1}$, in joules by kelvin. This formula is actually due to Max Planck in 1900, but it expresses Boltzmann's thought. In modern words, we may thus think that Boltzmann interpreted the physical entropy as measuring how much information is lost in a macroscopic observation. This interpretation was very bold since the community of physicists of his time was far from unanimously accepting the reality of atoms. Moreover, there was no known reason why the number $W$ of complexions should be finite. The answer to this last problem was provided later by Planck with the concept of energy quanta.

It turns out that the number $W$ of distinct microscopic configurations which can not be distinguished at the macroscopic scale is inconceivably huge. The entropy, in its physical meaning, which occurs in macroscopic phenomena is thus very large when expressed in binary information units (shannons). For instance, melting a gramme of ice increases the entropy by 1.2 JK$^{-1}$, which amounts to the enormous information quantity of $1.3 \times 10^{23}$ shannons (Balian 1995). Through freeing the water molecules from assuming a (comparatively) small number of configurations, melting hugely multiplies their number but at the same time prevents an observer from knowing the one which is actually realized. Maxwell's demon could diminish the thermodynamic entropy of a physical system only inasmuch as it is able, perceiving objects at the molecular scale, to acquire information at this very scale (Brillouin 1956).

The basic disorder at the microscopic level also results in thermal noise. It is why Boltzmann constant is met in the expression of the spectral density $N_0$ of thermal noise (see Sect. 3.3), which reads $N_0 = k_B T_{abs}/2$.

After Newton and the Enlightenment philosophers of the XVIII-th century like Voltaire, the universe was described as an immense clockwork. An extremely different vision of the world arose in the second half of the XIX-th century with the advent of statistical physics and, although it is not usual to speak of 'Boltzmannian revolution', it could well be qualified so. Schrödinger thought that it should have deeply changed the way humans perceive the world, and he wondered why it was not so for most people. Discussing the evolution of ideas about physical reality, he wrote in *Mind and Matter* (Schrödinger 1943, p. 161):

> Now between Kant and Einstein, about a generation before the latter, physical science has witnessed a momentus event which might have seemed calculated to stir the thoughts of philosophers, men-in-the-street and ladies in the drawing-room at least as much as the theory of relativity, if not more so. That this was not the case is, I believe, due to the fact that this turn of thought is even more difficult to understand and was therefore grasped by very few among the three categories of persons, at the best by one or another philosopher. This event is attached to the names of the American Willard Gibbs and the Austrian Ludwig Boltzmann.

---

[5] How this formula can be derived from Clausius' definition is summarized in Appendix A of the book by (Avery 2012, pp. 215–220). It is far from being straightforward.

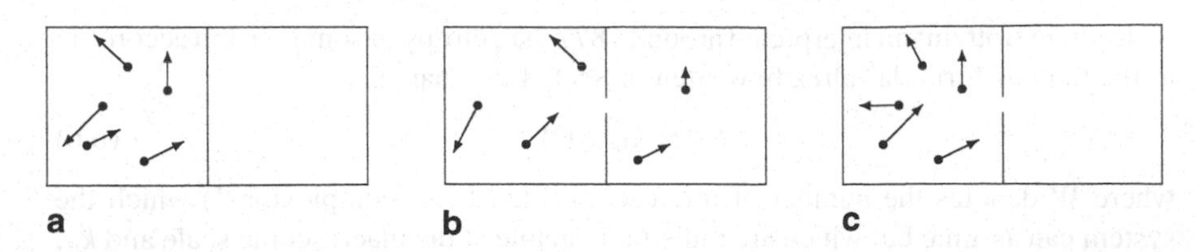


**Fig. 6.1** Gas molecules within an enclosure (a molecule is represented as a *black dot* and its instantaneous speed vector is shown). **a** Initial state, two separated compartments, all gas molecules are at *left*. **b** A hole is made in the partition which separates the compartments. The molecules then tend to be evenly distributed within the whole volume; the physical entropy increases. **c** All molecules are located again in the *left* compartment. It is highly improbable, although not strictly impossible, that this situation spontaneously occurs: a thermodynamic miracle

At least at the microscopic scale, the world could no longer be seen as a clockwork, but as a chaos of innumerable fast moving molecules hitting each others at random. Less than two decades later, the clockwork metaphor lost its relevance even at the cosmic scale, when in 1889 Poincaré realized that the three-body problem of gravitation has no analytical solutions and that the very long term behaviour of celestial bodies is actually chaotic. The world can now be described only in terms of probabilities. J.L. Borges wrote 'I belong to a vertiginous country where lottery is an essential part of reality'. His fiction actually matches Boltzmann's world.

Let us illustrate Boltzmann's vision of a gas in an enclosure by the simple example of Fig. 6.1. The distribution of the molecules of a gas within some enclosure is disordered and, at variance with the figure, the actual number of molecules is extremely large: for instance, there are about $3 \times 10^{23}$ molecules in one gramme of molecular (diatomic) hydrogen. If the volume of the enclosure is initially divided into two compartments, one of which containing gas molecules whereas the other is empty, the kinetic theory of gases tells that eventually the gas molecules will evenly occupy the whole volume of both compartments when a hole is made in the separating partition. The initial state is then more ordered than the final one, in accordance with the second law, and the increase of disorder is measured by that of the entropy. The system tends towards a state of maximal entropy, i.e., of maximal disorder. Notice that entropy is a mean quantity associated with a probabilistic system, hence its significance is statistical. In the above example, it may occur that at some instant all the gas molecules are located in one of the two previously separated compartments (see Fig. 6.1c). However, because of the huge number of molecules (unlike the figure), this event would be a 'thermodynamic miracle' of extremely low probability, never observed in practice although possible in principle.

The system just described conceals a paradox. The trajectories of individual molecules are perfectly reversible. How could then an overall irreversibility result when they are taken together? Loschmidt formulated in 1876 against Boltzmann's model the objection that it suffices to invert the speed of all molecules at a given instant to revert to the initial state. This objection illustrates how a certain mathematical abstraction fails to fit a physical reality. Changing the sign of the variable $t$ which denotes time in the equations which describe the movement of the molecules is mathematically trivial, of course. However, *implementing* the simultaneous inversion of the speed of all molecules would imply the *exact* measurement of their speed because the 3-dimentional billiard game played by molecules is chaotic, although deterministic according to classical mechanics, so the system evolution is highly sensitive to the initial conditions. Moreover, all molecules should be endowed with a speed *exactly* equal to minus their previously measured speed, which implies a perfect control on their speed and an infinite acceleration. Furthermore, less than 30 years later, Einstein's special relativity denied simultaneity and still later quantum physics denied the relevance of classical mechanics at the molecular scale. Boltzmann's vision of the world has thus won against Loschmidt's objection.

Although Boltzmann's interpretation may be deemed illuminating, the entropy has long been perceived as a rather strange entity. For instance, it is reported that before completing his seminal work (Shannon 1948) Shannon asked some renowned scientists how he could name the quantity he defined by Eq. (4.10). John von Neumann is said to have answered: "I suggest that you name it 'entropy'. First of all, because it is formally similar to Boltzmann's entropy. But also because nobody exactly knows what is entropy, so you'll always have the last word in discussions" (Avery 2012, p. 87). True or not, this anecdote is telling: in the middle of the XX-th century, entropy was still considered as somewhat mysterious by many scientists. In any way, Shannon did name 'entropy' his quantitative measure of information.

### *6.3.2 Boltzmann Constant as a Signal-to-Noise Ratio*

We already noticed in Sect. 5.2.3 that the signal-to-noise ratio determines the largest possible information quantity that an additive white Gaussian noise channel can transfer. The equalities (5.4) or (5.5) thus establish a correspondence between the signal-to-noise ratio, a dimensionless physical parameter, and an information quantity. We now try to interpret Boltzmann constant as a signal-to-noise ratio.

It has yet been met twofold in this book. First of all, we stated in Sect. 3.3 that the power spectral density of thermal noise, the most fundamental perturbation which affects any received signal, is given by

$$N_0 = k_B T_{abs}/2, \tag{6.7}$$

where $k_B$ denotes Boltzmann constant and $T_{abs}$ the absolute temperature, which measures the average kinetic energy of a particle in a gas or in any set of a large number of freely moving particles, like the electrons in a metal or the photons of the black-body radiation. Since their movements are completely disorderly, they jointly result is a 'noise' of average energy per physical dimension equal to $N_0$. The Boltzmann constant expresses the value of $N_0$ in physical units. In joules per kelvin, it equals $k_B = 1.38 \times 10^{-23}$ JK$^{-1}$.

We also found Boltzmann constant in Sect. 4.2.7 as the coefficient in the Boltzmann-Planck Eq. (6.6) which relates the thermodynamic entropy of a system

$S_{th}$ and the number of complexions $W$ which can be distinguished at the microscopic scale but correspond to a single macroscopic state of this system.

The most straightforward interpretation of Boltzmann constant is that it simply operates a conversion of units. Since the temperature measures the average kinetic energy of an atom or a molecule, Boltzmann constant just expresses this very small quantity in terms of units at the macroscopic scale, e.g., in joules per kelvin as above. The change of unit is anecdotic and its necessity comes from the measurement of temperatures having antedated by centuries the proof that atoms have a physical reality, which enabled the modern statistical interpretation of temperature. What really matters is the change of scale. The macroscopic quantities are actually averages over an extremely large number of objects having absolutely random locations and speeds. From this point of view, the Boltzmann constant appears as bridging the microscopic and the macroscopic scales, and it is far more interesting in this respect.

Let us consider $N_a$ atoms of a monatomic gas in some enclosure of volume $V$, at thermal equilibrium. Let us consider a single atom of this gas and let $v_x$ denote the component of its speed according to some given spatial direction $x$ at some time $t$. The thermal equilibrium entails that $v_x$ does not depend on the spatial location of the considered atom within the enclosure nor on the considered direction. Moreover, the parameters of its distribution do not vary with time. The mean $\mathrm{E}(v_x)$ is zero. Let $v \triangleq \mathrm{E}(v_x^2)$ denote its variance. Due to the enormous number $N_a$ of atoms in any enclosure at the macroscopic scale, $v_x$ has with an excellent approximation a Gaussian probability distribution (according to the 'central limit theorem', also referred to as Liapunov's). We consider the distribution of the random variable $K = v_x^2$ which, up to a proportionality factor, equals the kinetic energy of the atom in the considered direction. Now $K$ is essentially positive. Let $\kappa$ denote its mean. The variance $w$ of $K$ is, by definition, $w = \mathrm{E}(K-\kappa)^2 = \mathrm{E}(K^2) - \kappa^2 = \mathrm{E}(v_x^4) - \kappa^2$. Since for a Gaussian random variable $\mathrm{E}(v_x^4) = 3\kappa^2$, $w = 2\kappa^2$. Then the distribution of the kinetic energy of atoms according to any direction has a variance equal to twice its squared mean.

In order to interpret the Boltzmann constant as a signal-to-noise ratio, let $P$ denote the pressure and $T_{abs}$ the absolute temperature. Then, the equality $PV = N_a k_B T_{abs}$ entails

$$k_B = (PV/N_a)/T_{abs}.$$

$PV$ is a mechanical work, hence an energy, and $T_{abs}$ measures the average kinetic energy of a single atom. $PV$ is a macroscopic quantity since $P$ is a macroscopic average quantity, and $N_a$ is a constant number. In denominator, $T_{abs}$ is the macroscopic average of a quantity which is completely random at the microscopic scale and can thus be dealt with as noise. In this interpretation, the signal energy is $S = PV/N_a$ where all the quantities in this expression are macroscopic. Then $T_{abs}$ may be interpreted as measuring a noise energy, so it sets a limit to the precision of the measurement. The same units must of course be used for both signal and noise, so the Boltzmann constant should be equated to 1. Interpreted as a signal-to-noise ratio of 1, it corresponds according to Eq. (5.4) to a channel capacity of $C = 1/2$ shannon per sample. The measurement of the kinetic energy of a single atom thus cannot provide more than this information quantity. Its basically random character entails that it acts as 'self-noise', limiting the accuracy of its own observation.

This result meets the idea expressed by Gilles Cohen-Tannoudji of the Boltzmann constant being a quantum of information quantity (Cohen-Tannoudji 1998) and shows that its value in information units is 1/2 shannon. But what interpretation can be given to this result? Notice that this limitation of the capacity results from a perturbing noise intrinsic to the measured object itself, under the assumption that it has been perfectly singled out. This is very optimistic and any actual measurement at the atomic scale is likely to be perturbed by other noise sources so the limit of 1/2 Sh should be understood as an upper bound to the actual informational capacity of this measurement, most often grossly overestimating it. Moreover, the capacity is the largest information quantity which can be obtained. Only the optimized design of the measurement apparatus and of the experimental protocol would ideally enable eventually reaching it.

### *6.3.3 Exorcizing Laplace's Demon*

Laplace's and Maxwell's demons are two fictitious beings which were popular in the XIX-th century for illustrating physical problems. Let us try to exorcise them by means of informational concepts. We begin here with Laplace's demon. We defer to Sect. 10.2 below dealing with Maxwell's.

Laplace described the demon who bears his name as follows as quoted in (Postel-Vinay 2004):

> For a being as clever as to know at a given instant all the strengths which are exerted in nature and all the positions and relative speeds of the particles which constitute the Universe, [...], nothing would be uncertain, and the future as well as the past would be present to our eyes. (*Pour un être suffisamment intelligent pour connaître à un instant donné toutes les forces s'exerçant dans la nature et toutes les positions et vitesses relatives des particules qui composent l'Univers, [...], rien ne serait incertain, et l'avenir comme le passé serait présent à nos yeux.*)

Notice that Laplace refers to his 'demon' as clever (*intelligent*), although it must be able to exactly know a huge number of data, i.e., it should be immensely *informed*. There is moreover an inconsistency in the quoted sentence: it refers at its beginning to 'a being' (*un être*), but ends with 'our eyes' (*nos yeux*), meaning that Laplace actually likened the demon to himself or ourselves.

Was Laplace's omniscient demon a scientific substitute for God, or intended as a counter-example for illustrating the limits of pure determinism (remember that Laplace has been one of the founders of probability theory)? In any way, our interpretation of information implies that the demon should be able to memorize the measurements made on each of the particules of the universe, but the measurement apparatuses as well as the medium for storing this information could only be made of certain of these particles, which should themselves be similarly measured ... in an endless chain. The demon should thus be outside the universe, hence cannot exist as a physical object. Moreover, the precision of these measurements should be absolute because of the chaotic character of the particle interactions, which in principle would need more than all the available memory in the universe for any single measurement. How absurd is the existence of the demon becomes obvious once the necessity of a physical inscription of any information, as stated at the beginning of this book, is accepted. Implementing Loschmidt objection to Boltzmann would need the help of Laplace's demon and the impossibility of its existence is consistent with the refutation of Loschmidt we proposed above. The huge quantity of information acquired by the demon would not be tractable and could not be communicated. It could not be processed in order to obtain any usable prediction at the macroscopic scale, that of a human observer. We may thus interpret the physical entropy as measuring the information about a physical system that the demon would have acquired if it existed: as *a missing information*. Avery credits Szilard for having first interpreted thermodynamic entropy as measuring a missing information (Avery 2012, p. 84).

Besides being non-existent, Laplace's demon is 'overinformed'. It possesses an immense information quantity but no thinking being can use it. Replacing innumerable details by a few quantities which are actually statistical means is an absolute necessity. The information loss which is measured by the physical entropy is in fact highly beneficial to an intelligent being. Quoting Jean-Pierre Changeux: 'learning is eliminating (*apprendre, c'est éliminer*)' (Changeux 1983). Chaitin similarly considers the shortening of sequences by source coding as a necessary step towards understanding (Chaitin 2005).

### 6.3.4 Information is not a Physical Entity

Some prominent physicists tried to define information as the negative of the physical entropy. Schrödinger did so and named it 'negentropy' (Schrödinger 1943). Brillouin followed Schrödinger in this respect and designated by 'negentropy' the quantitative measure of information introduced by Shannon (Brillouin 1956, 1959). Many are convinced that this is right, others think that the identity of the expressions of the physical entropy and of Shannon's quantitative measure of information is purely formal and does not express any reality. For instance, Yockey nicely mocks the upholders of using negentropy in biology, writing: 'Life does not feed on *negentropy* as a cat laps up cream' (Yockey 2005).

We accept the concept of negentropy only as regards the quantitative measure of information because it corresponds to a *resolved* uncertainty, but we reject the identification of information, *as an entity*, with the negative of the physical entropy. This identification entails that the quantity of information borne by $n$ clones bearing each an information quantity of $H$ is measured by $nH$, as written by (Brillouin 1959; Eigen 1971) and others. We cannot accept this conclusion and we just stated the contrary in Sect. 6.3.3 above. We introduced information in Sect. 2.2 as an abstract entity intrinsically foreign to the physical world although it *resides* in it as necessarily borne by a physical medium, which forbids its derivation from any physical entity. As a consequence, the ability to proliferate is specific to information as we defined it. We show below that life, and only life, uses information, which leads to the conclusion that life thrives at the expense of the physical entropy. Although making $n$ clones bearing each an information quantity of $H$ actually diminishes the physical entropy by $nH$, the basic properties of information stated in Sect. 2.2 forbid that making clones can create any information, for lack of creating any novelty.

We already interpreted in Sect. 4.2.7 the entropy of physics as the information quantity associated with the knowledge of the microscopic state of a physical system. Since the number of equally possible states is huge and since no measurement can let know the actual state, the system is known only in the average, through physical quantities like temperature and pression. Taking information as a fundamental entity, we interpret the physical entropy as measuring an inevitably *missing* information. Then, the negentropy of Schrödinger and Brillouin becomes the lack of a lacking entity. But cancelling cancels differences. Multiplying by 0 cannot be inverted; 0 is a devouring entity, something like a mathematical black hole, so the lack of a lacking entity does not define anything. This situation reminds what Jared Diamond names 'the Anna Karenin principle' (Diamond 1997) according to the first sentence of Tolstoy's novel, which reads 'Happy families all look alike; unhappy families are each unhappy according to their own way.' Diamond's principle states that 'being successful actually demands that many failure causes are avoided'.

It must be accepted that information is a fundamental scientific entity of its own which up to now has been overlooked. The attempts of (Schrödinger 1943) or (Brillouin 1959) to derive it from the physical entropy have failed because information is basically an abstract entity, hence foreign to the physical world although it resides in it. Moreover, no other physical entity than entropy could enable such a derivation. The interpretation of the physical entropy as measuring a lack of information, which is rather commonplace and with which we fully agree, is meaningful only provided information is taken as a fundamental entity.

It should not be forgotten that science is a human activity. Any scientific discipline demands from its researchers a commitment to its bases, especially as regards the few entities which found it. They become to some extent prisoners of the fundamentals of their discipline, and it is one of the reasons why transdisciplinary research meets so many obstacles. It is why Schrödinger, Brillouin and many other physicists could not conceive information but as a physical entity. On the contrary, information is *the* fundamental entity of communication engineering and it is not related to any physical entity[6]. We assert that information is not physical, but why should a fundamental entity be necessarily so? Information is basically mathematical, hence abstract. Promoting it as a fundamental entity in physics and biology deserves being tried; it can and will open new horizons (Battail 2009).

To shortly conclude this section, we may state that

information *is not* negentropy, but
physical entropy *measures* neginformation.

[6] We already expressed our disagreement with Landauer's statement that 'information is physical' at the beginning of this book; see Sect. 2.2

The first statement contradicts Schrödinger and Brillouin; it expresses our denial of information being physical, as it would be if information could derive from a physical entity.

## 6.4 Information Bridges the Abstract and the Concrete

Although it is an abstract entity, an information necessarily resides in the physical world and can have some control on it. As a consequence of the second law, the physical world tends to randomly modify the representatives of information which dwell in it, but information itself can be made to some extent resilient to such changes, by means of some kind of error-correcting code, at the price of redundancy.

We stated in Sect. 4.3.5 that an information message can be endowed with semantics by associating with each of its bits the answer to a dichotomic question. These answers specify a path in a tree and the abstract information message then represents a linguistic statement. Information messages reside in the human brain and have their material support in neurons, although the mental mechanisms escape our understanding to a large extent; their written versions are contained in man-made memories and contribute in the human culture.

In other instances, however, the bits are not intended to answer questions but consist of instructions: each information bit tells what of two possible alternative actions should be performed. In the case of the algorithmic information theory, for instance, a bit of the program tells the computer what elementary operation should be executed at the next step of the computing process. It thus triggers physical actions, the cumulated result of which is eventually the specified computer output. When one among $2^k$ possible actions is performed depending on some $n$-bit sequence representing an information of $k$ shannons, this abstract entity exerts a control on physical objects. For instance, in genetics, reading a codon in the messenger RNA (a 6-bit sequence) results in the ribosomal machinery appending one more specified amino-acid to a polypeptidic chain which eventually folds into a protein, or stopping the translation process. Then an abstract information has concrete physical results.

Let us emphasize that the correspondence of bits with dichotomic questions or elementary instructions to a computer only concerns the bits of an information message, i.e., devoid of redundancy. The information borne by such a message is destroyed by any casual error and Chaps. 3 and 5 have shown that the use of an error-correcting code is mandatory for ensuring the conservation of informations. The concept of 'soft code' as discussed in Sect. 5.5.11 explains why genomic error-correcting codes may pass unnoticed: they are (or seem to be) mere by-products of more visible functions.

The conservation of human culture, similarly to that of biological information, heavily relies on soft codes. That linguistic constraints result in efficient protection of information against casual errors is indeed a matter of daily experience. As an example, the recognition of phonemes in spoken language is quite poor, so merely spelling a word does not suffice in a noisy surrounding for its letters to be correctly perceived. It is often necessary to designate each letter by a spoken word, e.g., to use Alpha, Bravo, Charlie, ... for A, B, C, ..., thus providing redundancy, for ensuring the correct reception of a spelled word. Yet, the spoken language is literally understood even in very noisy surroundings like vehicules, crowded streets or cocktail parties, which clearly shows that the many linguistic constraints (morphological, syntactic, semantic, ...) act as a powerful error-correcting soft code (having, moreover, a nested structure similarly to the genomic error-correcting codes to be hypothesized in Sect. 8.1.4). How its decoding is performed is unknown, but a highly complex machine clearly succeeds in doing so: the human brain. Would the human language lack error-correcting ability, any conversation would be impossible and our social life would be very different, if even possible.

The necessity of redundancy entails that sequences of natural or cultural origin are long. However, a semantically significant information message is necessarily short because it specifies one among many instances, the number of which exponentially increases in terms of its length. This number soon becomes so huge that it loses any meaning. A sequence of $k$ bits enables discriminating between $2^k$ instances. This number exceeds $10^{80}$, the estimated number of atoms in the visible universe, if $k$ exceeds 266. If far longer messages are met in both natural and cultural sequences, it is not because very many binary statements are necessary, but because a very redundant encoding is needed for conserving short information messages.

In the case of genomes, this order of magnitude leaves much room for redundancy, meaning that they possibly involve very efficient error-correcting codes. The length of the shortest virus genome is of about thousand base pairs. In the absence of redundancy, thus, it would enable selecting one instance among much more than the atoms in the universe. The genomes of animals and plants are still much longer, of at least hundreds of thousand base pairs. As an example, the size of the human genome, far from being the longest, is of about $3.2 \times 10^9$ base pairs, which corresponds to a binary length twice as large. Writing the number of distinct binary choices which could be thus specified if no redundancy were present would require about 1.92 billion decimal digits! This number is inconceivably huge. It would be absurd to interpret it as a number of possible independent choices. What may be concluded is simply that genomes are immensely redundant. Similarly to genomes, any sequence originating in human culture is extremely redundant: the memory size of a computer, the binary length of a book or even of a newspaper article, all exceed by far the modest length of 266 binary digits which suffices for counting the atoms of the visible universe. A *huge redundancy* is thus the rule in symbolic sequences of natural as well as cultural origin. It is the price that must be paid for conserving informations.

## References

Avery, J. S. (2012). *Information theory and evolution, 2nd edition*. Singapore: World Scientific.

Balian, R. (1995). Le temps macroscopique. In E. Klein & M. Spiro (Eds.), *Le temps et sa flèche* (2nd edn.). Gif sur Yvette: Frontières.

Battail, G. (2009). Living versus inanimate: The information border. *Biosemiotics, 2*(3), 321–341. doi:10.1007/s12304-009-9059-z.

Battail, G. (2011). An answer to Schrödinger's *What is life?*. *Biosemiotics, 4*(1), 55–67. doi:10.1007/s12304-010-9102-0.

Brillouin, L. (1956). *Science and information theory*. New York: Academic Press.
Brillouin, L. (1959). *Vie, matière et observation*. Paris: Albin Michel.
Carnot, S. (1824). Réflexions sur la puissance motrice du feu. Paris: Bachelier.
Chaitin, G. J. (1988). *Algorithmic information theory* (2nd revised edn.). Cambridge University Press.
Chaitin, G. J. (2005). *Meta Math!* New York: Pantheon Books.
Changeux, J.-P. (1983). *L'homme neuronal*. Paris: Fayard.
Cohen-Tannoudji, G. (1998). *Les constantes universelles, new edition*. Paris: Hachette.
Diamond, J. (1997). *Guns, germs, and steel. The fates of human societies*.
Eigen, M. (1971). Self organization of matter and the evolution of biological macromolecules. *Naturwissenschaften, 58*, 465–523.
Gàcs, P., & Vitànyi, P. M. B. (2011). Raymond J. Solomonoff 1926–2009. *IEEE Information Theory Society Newsletter, 61*(1), pp. 11–16.
Postel-Vinay, O. (2004). *La Recherche*, No. 381, p. 34.
Schrödinger, E. (1943). *In what is life? and mind and matter*. Cambridge: Cambridge University Press.
Shannon, C. E. (1948). A mathematical theory of communication. *The Bell System Technical Journal, 27*, 379–457, 623–656. (Reprinted in Shannon and Weaver 1949, Sloane and Wyner 1993, pp. 5–83 and in Slepian 1974, pp. 5–29).
Shannon, C. E., & Weaver, W. (1949). *The mathematical theory of communication*. Urbana: University of Illinois Press.
Slepian, D. (ed.). (1974). *Key papers in the development of Information Theory*. Piscataway: IEEE Press.
Sloane, N. J. A., & Wyner, A. D. (eds). (1993). *Claude Elwood Shannon, collected papers*. Piscataway: IEEE Press.
Yockey, H. P. (2005). *Information theory, evolution, and the origin of life*. Cambridge: Cambridge University Press.

# Part II
# Information is Coextensive with Life

# Chapter 7
# An Introduction to the Second Part

**Abstract** Chapter 7 introduces the second part devoted to the application of information theory to life. It first emphasizes its difference with biosemiotics, a discipline which applies semiotics to biology. Semiotics is restricted to semantic communication, implicitly assuming that literal communication is trivially secured. The necessity of literal communication makes information theory a prerequisite to semantics. As a mathematical discipline, information theory needs a precise vocabulary and methods hopefully introducing a rigourous theoretical framework in biology. Information is thus considered as a fundamental entity for dealing with life, as important as the physical entities of matter and energy.

## 7.1 Relationship with Biosemiotics

The title of this second part, 'Information is coextensive with life', is a paraphrase of Thomas Sebeok's statement: 'Semiosis is coextensive to life'. 'Semiosis' means the use of signs. Signs may be understood as events which are intended to refer, according to some conventions, to objects otherwise unrelated with them. They 'stand for' something else. The signs are unimportant by themselves, only the *meaning* they are given according to the conventions is relevant. The human language is an example of a system of signs indissolubly linked to humankind. Sebeok's sentence boldly extends to life in general a statement which obviously applies to human life. The science of semiosis, *semiotics*, is primarily concerned with semantic communication and has been mainly devoted to the understanding of linguistic phenomena. *Biosemiotics* is the scientific trend initiated by Sebeok which uses semiotics for analyzing the many instances where communication takes place in the living world (Barbieri 2008; Favareau 2010). Biosemiotics mainly relies on the semiotic work of the American philosopher Charles Peirce (1839–1914).

In the first part of this book, we developed concepts associated with communication engineering. We stated that information theory, the science of *literal* communication, was elaborated without any consideration to semantics and thanks to this divide. However, literal communication is an obvious prerequisite to semantic communication, so information theory should precede semiotics (Battail 2009). It turns out that, just like mainstream biologists, many biosemioticians ignore or reject it. We try in the sequel to stress the importance of information in life sciences. We

G. Battail, *Information and Life,* DOI 10.1007/978-94-007-7040-9_7,

may replace ‘semiosis’ in Sebeok’s statement with ‘information’ insofar as information is the scientific entity necessarily associated with communication. This change of vocabulary expresses our intention of opening biology to the science of information as expounded in the first part of this book. The reformulated statement does not replace the original one, of course, but lays emphasis on the necessity of *first* taking into account *information*, yet almost completely ignored as a scientific entity despite its paramount importance.

Communication occurs in the living world in very numerous instances and we do not attempt to list them. They range from communication between molecules up to communication within ecosystems and advanced animal societies, including the human ones. The physical means which implement these communications are extremely diverse, but information theory holds for all communication means and does not need the details of their operation. Indeed, it draws much of its synthesis ability from its blindness to such details. Communication in the living world has been studied from the semiotic point of view by biosemioticians, e.g., (Witzany 2006/2007). The reader will also find in (Mian and Rose 2011) a comprehensive discussion of many biological communication problems at the cellular level which are relevant to information theory.

## 7.2 Content and Spirit of the Second Part

Among the numerous communication engineering problems solved by Nature, we restrict ourselves in Chap. 8 to communication of genomes over time at the geological scale, as a single paradigmatic example which concerns a major feature of life, *heredity*. Nature solved there a very hard communication problem but biologists were unaware of the solution for lack of perceiving the problem. We show in Sect. 8.1 that its solution necessarily implies the existence of genomic error-correcting codes. Assuming their existence, in turn, explains many basic facts of life left unexplained by mainstream biology (Sect. 8.2). Section 8.3 shows how the properties of information and of error-correcting codes enable roughly simulating heredity. Section 8.4 is devoted to the yet unsolved problem of identifying the genomic error-correcting codes.

More generally, recognizing the prominent role of information in life further suggests that transmitting, receiving, recording, processing, and using information is as specific to the living world as to delineate its border with the inanimate world (Chap. 9, and especially Sect. 9.1). Information then appears as the yet missing fundamental entity that science needs for properly dealing with life (Sect. 9.3). Chapter 10 continues Chap. 6 as regards the place of life within the physical world. Chapter 11 concludes the whole book.

Our approach will be better understood in contrast with the following quotation from the biologist and historian of biology (Morange 1994, pp. 223–224):

> If [the schemes of molecular biology] are so attractive, if they look obvious and in accordance with 'reality', it is because their underlying logic and the image of the biological world they uncover are in harmony, in resonance with the image of the surrounding world, delivered as well by the media as by the other scientific disciplines. Explaining the operation of living beings in terms of information, memory, code, message, feedback regulation, amounts to use a language and images that everybody knows. (*Si [les schémas de la biologie moléculaire] sont si attirants, s'ils semblent évidents, correspondre à la 'réalité', c'est que la logique qui les inspire et l'image du monde biologique qu'ils révèlent sont en harmonie, en résonance avec l'image du monde environnant, livrée tant par les médias que par les autres discipline scientifiques. Expliquer le fonctionnement des êtres vivants en termes d'information, de mémoire, de code, de message, de régulation par rétroaction, c'est utiliser un langage et des images connus de tous.*)

This text illustrates how the vocabulary of communication and information impregnates the current biological thoughts and literature, but also that it lacks scientific relevance. The media are indeed replete with *words* which belong to this vocabulary, but are the corresponding *concepts* understood? For instance, one has read in newspapers, even of good standing, that biologists discovered the genes of homosexuality or of belief in God. Such twaddles should have convinced Morange that using 'a language and images that everybody knows'[1] is not sufficient (nor necessary) for making science! Prominent philosophers and scientists like Henri Poincaré state on the contrary that science must use a language distinct from that of everybody where the extant words, when they are used, are endowed with a precisely defined and unambiguous meaning (Poincaré 1911). Precise concepts are actually needed, and mathematics is for Poincaré the best example of the needed 'language of science'. Morange rightfully places information at the heart of the modern vision of life, but in the looser possible meaning of the word. In his later book (Morange 2005), he moreover discards information theory with the back of one hand after giving a caricatural account of it. He clearly does not think of information as a scientific concept on the same footing as those of physics or chemistry. We take the opposite of his opinion in what follows: as stated in Chap. 1, we try to use information theory and the principles of communication engineering to deal with biological problems, and we show that information is a needed fundamental scientific entity yet generally overlooked.

Suggesting that biology should explain the reality of the world by the daily experience, Morange goes against the general trend of scientific history. Did physics proceed this way? Even classical physics as initiated by Galileo and Newton describes the reality of the world in terms which deeply depart from the daily experience. Alexandre Koyré wrote that Galileo's decision to deal with mechanics as a branch of mathematics is paradoxical as intended to 'explain the real by the impossible'. One may interpret his celebrated sentence 'The book of nature is written in mathematical language (*Il libro della natura è scritto nella lingua matematica*)' as stating the need of a precise language for describing reality. Since the revolution brought to physics at the beginning of the XX-th century by the relativity and quantum theories, the gap between the daily experience and the scientific description of things has moreover

[1] Moreover, the *precise* meanings of information, memory, code, ... are actually known by few people. It would be closer to the truth to say that everybody *believes* he/she knows these meanings.

become so wide that not only the former is of no use for understanding the latter, but practicing physics needs an effort against sensible intuition. Far from being useful, the daily experience has become a burden.

The extremely complex reality of life demands, at the opposite of Morange's opinion, abstract and general enough scientific concepts. We think that the concept of information, due to the ubiquiteness and crucial importance of communication in the living world, is the most adequate one. We look forward to biology becoming a great extension of information theory. Information theory may be thought of as well as an integral part of biology.

Some thinkers foreign to mainstream biology actually stated that information is an unescapable concept for understanding life. It is especially pleasant to ackowledge the opinion of Kenneth Boulding (1910–1993) in this respect (Boulding 1956).

Morange is far from being the sole biologist who discarded information theory without seriously examining it: this prejudice is shared by a vast majority of biologists. Many of them, however, clearly understood that information is a crucial actor in life phenomena. Maynard Smith quotes Weismann as having first realized it was so as regards heredity. He thought himself that information is very important in biology, but he quickly discarded information *theory*, briefly arguing that it could be of no use in biology because it ignores semantics (Maynard Smith and Szathmáry 1997, 1999). This negative opinion was shared without further examination by most biologists, all the more it avoided any effort for acquiring a mathematical science often foreign to their culture.

At the beginning of the nineteenth century, biologists believed that organic compounds were the product of a specific 'vital force' so they could not be synthesized but by living beings. According to this belief, chemistry was divided into the strictly separated domains of 'inorganic' and 'organic' chemistry. This belief was ruined in 1828 when Wöhler performed the synthesis of an organic molecule: urea, $H_2N-CO-NH_2$, from inorganic elements. At the beginning of the twenty-first century, biologists think that biological information is foreign to the concept of information introduced by Shannon and proved to be highly successful by the engineers' experience. Similarly to the chemists of the past, they believe that there should exist a science of 'organic' information, which moreover—at variance with organic chemistry at the beginning of the nineteenth century—has to be created *ex nihilo*. For instance, Judson explicitly refers to such a duality, setting information in Shannon's sense against Crick's opinion about information (Judson 2001). He also quotes François Jacob: 'Language studies the message transmitted from an emitter to a recipient. Now there is nothing of the kind in biology: no emitter, no recipient. The famous message of heredity transmitted from one generation to the other, no one has ever written it; it is constituted by itself, slowly, painfully traversing the vicissitudes of reproductions subtended by evolution.' Obviously, a language has a human emitter and a human recipient. Precisely *because it ignores semantics*, information theory is relevant to life. Jacob's objection does not apply to information theory, although Judson interprets it as ruling out the use of Shannon's information. Almost all biologists similarly believe it is irrelevant to their field. When and how will the belief in a still unformulated organic information theory be ruined?

# References

Barbieri, M. (2008). Biosemiotics: A new understanding of life. *Naturwissenschaften, 95,* 577–599.
Battail, G. (2009). Applying semiotics and information theory to biology: A critical comparison. *Biosemiotics, 2*(3), 303–320. doi:10.1007/s12304-009-9062-4.
Boulding, K. E. (1956). *The image: Knowledge in life and society*. Ann Arbor: University of Michigan Press.
Favareau, D. (2010). *Essential readings in biosemiotics*. Dordrecht: Springer.
Judson, H. F. (2001). Subtended by evolution. *Nature, 410*(6825), 146–147. (Review of the book by Lily E. Kay: *Who wrote the book of life? A history of the genetic code*, Stanford University Press, 2000).
Maynard Smith, J., & Szathmáry, E. (1997). *The major transitions in evolution*. Oxford: Oxford University Press.
Maynard Smith, J., & Szathmáry, E. (1999). *The origins of life: from the birth of life to the origins of language*. Oxford: Oxford University Press.
Mian, I., & Rose, C. (2011). Communication theory and multicellular biology. *Integrative Biology, 3,* 350–367.
Morange, M. (1994). *Histoire de la biologie moléculaire*. Paris: La Découverte.
Morange, M. (2005). *Les secrets du vivant. Contre la pensée unique en biologie*. Paris: La Découverte.
Poincaré, H. (1911). *La valeur de la science*. Paris: Flammarion.
Witzany, G. (2006/2007). *The logos of the bios, I, and II*. Helsinki: Umweb Publications.

# Chapter 8
# Heredity as a Communication Problem

**Abstract** Chapter 8 considers the capital problem of heredity. The conservation of genomes over the ages is shown to blatantly contradict the existence of frequent mutations in genomes of somatic cells. This contradiction can be solved only by hypothesizing that genomes are endowed with error-correcting codes. Moreover, the better conservation of very old part of the genome like the *HOX* genes suggests that the genomic error-correcting code should combine a number of nested component codes which appeared progressively during the evolution process, and can be likened to Barbieri's organic codes. It is shown that these hypotheses meet reality and explain a number of facts left unexplained by mainstream biology: genomes are immensely redundant, discrete species exist which moreover can be ordered according to a hierarchical taxonomy, and nature proceeds with successive generations. Evolution appears as contingent and saltationist because the genetic information originates in regeneration errors. Even the trend of evolution towards increased complexity, hence towards longer genomes, is explained by the better performance of long error-correcting codes, which thus have been favoured by Darwinian selection. We introduce a very simple model of 'genomes' consisting of binary sequences subjected to random errors. They are either left uncoded, or encoded by means of a simple error-correcting code. They are periodically replicated, and regenerated if they have been encoded. The resulting population of 'genomes' is referred to as 'toy living world'. We examine the permanence of such 'genomes' in the presence of random symbol errors, with and without error correction. If the genomes are not coded, a chaotic set of 'genomes' results where 'species', i.e., populations of identical 'genomes', are the more fleeting, the longer they are. Only the existence of error correction can account for the permanence of 'genomes' of lengths compatible with those of actual genomes and the existence of discrete species. The hypothesized genomic error-correcting codes remain unidentified although some guesses can be made about them, and the means of their regeneration remain unknown.

## 8.1 The Enduring Genome

As quoted by Etienne Klein (2010, p. 153), Ludwig Wittgenstein wrote:

> Strangely, it is said that God created the world and not: God continuously creates the world. Why would the fact that the world once began to be be a greater miracle than the fact that it continued to be?

G. Battail, *Information and Life*, DOI 10.1007/978-94-007-7040-9_8, 

Similarly, biologists seem to think of the fact that genomes once began to be what they are is much more important than the fact that they continued to be so. That this is wrong is clear once it is realized that the integrity of genomes is continuously threatened by mutations, a biological fact as well as the existence of genomes.

Richard Dawkins wrote in *The selfish gene* (Dawkins 1976):

> We do not know how accurately the original replicator molecules made their copies. Their modern descendants, the DNA molecules, are astonishingly faithful compared with the most high-fidelity human copying processes.

Copying, however faithful it may be, is actually not sufficient for ensuring the genome conservation. Would the copying process be absolutely faithful, the integrity of genomes would still be threatened by casual mutations occurring outside the replication process. Only error-correcting codes can secure the permanency of genomes by *intrinsically* endowing them with *resilience* to changes.

### 8.1.1 A Blatant Contradiction

The powerful theorems of information theory are not even needed to perceive a blatant contradiction in the usual accounting of how genomes are communicated over time; common sense suffices. A genome[1] can be replicated because the double-helix structure of DNA enables its copy. Each of its two strands is used as a template for reconstructing the complementary one. Then a single double-strand DNA molecule gives rise to two identical molecules. A faithful copy can thus result from DNA replication, all the more several 'proof-reading' mechanisms ensure that the copied molecule is identical to the original one. Dawkins, as quoted above, is thus right in underlining the extreme faithfulness of the replication process.

The genomes incur successive replications performed at time intervals of the same order of magnitude as the lifetimes of living beings, but they are faithfully conserved during the inconceivably long time intervals of the geological scale, thus after a huge number of replications. It is now established that the origin of life has been a single event which occurred at least 3.5 (maybe 3.8) billion years ago. By the mere strength of his thought, Darwin (1809–1882) may be credited for having realized how deep were geological times, as well as the unicity of life on Earth despite the astonishing variety of its manifestations. The evidence that DNA is the support of heredity for all living beings, with moreover a common rule for converting DNA into proteins: the 'genetic code' (which we prefer naming 'genetic mapping'; see Fig. C.3 in Appendix C), came not before a century later as a dazzling confirmation of the unicity of life. As regards the depth of time, absolute datation tools were not available to geologists before the XX-th century. Around 1900, lord Kelvin tried to estimate the Earth's age using arguments based on its cooling. He found a few tens

[1] 'Genome' is intended to mean the whole sequence of base pairs borne by DNA molecules, regardless of their possible function, and especially not restricted to genes.

of million years, at most, to be compared with the modern estimate of 4.6 billion years. The subsequent discovery of radioactivity provided both an explanation for the Earth's excess of internal heat and means for the absolute datation of geological material.

However, at a timescale shorter by far than that of geology, say that of a human life, random errors referred to as mutations are observed to occur in genomes of somatic cells. Their effects on individual phenotypes are as dramatic as ageing or cancer. They are due to many permanent causes, especially chemical reactants and radiations of terrestrial, solar and cosmic origin. Cell membranes can efficiently shield DNA molecules against chemical agents, but much less so against radiations. The existence of such mutations blatantly contradicts the postulated conservation of genomes at the timescale of geology by the agency of the genomes of germinal cells. If mutations are as frequent as to have observable effects at a short timescale, merely copying genomes, however faithfully, cannot secure their long-term conservation, in contradiction with Dawkins' statement quoted above. This obvious remark is fully confirmed by information theory, which shows that the channel capacity of DNA vanishes as time passes. We present below (Sect. 8.1.2) the computation of an upper bound on the capacity of the genomic channel, based for lack of precise data on unrealistic assumptions which nevertheless result in greatly overestimating the capacity. It turns out that this upper bound exponentially decreases with time. This bound depends on the error rate in genomes. An attempt to estimate it is made in Sect. 8.1.2, too. It can but be rough and it relies in part on our own hypotheses, yet using it in the upper bound expression gives the expected order of magnitude for the decrease of DNA capacity.

In order to illustrate by a simple example how comparatively infrequent but cumulative errors quickly destroy the information contained in a replicated text, a 72-letter sentence has been successively replicated 42 times with a single letter, chosen at random, replaced by an arbitrary one of the same alphabet at each replication (the space which separates words is dealt with as a letter of the alphabet, thus assumed to contain 27 symbols; it has been represented as a tilde in the example). The result is shown in Fig. 8.1. The number of letters in the sentence, 72, and the number of replications, 42, are both arbitrary and were chosen so as to fill in a whole page.

Of course, the replicated sentence is much shorter than actual genomes and the assumed error frequency is much larger than in reality, but the brevity of the former (very roughly) compensates for the excess of the latter, and the number of successive genome replications to be considered in heredity is much larger. This example is just intended to show the progressive degradation of an initial message. The first steps seem to have little impact on the message but the accumulation of errors soon makes it impossible to guess what the initial message was. A similar final result could of course be obtained through drawing at random 42 letters of the initial message and substituting for each of them an arbitrary letter. No wonder then that the initial message becomes incomprehensible.

As the only possible way for conciliating the long-term conservation of genomes and the existence of comparatively frequent mutations, we are led to assume that genomes are endowed with an intrinsic error-correction ability which, moreover,

```
genomes~are~subjected~to~random~errors~and~incur~successive~replications
genomes~arm~subjected~to~random~errors~and~incur~successive~replications
genomes~arm~subyected~to~random~errors~and~incur~successive~replications
genomes~arm~subyected~to~ranjom~errors~and~incur~successive~replications
genomes~arm~subyectedoto~ranjom~errors~and~incur~successive~replications
genomes~arm~subyectedoto~ranjom~errors~anz~incur~successive~replications
genomes~arm~subyectedoto~ranjom~earors~anz~incur~successive~replications
genomes~arm~subyectedoto~ranjomjearors~anz~incur~successive~replications
genomes~arm~subyecqedoto~ranjomjearors~anz~incur~successive~replications
genomes~arm~subyecqedoto~ranjomjearors~anz~incur~suciessive~replications
genomes~arm~subyecqedotv~ranjomjearors~anz~incur~suciessive~replications
genomys~arm~subyecqedotv~ranjomjearors~anz~incur~suciessive~replications
genomys~arm~subyecqedotv~ranjomjearors~anz~incur~suciessivevreplications
genomys~grm~subyecqedotv~ranjomjearors~anz~incur~suciessivevreplications
genomys~grm~subyecqedotv~ranjomjearors~anz~incur~suciessivenreplications
genomys~grm~subyecqedotv~ranjomjearors~anz~incur~suciessivenreplxcations
genomys~grm~subyecqedotv~ranjomjearorsvanz~incur~suciessivenreplxcations
genomys~grm~subyecqedotv~ranjomjearorsvanz~incur~suciessavenreplxcations
genomys~grm~subyecqedotv~rpnjomjearorsvanz~incur~suciessavenreplxcations
genomys~grm~subyecqedotv~rmnjomjearorsvanz~incur~suciesskvenreplxcations
genomys~grm~subyecqedotv~rmnjomjearorshanz~incur~suciesskvenreplxcations
geuomys~grm~subyecqedotv~rmnjomjearorshanz~incur~suciesskvenreplxcations
geuomys~grm~subyecqebotv~rmnjomjearorshanz~incur~suciesskvenreplxcations
geuomds~grm~subyecqebotv~rmnjomjearorshanz~incur~suciesskvenreplxcations
geuomds~grm~subyecqebotv~rmnjomjearorshanz~incur~suciesskvenreplxcatioms
geuomds~grm~subyecqejotv~rmnjomjearorshanz~incur~suciesskvenreplxcatioms
geuomds~grm~subyecqejot~~rmnjomjearorshanz~incur~suciesskvenreplxcatioms
geuomds~grm~subyecqejot~~rmmjomjearorshanz~incur~suciesskvenreplxcatioms
geuomds~grm~subypcqejot~~rmmjomjearorshanz~incur~suciesskvenreplxcatioms
geuomys~grm~subypcqejot~~rmmjomjearorshanz~incur~suciesskvenreplxcatioms
geuomys~grm~subypcqejot~~rmmjomjearorshanz~incur~suciesskvenrdplxcatioms
geuomys~grm~subypcqejot~~rmmjomjearorshanz~incur~suciesskvynrdplxcatioms
geuomys~grm~subypcqejot~~jmmjomjearorshanz~incur~suciesskvynrdplxcatioms
geuomys~grm~subypcqejot~~jmmjomjearvrshanz~incur~suciesskvynrdplxcatioms
geuomys~grm~subypcqejot~~jmmjomjealvrshanz~incur~suciesskvynrdplxcatioms
geuomys~grm~subjpcqejot~~jmmjomjealvrshanz~incur~suciesskvynrdplxcatioms
geuomys~grm~subypcqrjot~~jmmjomjealvrshanz~incur~suciesskvynrdplxcatioms
geuomys~grm~subypcqrjot~~jmmjomjenlvrshanz~incur~suciesskvynrdplxcatioms
geuomys~grm~subypcqrjot~~jmmjomjenlvrshanz~incur~suciesskvynrdplxoatioms
geuomys~grm~subypcqrjot~~jmmjomjenlvrshanzlincur~suciesskvynrdplxoatioms
geuombs~grm~subypcqrjot~~jmmjomjenlvrshanzlincur~suciesskvynrdplxoatioms
geuombsqgrm~subypcqrjot~~jmmjomjenlvrshanzlincur~suciesskvynrdplxoatioms
geuombsqhrm~subypcqrjot~~jmmjomjenlvrshanzlincur~suciesskvynrdplxoatioms
```

**Fig. 8.1** An example of replications with symbol errors

unequally protects the different parts of the genome. We formulate these assumptions with more detail in Sect. 8.1.3. We may already notice that they are actually not entirely foreign to the biological literature. Rather, they make explicit more or less implicit ideas it contains. Biologists often assume that mutations are not evenly distributed within genomes, or that organisms have some control on the mutation rate. Such assumptions actually imply that genomic error-correcting codes exist. Insofar as this assumption remained implicit, biologists did not realize its importance, especially as regards its consequences for the living world as a whole, which will be briefly expounded in Sect. 8.2 below.

As far as I know, the hypothesis that genomes are endowed with error-correction ability was first formulated by Donald Forsdyke (1981). He suggested that the genes of eukaryotes are similar to words of a code in systematic form, where the introns are made of check symbols associated with an information message borne by the exons. We present below (Sect. 8.4) arguments in favour to this interesting hypothesis.

Other attempts were made later, for instance (Rzeszowska-Wolny 1983; Liebovitch et al. 1996). Neither the theory and practice of error-correcting codes, nor the knowledge of genomes, were advanced enough when these studies were made to enable reaching clear conclusions as regards the existence of genomic error-correcting codes. Moreover, the researches on such topics are necessarily trans-disciplinary hence difficult. It was especially hard for biologists to get a clear idea of error-correcting codes when most of the literature published about them was basically algebraic and rather opaque for non-mathematicians. Since these studies failed to uncover genomic error-correcting codes, hasty or lay readers understood that such codes do not exist although their failure was due to assuming simplistic and inadequate coding schemes. The prejudices thus created, as usual, are unfortunately long-lasting. Our statement that genomic error-correcting codes exist results from a logical necessity, and guessing what they are and how they work needs matching Nature's inventiveness: an impossible challenge since, according to Jerome Wiesner, 'no one is visionary enough to match reality'. The identification of genomic error-correcting codes remains indeed an open problem for lack of experimental data although some guesses can be made (see Sect. 8.4).

### *8.1.2 An Upper Bound on the DNA Channel Capacity*

Let us first of all notice that the symbol errors are cumulative: regardless of the possible variation of their frequency of occurrence, the number of such errors which affect the genome is an increasing function of time. Precisely computing the DNA capacity is impossible for lack of several data which are not available. It is possible however to compute an upper bound of this capacity, which is easily shown to vanish with time. We may thus assert that, *a fortiori*, the channel capacity of DNA, in the information-theoretic meaning, does so, which dramatically confirms that it is actually unable to communicate the genetic information through geological time intervals.

The symbol errors may consist of erasures, substitutions, deletions, or insertions. 'Erasure' means that a symbol foreign to the alphabet is substituted for the correct one; 'substitution' means that a symbol of the alphabet different from the correct one is substituted for it; 'deletion' means that that one or several symbols are removed from the given sequence, and 'insertion' that spurious symbol(s) are appended somewhere in it. The proportion of each type is unknown, although the substitutions are probably the most frequent. However, we can compute an upper bound of the genome capacity if we assume that all errors are of the mildest type, i.e., consist of erasures. In the case of an erasure the received symbol is not recognized as one of the alphabet symbols but is not mistaken for one of them (see Sect. 5.1). The decrease of capacity that erasures entail is significantly less than that resulting from substitutions (roughly speaking, a single substitution is as harmful as two erasures in the binary case; see Sect. 5.2). We moreover take account of the redundancy provided by the availability of two complementary strands in the capacity computation. The true capacity is thus grossly overestimated. It turns out that the computed upper bound is a vanishing function of time, so *a fortiori* the true capacity has the same behaviour.

As regards the alphabet size, it would seem self-obvious that it should be 4, i.e., equal to the number of nucleotides, but it is possible that other alphabet sizes are relevant. For example, a binary alphabet results in distinguishing only the chemical structure, purine (**R**) or pyrimidine (**Y**), of the nucleic bases. It may be also that blocks of three nucleotides (as in the 'genetic code') are relevant, then defining a 64-symbol alphabet. It is why we denote the alphabet size by $\alpha$, leaving indeterminate its actual value.

For computing an upper bound on the actual capacity of DNA, we consider the capacity of an $\alpha$-ary erasure channel such that during the infinitesimal time interval $dt$, the probability of an erasure is $\nu dt$, where $\nu$ denotes the erasure frequency. For simplicity's sake we assume that the frequency $\nu$ is constant and that the symbol erasures are independent events. Then the probability of erasure as a function of time, $e(t)$, is easily shown to be:

$$e(t) = 1 - \exp(-\nu t). \tag{8.1}$$

In order to take into account the availability of two complementary strands, we consider a pair of complementary nucleotides as erased only if both are erased. Assuming the erasure of a symbol to occur independently in the two strands, the erasure of a complementary pair occurs with a probability which equals the square of the probability of erasure in a single strand, namely $e^2(t)$, where $e(t)$ is given by Eq. (8.1).

The capacity of single-strand DNA in the presence of erasures only is easily shown to be $C_{\alpha,\text{er}} = (1 - \delta)\log_2(\alpha)$ binary units, where $\alpha$ is the alphabet size and $\delta$ is the erasure probability (see Sect. 5.2). For double-strand DNA we have $\delta = e^2(t)$, where $e(t)$ is given by Eq. (8.1), so the capacity $C_{\alpha,\text{er,ds}}$ of a complementary pair is:

$$C_{\alpha,\text{er,ds}} = \exp(-\nu t)[2 - \exp(-\nu t)]\log_2(\alpha) \tag{8.2}$$

binary units. This function is represented in the figure below assuming $\alpha = 2$.

The hypotheses which were made entail that this expression grossly overestimates the actual DNA capacity. This upper bound is an exponentially decreasing function of the product $\tau = \nu t$, which can be interpreted as a measure of time using as unit the

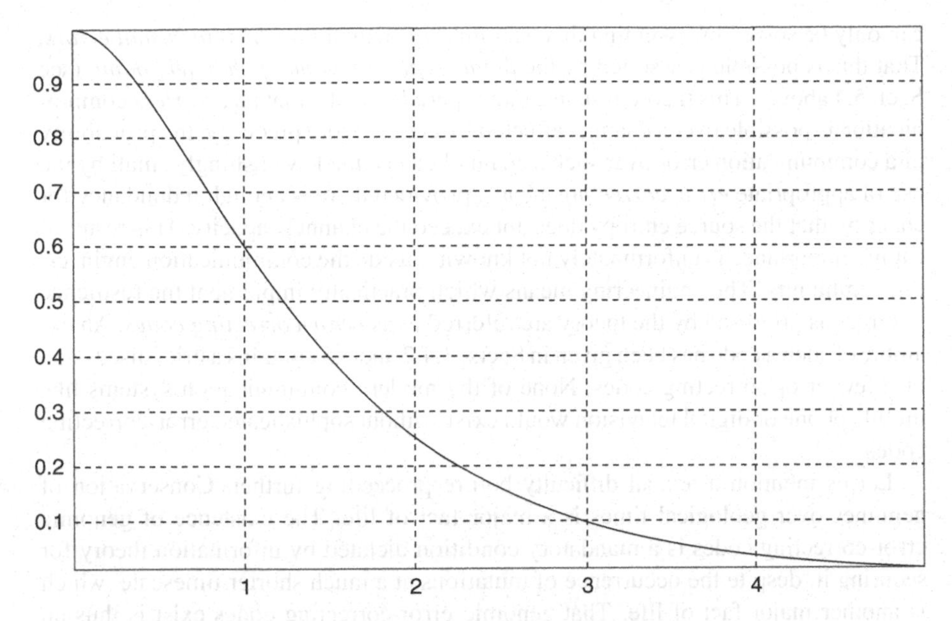


**Fig. 8.2** Upper bound on the DNA capacity as a function of time for $\alpha = 2$. The unit in the horizontal (time) axis is $\tau = 1/\nu$, where $\nu$ denotes the erasure frequency

reciprocal of the error frequency, $1/\nu$. The channel degrades with time and becomes useless after a time interval much less than $\tau$ if no means is available for correcting errors. (Fig. 8.2)

We try now to roughly estimate the error frequency $\nu$. To this end we rely on published data on the mutation rate in humans, assuming that it does not significantly differ from a species to another. This rate is estimated to about $2 \times 10^{-8}$ mutation per nucleotide and per generation (Nachman 2004; Kondrashov 2003), or about $10^{-9}$ mutation per nucleotide and per year. According to an assumption intended to complement our subsidiary hypothesis (see Sect. 8.1.4), a fraction $\mu$ of the nucleotides remains uncoded. We may thus think that only the errors occurring in the uncoded fraction significantly contribute in the observed error rate, so the mutation rate before correction is the observed rate divided by $\mu$. Taking $\mu = 10^{-3}$ results in $\nu \approx 10^{-6}$ mutation per nucleotide and per year. The time unit above, $\tau = 1/\nu$, is thus about a million years (we refer to this result as the 'Haldane-Kondrashov estimate'). This is a short time interval at the geological scale (although it exceeds by far any experience humans can have of time), so DNA is actually an *ephemeral memory*. Notice that a similar conclusion obtains for any 'permanent' memory incurring random errors at a nonzero frequency: in adamant words, no memory is permanent.

### 8.1.3 Main Hypothesis: Genomic Error-Correcting Codes must Exist

The engineering experience, confirmed by information theory, tells that the contradiction between frequent mutations and very faithful conservation just pointed out can only be solved by assuming that genomes are actually *resilient to casual errors*. That this is possible is asserted by the *fundamental theorem of channel coding* (see Sect. 5.4 above). This theorem states, rather paradoxically, that an *errorless* communication is possible over a channel affected by errors. More precisely, the probability of a communication error over such a channel can be made vanishingly small by the use of appropriate *error-correcting means*, provided there is enough redundancy for ensuring that the source entropy does not exceed the channel capacity. This result of capital importance is unfortunately not known outside the communication engineering community. The engineering means which practically implement the resilience to errors as promised by the theory are referred to as *error-correcting codes*. An example of such a code has been given in Sects. 3.4.2 and 5.5 contains a brief discussion of a few error-correcting codes. None of the modern communication systems like mobile phone or digital television would exist without sophisticated error-correcting codes.

Let us mention a lexical difficulty before proceeding further. Conservation of genomes over geological times is a major fact of life. The existence of genomic error-correcting codes is a mandatory condition dictated by information theory for securing it, despite the occurrence of mutations at a much shorter timescale, which is another major fact of life. That genomic error-correcting codes exist is thus an unescapable conclusion. We nevertheless refer to it as a *hypothesis*, despite the speculative connotation of the word, since no one can yet exhibit means of genomic error correction. More precisely, although many encoding processes can be identified as potentially error-correcting, the mechanisms which perform the regeneration of genomes are still unknown. Their experimental identification is obviously a difficult task which needs that molecular biologists and communication engineers closely collaborate. We think however that the word 'hypothesis' fails to express the compelling character of the statement since information theory leaves no alternative to it. We nevertheless use it for lack of a more appropriated one. 'Mandatory hypothesis' would perhaps be better, but looks too much like an oxymoron.

### 8.1.4 Subsidiary Hypothesis: Nested Codes

The existence of genomic error-correcting codes, our *main hypothesis*, does not suffice to fully account for the biological reality. It turns out that some very old parts of genomes are faithfully conserved, in contradiction with the decrease of DNA capacity with time as already discussed[2]. For instance, the *HOX* genes are shared by beings as far from each other in the evolution tree as flies and humans (whose last common ancestor lived more than half a billion years ago). Interestingly, some parts of the genome outside the genes are also known to be very faithfully conserved. It is thus necessary to further assume, as a *subsidiary hypothesis*, that a genomic code unequally protects the genetic data, so that the older is an information, the

[2] This remark alone entails that genomic error-correcting codes are needed.

more faithfully it is conserved. We assume it is organized as a system of 'nested codes', made of successive component codes encoding messages made of already encoded sequences to which new information symbols have been appended. Then the assumed error-correcting nested system may be described according to the fortress metaphor, where a component code is depicted as a wall which protects what is inside it against outside attackers. The more numerous walls enclose an information message, the better it is protected. Notice that a very efficient protection of the most central information does not demand very efficient individual codes: the multiplicity of enclosing walls is much safer than each of them separately. Assuming they were built successively during geological ages accounts for the better protection of older information. This scheme is made compatible with crossing-over, hence with sexual reproduction, by further assuming that the most lately appended information message has been left uncoded and that it concerns only the small fraction of the total genome which accounts for individual differences among the members of a species.

Any symbol in the genome which is not left uncoded belongs to at least one component code. The deepest (and oldest) component code of an $l$-layer nested system will be denoted by $C_1$ and the following ones by $C_2, \ldots, C_l$. A symbol which belongs to the component code $C_i$ also belongs to $C_{i+1}, \ldots, C_l$. Roughly speaking, we may think of the words of $C_i$ as distant of at least $d_i$ in the Hamming space, where $d_i$ is the *cumulated* minimum distance of codes $C_1, C_2, \ldots, C_l$. These distances form a decreasing sequence: $d_1 > d_2 > \ldots > d_l > 1$.

As a typical example, we may assume that all component codes are binary and have the same rate $R$. We define their common expansion factor as $\lambda = 1/R$. We also assume that the information messages appended to the result of the previous encoding at each coding step have all the same length, say $k$ bits. At the deepest coding level (step 1) a first $k$-bit information message is encoded into a $\lambda k$-bit word. At the next step, a new $k$-bit information message is appended to the word which results from the previous encoding, and this whole message is encoded with the expansion factor $\lambda$, resulting in a word of length $\lambda(\lambda + 1)k$. The codeword length at the encoding step $i$, similarly, is $\lambda(\sum_0^i \lambda^i)k = \lambda(\lambda^{i+1} - 1)k/(\lambda - 1)$. The total number of information bits at step $i$ is merely $(i+1)k$ since $k$ new information bits are appended at each step, hence the cumulated expansion factor at step $i$ is $\Lambda_i = \lambda(\lambda^{i+1} - 1)/(\lambda - 1)(i + 1)$, an increasing function of $i$.

Increasing the number of component codes of the nested system considered here thus results not only in an increase of the overall code length, but in an increase of the overall redundancy since $\Lambda_i$ increases with $i$. One easily checks that avoiding to increase the overall code redundancy necessarily implies the use of component codes with an expansion factor $\lambda_i$ which tends to 1 as $i$ increases. Then the component codes become the less efficient, the larger $i$, at the expense of the overall efficiency. We may thus think of the above example as typical, and consider that the simultaneous increase in length and redundancy as the number $i$ of component codes increases generally occurs in a nested system.

None of the assumptions which we made to illustrate the nested system in Fig. 8.3 is mandatory. It is not necessary to assume that the alphabet is binary in all component codes of the nested system, not even that it is the same for all. Its size can be different

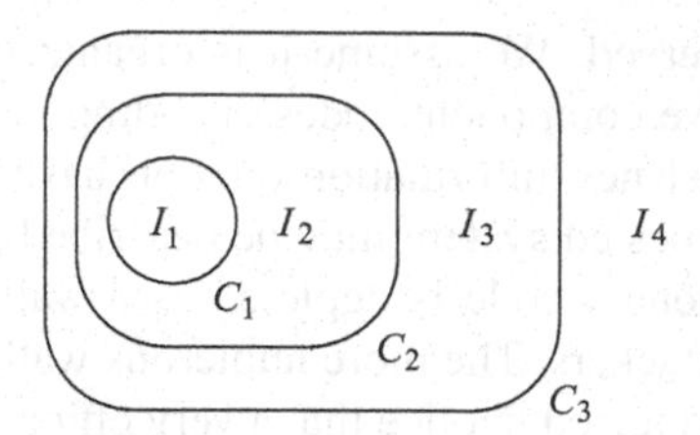


**Fig. 8.3** The proposed system of nested codes represented according to the fortress metaphor. A code $C_j$, $j = 1, 2, 3$, is represented as a closed wall which protects what is inside it. $I_1$, $I_2$, $I_3$ and $I_4$ are successive information messages. $I_1$ is protected by 3 codes, $I_2$ by 2 codes, $I_3$ by a single code and $I_4$ is left uncoded

from a coding layer to another one: a newly performed encoding may well deal with some extension of the information message, i.e., consider blocks of $m$ successive $\alpha$-ary symbols as symbols of an alphabet of size $\alpha^m$. The codes do not need to be in systematic form and, if they are, the location of the information bits is arbitrary. Even the condition that the dimension $k_i$ of each code is larger than the codeword length of the previous coding layer has not to be strictly satisfied. It suffices that the encoding performed at the $j$-th step brings some further protection to the message which has already been encoded at the previous steps. Moreover, the codes actually used by nature do not result from an engineering design but likely belong to the broader class of 'soft codes', as introduced above (Sect. 5.5.11). (Fig. 8.3)

## 8.2 Consequences of the Hypotheses Meet Biological Reality

It turns out that the main hypothesis (genomic error-correcting codes exist) and the subsidiary one (these codes are organized as a nested system) suffice to account for many basic and conspicuous features of the living world, which may be considered as a strong, although indirect, proof of the hypotheses. Most of these features are left unexplained by biology, being often taken as postulates. The consequences of the hypotheses also suggest answers to still debated questions. Among features of the living world to which the above hypotheses provide an explanation, let us briefly mention the following ones.

### *8.2.1 Genomes are Redundant*

A rough estimate of the number of past and extant species is $10^9$. A genome length of 15 base pairs would thus suffice to endow each of these species with its own unique genome since $4^{15} = 1,073,741,824 > 10^9$. Of course, genomes are not mere labels and act as recipes for building phenotypes. We already noticed in Sec. 6.4 that a genome 133 base-pairs long would suffice to label each of the $10^{80}$ or so atoms of the visible universe, since $4^{133} \approx 10^{80}$. In contrast, the shortest genomes of viruses are a few thousands base-pairs (b.p.) long. That of bacterial genomes is of a few millions b.p., that of animals and plants range from $10^8$ to more than $10^{11}$ b.p. (the human genome, for instance, is approximately $3.2 \times 10^9$ b.p. long and is far from being the longest). Even the shortest genomes are by far much longer than the strict minimum. There is thus ample room for redundancy, a necessary feature of error-correcting codes.

We may indeed think of a genome as made of three parts: (1) an information message, (2) redundant symbols associated with this message, and possibly (3) symbols unrelated to the information message and not involved in the molecular processes which eventually result in the construction of a phenotype. Parts (1) and (2) would constitute together according to our main hypothesis a word of an error-correcting code in systematic form. Although it may reasonably be conjectured that some parts of eukaryotic genomes are indeed codewords in systematic form (see Sect. 8.4), it would be utterly restrictive to assume that a genomic error-correcting code is always so. It needs only to be redundant. Parts (1) and (2) thus need not be distinct. It suffices that they constitute together the output sequence of a redundant source, i.e., of entropy per symbol, expressed using the size $\alpha$ of the alphabet as unit, less than unity. As regards part (3), biologists have long believed that, in many species, much of the DNA is devoid of any function, being parasitic or fossil, and dubbed it 'junk DNA'. However, 'of unknown use' does not mean 'useless', and an increasing number of parts of DNA believed to be non-functional are actually found to participate in molecular processes, and especially to be transcribed into short RNA molecules. In any way, one should keep in mind that all the symbols of a codeword both participate in, and benefit from, the resilience to casual errors that encoding provides: sequences are conserved only insofar as they contribute in their own conservation. Maybe there is no junk DNA at all in genomes, although DNA seems to be very 'cheap'. It is possible that some DNA sequences, e.g., made of short motifs repeated many times, merely act as 'space-keepers', ensuring a separation along the sequence of nucleotides which is needed by, or useful to, the encoding process.

The existence of constraints incurred by nucleic-base sequences in DNA limits their possible number, thereby introducing redundancy. For instance, in eukaryotic cells double-strand DNA is closely packed in nucleosomes, wrapped around histone octamers acting as spools. Bending constraints result. We suggest that genomic codes are actually defined by such kind of constraints, not by mathematical equalities as are engineering error-correcting codes; in other words, that they are soft codes in the meaning of Sect. 5.5.11. Section 8.4 below contains a tentative, and presumably very partial, list of such constraints. The autocorrelation and spectral properties of genomes, too, confirm that they are redundant (Voss 1992; Arnéodo et al. 1998; Audit et al. 2002).

### 8.2.2 *Discrete Species Exist with a Hierarchical Taxonomy*

As a collection of $n$-symbol words, an error-correcting code is a $k$-dimensional subset of the $n$-dimensional Hamming space, with $k < n$. For an efficient protection against errors, these words need to be as distant as possible to each other according to the Hamming metric. Associating a word of an error-correcting code with the genome of some species results in species necessarily being discrete in the sense that their genomes are separated from each other by large Hamming distances.

The subsidiary hypothesis moreover implies that the distance between the codewords depends on the coding layer to which they belong in the nested system. The deeper the layer, the more distant they are. Then species can be ordered according to a hierarchical taxonomy. The further assumption that a small fraction of the genome remains uncoded suffices to account for the difference between individuals inside a species.

The main and subsidiary hypotheses then explain one of the most basic biological facts, namely, the existence of distinct species and the possibility of their ordering according to a hierarchical taxonomy. These properties appear as consequences of both the necessity of conservation in the presence of errors and the effect of residual errors. The concepts of species and of taxonomy have thus objective roots in the error-correcting structure of genomes.

### 8.2.3 *Nature Proceeds with Successive Generations*

**Constraints on regeneration timing** We now assume, contrary to Sect. 8.1.2 which was intended to obtain an upper bound of the DNA information-theoretic capacity, that all errors consist of substitutions. Given a constant error frequency $\nu$, the average number of symbol errors at time $\Delta t$ with respect to the genome at instant 0, referred to as the initial genome, is then $\bar{n}_e = \nu \Delta t$. The initial genome can be recovered with certainty only provided the number of errors which actually occurred, $n_e$, is less than $d/2$ where $d$ denotes the minimum Hamming distance of the genomic error-correcting code. The correct recovery of the initial genome will thus be secured only if the average number of errors $\bar{n}_e = \nu \Delta t$ is significantly smaller than $d/2$. Then successful decoding, resulting in the genome *regeneration*, will occur with high probability. Failure to properly regenerate the genome would result in a genome at a Hamming distance with respect to the initial one of at least $d$, to be referred to as the 'amplitude' of the regeneration error. If the average number of errors $\nu \Delta t$ is large, its distribution around its mean $\bar{n}_e$ is narrow as compared with $\bar{n}_e$. Then, the probability of a regeneration error is very low when $\Delta t$ is small, but increasing $\Delta t$ up to the vicinity of $d/2\nu$ results in a large increase of the regeneration error probability, which tends to 1 if $\Delta t$ increases beyond this threshold. $\Delta t$ thus appears as a crucial parameter which needs to be small enough for securing the genome regeneration with high probability. Nature must indeed proceed with successive generations, actually interpreted as **re**generations, separated by a sufficiently small time interval. For values of $\Delta t$ larger than the threshold $d/2v$, no stable species can exist.

The observed persistency of species shows that $\Delta t$ is actually small enough to secure the conservation of their genomes. The cellular mechanisms which implement the regeneration of the genome of some species belong to its phenotype, so we may think of the proper adjustment of $\Delta t$ as a specific result of natural selection. Too short this time interval would result in a very well conserved genome, with the drawback of a lack of flexibility which may prevent the species to adapt itself to environmental changes and lead it to extinction. If on the contrary this time interval is too long, many comparatively short-lived species will appear but cannot last. Maybe the Cambrian explosion resulted from too long a time interval between regenerations not yet stabilized by natural selection to a shorter, more adequate, value.

The above analysis only relies on the main hypothesis that a genomic error-correcting code exists. It has to be extended so as to take the subsidiary hypothesis into account, i.e., the existence of a nested system of codes. First of all, a fraction of symbols is assumed not to belong to any error-correcting code and thus to escape regeneration. The remainder of the genome belongs to an $l$-layer nested system as described in Sect. 8.1.4 which has as minimum distance the smallest of the cumulated distances, denoted by $d_{l-1}$. Instead of comparing the number $n_e$ of occurring symbol errors with the single threshold $d_{l-1}/2$, we locate it within the sequence of half the cumulated minimum distances of the component codes. We assume that $d_i/2 < n_e < d_{i-1}/2$, an event the less probable, the smaller is $i$. The genome then incurs a regeneration error of amplitude equal to, or slightly larger than, $d_{i-1}$. Thus the larger the number of symbol errors $n_e$, the lower the probability of a regeneration error, but the larger its amplitude when it occurs.

**Regeneration and replication** On the one hand, the conservation of a genome endowed with an error-correcting code needs its (almost) periodic regeneration in order to avoid that the cumulated number of symbol errors exceeds the correction ability of the code. On the other hand, it is a trivial fact that Nature proceeds with successive generations at an (almost) periodic pace for a given species. Assuming that genomes are endowed with error-correcting codes, the need expressed by the first statement may explain the fact asserted by the second one. In other words, Nature proceeds with successive regenerations made possible by the existence of a genomic error-correcting code and needed to secure the genome conservation.

The existence of successive generations is generally thought of as resulting from the tendency of living beings to proliferate. Then, genome replication is the main function which has to be performed. According to the point of view defended here, the mere template-replication made possible by the double helix structure of DNA does not suffice to ensure the long-term conservation of genomes but should be complemented with the regeneration process. It is thus plausible to assume that regeneration and template-replication are jointly performed and, for instance, this has been assumed in the 'toy living world' analyzed for the purpose of illustration in Sect. 8.3 below. However, it is by no means necessary that they are jointly performed.

The functions of template-replication and regeneration are indeed conceptually distinct. Moreover, regeneration is much more complex than mere replication hence more costly in some sense, and is necessary but for long-term conservation. It is thus possible, for instance, that the template-replication is frequently, almost periodically, performed while regeneration is more infrequent, possibly triggered by external events in some instances. Assuming it is so could maybe explain epigenetic phenomena, as those reported in (Lolle et al. 2005). We now examine how these functions could be separately performed.

Both the functions of regeneration and template-replication are necessary for a unicellular being and may be jointly performed or not. Their dissociation is more plausible in multicellular beings. We may think of the comparatively few germinal cells as specialized in the genome conservation, at variance with the somatic cells. Genome regeneration is performed in germinal cells only and it may be presumed that it is effected during meiosis (or, maybe, at least in part during fertilization). Performing it for all somatic cells would be costly and unnecessary since the conservation of genomes can be secured by the regeneration of the genome in germinal cells only. Then accumulating mutations in somatic cells eventually results in the death of individual phenotypes.

### 8.2.4 Evolution is Contingent and Saltationist

Assuming that the encoded part of the genome uniquely corresponds to a species entails that a new species can originate in a regeneration error. In this case, the new genome is at least at the minimum Hamming distance of the code apart from the original one. A regeneration error is thus a potential speciation mechanism, besides the already known ones. More precisely, such an error is at the origin of the genome variation which results in a target of Darwinian selection. We may deduce from this that evolution is contingent and saltationist, at least as far as it results from regeneration errors, since then it depends on chance events and proceeds by jumps.

Assuming that species originate in regeneration errors hints at their contingency. A regeneration error is indeed a chance event, namely, the wrong recovery of a codeword, but each of its symbols is not chosen at random. This suffices to refute arguments from the 'intelligent design' upholders on the improbability of errors simultaneously affecting two different nucleotides. The probability of two simultaneous errors equals the square of the probability of error in a single symbol only provided the symbol errors are independent events, and they are not so.

Species do not solely originate in regeneration errors. Transpositions, chromosome rearrangements, and integration of genetic material of outer origin, especially viral, are already known mechanisms which deeply modify genomes. The setting up of a new code in the nested system assumed according to the subsidiary hypothesis necessarily results from an event of this kind, referred to as *horizontal genomic transfer*. A better knowledge of the genomic error-correcting codes will be necessary in order to understand their possible connection with such phenomena.

As regards saltationism, let again $n_e$ denote the number of symbol errors which affected the genome before its regeneration. The event $d_i/2 < n_e < d_{i-1}/2$ results in a regeneration error of amplitude at least $d_{i-1}$, where $d_i$ denotes as above the cumulated Hamming distance associated with the $i$-th component code in the nested system (with $1 \leq i \leq l$). The probability of its occurrence is the lower, the smaller $i$. The frequency and amplitude of evolutive jumps thus depend on the layer depth $i$ in the nested system. The deeper this layer, the less frequent but of larger amplitude is a jump. Peripheral layers are thus concerned with *micro*evolution, deep layers with *macro*evolution. Beyond the dichotomy micro/macro, intermediate layers moreover provide gradual steps. The subsidiary hypothesis moreover implies that the younger a species, the more numerous coding layers in the nested system encode its genome. Then the complexity, the longevity (more precisely, the time interval between generations) and the diversity of individuals increase as the geological age of their species diminishes. Compare in this respect nematode worms and vertebrates.

Many biologists have an almost religious faith in the power of Darwinian selection and tend to use it in order to explain everything in the living world. It should however be underlined that its targets result from variations in genomes. Assuming that genomic error-correcting codes exist, these variations take the form of error patterns due to wrong regenerations. The weight of such error patterns is the largest, the deeper the coding layer involved; it reduces to a single nucleotide only if it has been left uncoded. This may explain phenomena which contradict Darwin's hypothesis of gradual evolution. To give a single example, Jean-Henri Fabre observed in his *Souvenirs entomologiques* (Fabre 1880) a female insect (a solitary wasp) that stings a spider much bigger than her (a tarentula) at a very precise point of its mouth so as to disable its venomous hooks. She then paralyses the spider by stinging nervous ganglia and lays her single egg inside it for feeding the larva after its hatching out. Were she not immediately successful in disabling the venomous hooks, she would be killed by the spider and her offspring would be lost. Could such an inherited behaviour gradually appear?

Another striking fact is the existence of 'hereditary invariants', an example of which is now given. The limbs of many vertebrates have 5-finger extremities. That this fact does not result from an adaptation but from the persistency of a hereditary information is rather obvious since the 5-finger pattern does not seem to have a high selective value if compared with 4- or 6-finger ones. The limb extremities widely differ from a species to another: think of the hand and the foot of a human, the foot of a lizard or a mole, the fin of a dolphin or a whale, the wing of a bat ... All these 5-finger patterns are fitted to very diverse functions, namely prehension, walking, running, digging, swimming, flying, ... This adaptation results in considerable variations in the form, size and mutual connection of bones, tendons, muscles, skin ... but *within* the 5-finger pattern. The 5-finger pattern could well result from a single 'regeneration error' of the hypothesized genome channel-coding system and thus appears as a hereditary invariant; the adaptation to specific functions involves variations of less basic characters only. The first tetrapods had different patterns (7- and 8-finger extremities) (Gould 1993). In 'modern' vertebrates, when the evolution results in changing the number of fingers it only diminishes it (e.g., a horse has a single-finger leg), but both the paleontological ancestors and early embryonic forms kept the 5-finger pattern.

There is an even deeper reason why selection alone cannot explain the conservation of genomes: selection is a process of elimination so it would be very strange that it could result in conservation, its exact contrary. A simple calculation shows in Sect. 8.3 that, in the absence of error correction, the lifetime of genomes varies as the inverse of their length. Selection can only further shorten it. This conclusion has been established for the simplified model we refer to as 'toy living world', but it obviously extends to the actual living world. As a consequence, long genomes would not be conserved for long time intervals, which blatantly contradicts the existence of very old species with an extremely long genome (e.g., the lungfish). The striking fact here is that their conservation has actually been ensured *against* natural selection, since their environment inevitably incurred many changes during the hundreds of million years of the species life.

### 8.2.5 Evolution Trends Towards Increasing Complexity

No concept is more difficult and controversial in biology than complexity. Let us first notice that the information-theoretic entropy can be used for objectively measuring the biological complexity. Consider the nested system of error-correcting codes used for encoding some genome. If we assume that an information message of a same length $k$ is inserted and encoded every time a new layer has set up, as in the simple example described as typical in Sect. 8.1.4, then the genomic entropy is $lk \log_2(\alpha)$ binary units, where $\alpha$ denotes as above the alphabet size, so the number $l$ of layers in the nested system is proportional to the actual entropy of the genome (Battail 2010). It may thus be thought of as a rough measure of the complexity of a species even if the conditions of the example are not exactly satisfied.

It has long been rather naively believed that a trend towards increased complexity, considered as a progress, was an intrinsic property of evolution, resulting in a hierarchical scale in which the more complex living things were considered as higher. This simplistic view has progressively been challenged and is no longer that of contemporary biologists. A serious objection to it is that, since all extant species descend from a common ancestor, the time needed to produce them is the same, regardless of their actual complexity. Some species evolved towards extremely complex forms but much others are simpler.

That evolution trends towards increased complexity can however be asserted in the more precise and limited meaning that beings more complex than those previously existing appeared during geological times (as a striking example, an estimated time interval of almost two billion years separates the origin of life and the emergence of the first multicellular being). Their survival implies that an increased complexity resulted in selective advantages. Information theory provides a very general argument in support to this statement. That the performance of codes in terms of error correction can be improved by lengthening them (a proven although paradoxical property) entails that increasing the genome length has in itself the evolutive advantage of enabling a smaller regeneration error probability. Darwinian selection operating on genomic error-correcting codes then results in the trend of evolution towards longer genomes even if the redundancy rate remains constant. Besides the immediate advantage of possibly decreasing the regeneration error rate, longer genomes give room for specifying more complex beings, which in turn can implement more sophisticated means for coping with the evolutive pressure. Moreover, the improvement of the error correction ability of longer codes is enhanced if their redundancy is simultaneously increased, as it occurs in the nested system described as typical in Sect. 8.1.4.

The co-evolution of all related species progressively appeared as more relevant to the history of life than the separate evolution of species since parasitism, commensalism and symbiosis played an important role in it. A species which 'chose' such a way of life could advantageously incur simplifications favoured by natural selection. A very tough problem posed to biologists is thus to distinguish, among 'simple' living beings, if they are actually close to an ancestral form, or if they resulted from simplifying a more complex one.

If lengthening a genome, for instance by setting up one more encoding layer in the nested system, has the evolutive advantage of improving its conservation by decreasing the probability of a regeneration error, it has also the drawback of diminishing the replication speed. The ability to survive thanks to a low probability of regeneration errors is an evolutive advantage, just as the ability of fast reproduction. The ability of fast reproduction is limited by the genome length, which should be as large as to specify the replication machinery (if we except viruses which use the cellular machinery of their hosts for their own reproduction), so there is a lower limit to this length. On the other hand, there is no limit to the evolutive advantage which can result from the lengthening of genomes since the regeneration error probability can be made arbitrarily small.

It is why, although successful genome lengthening occurred in the course of evolution, shorter genomes also survived as being able to faster proliferate. Increasing the genome length may be used to increase its information content, or the redundancy, or both. Mainly increasing the redundancy is a conservative strategy which makes the genome more resilient to errors at the expense of a decreased flexibility. Mainly increasing the information content is a more risky strategy which, if successful, can for instance enable conquering new ecological niches. We may think of any surviving species as having realized some compromise between the replication speed which decreases as the genome length increases, the resilience to casual errors provided by redundancy, and the ability to increase its phenotypic complexity. In the typical example of Sect. 8.1.4 the increase in genomic information quantity per encoding layer is $k$, the redundancy can be measured by $\lambda$, and the whole length $n = \lambda(\lambda^{l+1} - 1)k/(\lambda - 1)$ depends on these parameters and on the total number $l$ of encoding layers.

Notice that the encoding of genomes according to the nested scheme described above entails that they are very inhomogeneous. We noticed that some of the nucleotides must remain uncoded. We also noticed that the oldest informations are encoded by several code layers. The orders of magnitude found in Sect. 6.4 have moreover shown that these oldest informations are encoded with a very large average redundancy. A nucleotide chosen at random within a large genome (say that of an animal or a plant) can thus bear either a large information quantity if it is uncoded, or jointly with very many others contribute to the encoding of old and very important information. The information is thus in general very 'diluted' within a genome, but very unequally so and all the more it is long. Incidentally, this shows that it is meaningless to compare genomes as regards the percentage of nucleotides they have in common.

### 8.2.6 Some Comments About the Consequence of the Hypotheses

It happens that certain scientific hypotheses, even among those which were accepted after a slow maturation process, become *a posteriori* almost obvious. For instance, the existence of atoms was still questioned at the beginning of the XX-th century. We may think however that it is validated by the experience of daily life: how, else, could be explained that separated, well delimited solid bodies exist? I do think that the existence of genomic error-correcting codes is of this kind. Remember that, in the absence of redundant coding, the number of different genomes of length $n$ would be $4^n$, with $n$ of at least thousand and possibly equal to millions and even billions, although a DNA molecule of length $n = 133$ suffices to count the atoms of the visible universe! The genomes which correspond to viable beings are probably a tiny fraction of all conceivable DNA sequences, but a random choice among them would not result in definite species. Any living being would be a kind of chimera uniquely combining disparate features. Moreover, the longest of these genomes would be very short-lived at the evolutionary timescale (see Sect. 8.3). The living world would then be an incomprehensible chaos.

To summarize, although we do not yet know how the genomic error-correcting codes are implemented, we have two good reasons for accepting their existence. First of all, because we have no other means for solving the blatant contradiction stated above between the long-term conservation of genomes and the occurrence of comparatively frequent mutations. Second but not least, doing so enables deriving very basic but yet unexplained properties of the living world. Maybe the most dramatic one is the possibility of a taxonomy: any living being can be thought of as belonging to one of comparatively few *discrete species*. The chaos if it were not so is hard to imagine, and we would not exist for observing it. This is maybe the most convincing argument in favour of genomic error-correcting codes.

A genome is intrinsically enduring. First of all, because it possesses literal means which make it resilient to casual errors. Secondly, and now by the agency of semantics, because it can specify the assembly of phenotypic machines which implement its own regeneration by exploiting its intrinsic resilience properties.

## 8.3 A Toy Living World

### *8.3.1 A Toy Living World in Order to Mimic the Real World*

The necessity of a genomic error-correcting code has been asserted above as if a single genome had to be conserved during geological time intervals. The many successive replications of an ancestral genome and of its replicas actually generate a population of individuals among which mutations result in differences. As we already did in (Battail 2008), we now describe a world of sequences which mimics to some extent heredity in the actual living world although it consists only of words of an error-correcting code randomly subjected to symbol errors and then regenerated. Starting with some ancestor, this 'toy living world' proliferates and evolves. It is made of binary sequences of some given length $n$, to be referred to as 'genomes' (using quotes throughout reminds their fictitious character). It exhibits more and more distinct 'species' as time passes and can be endowed with natural selection. We begin with calculating its most significant parameters. Variants of this scheme could easily be designed so as to endow these species with more realistic features, e.g., with a hierarchical taxonomy thanks to nested codes, but we only deal with the simplest case of a single code.

This model is extremely simplified, so as to eliminate all details which could made its analysis difficult. (We think more generally that similarly considering 'toy living worlds' would be very useful in many biological problems as enabling simple computations of main parameters without being bothered with contingent details.) Despite its simplicity, it possesses the main properties of the living world as regards heredity. It is intended to mimic the successive replications of genomes subjected to random mutations. Since we have shown that the genomes must be endowed with error-correcting properties which ensure their conservation at the timescale of geology, the sequences we consider are words of an $(n, k)$ error-correcting code of minimum distance $d$. For comparison, we also deal with uncoded sequences of the same length, i.e., made of $n$ arbitrary symbols. Errors affect the symbols the same way in both cases. We assume that the bits of the 'genomes' incur independent errors with a constant probability $p_{su}$. Duplication of a 'genome' will mean its regeneration followed with its copying if it is encoded, and its mere copying if it is not. Duplication of each 'genome' originates in two identical ones which, except for infrequent errors, are identical to the initial one. The toy living world results from successive duplications of an ancestral 'genome' and of its descendants, and for simplicity's sake the duplications are assumed to occur periodically, simultaneously for all 'genomes'. The errors in the duplicated 'genomes' entail that differences between them occur. All identical 'genomes' in the toy living world are said to belong to a same 'species'and they constitute its 'population'. We intend to statistically analyze the toy living world as regards the number of species it contains. In a first step, we assume that no limit is set to the population growth.

When a 'genome' incurred a mutation, its duplication results in two 'genomes' different from the initial one. If no error-correcting code is used, this means that at least one of the $n$ bits of the original 'genome' is in error. The probability of a mutation, $P_{\text{unc}}$, is then the complement to 1 of the probability that no bit error occurred in the 'genome': $P_{\text{unc}} = 1 - (1 - p_{\text{su}})^n$, an increasing function of the 'genome' length $n$. If the error-correcting code is used, a 'genome' at a Hamming distance of at least $d$ from the original one results from a regeneration error. Let $P_{\text{cod}}$ denote the probability of such a mutation. Contrary to the uncoded case, it can be made *arbitrarily small* through the choice of a long and efficient enough code. Moreover, $P_{\text{cod}}$ can be made the smaller, the larger the codeword length $n$. No simple expression of $P_{\text{cod}}$ can be given. A plausible approximation of it for $p_{\text{su}}$ small enough is $Kp_{\text{su}}^{d/2}$ where $K$ depends on the code. An efficient error-correcting code results in $P_{\text{cod}}$ being much smaller than $P_{\text{unc}}$. $P$ stands below for $P_{\text{cod}}$ or $P_{\text{unc}}$, depending on an error-correcting code being used or not. Then the following calculations are valid in both cases, which helps exhibiting the differences brought by encoding.

### 8.3.2 Permanence of a 'Genome'

The probability that a 'genome' remains identical to itself after $i$ successive duplications is $(1 - P)^i$. In the uncoded case, this probability equals $(1 - p_{\text{su}})^{ni}$. The probability that a 'genome' incurs a mutation after exactly $i$ successive errorless duplications is $P(1 - P)^i$. This is also the probability that a 'genome' lasts exactly $i\Delta t$, where $\Delta t$ is the time interval between two successive duplications. The average lifetime $L(P)$ of a 'genome', expressed using $\Delta t$ as time unit, to be referred to in the sequel as its *permanence*, is the expectation of its lifetime $i$ affected with its probability $P(1 - P)^i$, namely

$$L(P) = \sum_{i=1}^{\infty} iP(1 - P)^i = (1 - P)/P \tag{8.3}$$

where the second equality results from deriving with respect to $x$ the identity $\sum_{i=0}^{\infty} x^i = 1/(1 - x)$, which is true for $0 < x < 1$, and letting $x = 1 - P$.

When an adequate error-correcting code is employed, the population contains only codewords, the total number of which is $2^k$. Then $P = P_{\text{cod}}$, the probability of a regeneration error, so the permanence of 'genomes' in this case is expressed according to Eq. (8.3) as

$$L_{\text{cod}} = (1 - P_{\text{cod}})/P_{\text{cod}}.$$

The probability $P_{\text{cod}}$ can be made very small so $1/P_{\text{cod}}$ is a simple approximation to the 'genome' permanence. It increases without limit if the error-correcting code is as efficient as to make $P_{\text{cod}}$ approach 0 as $n$ approaches infinity. This is possible according to the fundamental channel coding theorem provided the source entropy $k/n$ is less than the channel capacity, i.e., $k/n < 1 - \mathcal{H}_2(p_{\text{su}})$. Then the average lifetime of a 'genome' can be made arbitrarily large. The toy living world thus exhibits some of the most important properties of the living world which were shown in Sect. 8.2 to be consequences of the main hypothesis that genomic error-correcting codes exist: the existence of discrete species originating in infrequent but large regeneration errors, and that their permanence is improved by increasing the 'genome' length.

On the contrary, no sharply defined species exist in the absence of error correction, and their permanence decreases when the 'genome' length increases. Indeed, $P = P_{unc} = 1 - (1 - p_{su})^n$ in this case so the permanence reads in this case

$$L_{unc} = \frac{(1 - p_{su})^n}{1 - (1 - p_{su})^n}. \tag{8.4}$$

It is a decreasing function of the 'genome' length $n$, in sharp contrast with the encoded case. For $n = 1/p_{su}$, it equals $1/(e - 1) \approx 0.582\ldots$, where e denotes the base to the natural logarithms, when $p_{su}$ approaches 0.

### 8.3.3 *Populations of Individuals Within Species*

The total number of 'genomes' which remain identical to the ancestral one after $i$ duplications, hence the population at that time of the species it generates, is $[2(1 - P)]^i$ in the average since we assume that each successful duplication results in two identical 'genomes'. In the absence of any limiting factor, this number exponentially increases only if $P < 1/2$. If no error-correcting code is used this inequality is satisfied only for short enough 'genomes', i.e., such that $P_{unc} < 1/2$, which implies the inequality $(1 - p_{su})^n > 1/2$, or $n < -1/\log_2(1 - p_{su})$. An approximation of $-\log_2(1 - p_{su})$ for $p_{su}$ small enough is $p_{su}/\ln(2)$, so stable species can exist only if the 'genome' length $n$ satisfies the inequality

$$n < \ln(2)/p_{su} \approx 0.69/p_{su}. \tag{8.5}$$

For larger values of $n$, no definite species can exist. Moreover the larger $n$, the slower the population grows.

If on the contrary an error-correcting code is used, $P_{cod}$ can be made much smaller than 1/2 and the persistence of a species is ensured by the exponential increase of the population of individuals which share the same 'genome'. We shall see later that, for a plausible value of the symbol error probability $p_{su}$, the upper limit (8.5) to the length of 'genomes' is much too small to be compatible with that of actual genomes. In the uncoded case, all the $2^n$ possible $n$-bit 'genomes' eventually belong to the population. The ancestral 'genome' is progressively forgotten in the sense that its frequency in the population tends to become equal to that of any other $n$-tuple.

### 8.3.4 *An Illustrative Simulation*

For the purpose of illustration, we present here an example of a 'toy living world' using a very simple code and show some results of its simulation. Since the regeneration performance depends on the total number of erroneous symbols in a word, we may assume a very short 'genome' and a very high error rate. We ran simulations assuming a binary 'genome' of length $n = 7$ and a symbol error rate of $p = 0.1$. (In a realistic situation, a given regeneration error rate would be obtained by properly adjusting the time interval $\Delta t$ between successive regenerations, for a given symbol error frequency $p_{su}$.) We used either no error-correcting code, or the very simple (7,4) Hamming code which can correct all single errors. The probability of a regeneration error is then $P_{cod} = 1 - (1 - p_{su})^7 - 7p_{su}(1 - p_{su})^6$ which approximately equals 0.16 for $p_{su} = 0.1$. We assume that no other events than replications-regenerations occur. Especially, we assume that no natural selection limits the number of replicated 'genomes'. Despite the extreme simplicity of this model and the lack of natural selection, i.e., the factor which is believed the most important for shaping the living world, the obtained results mimic it rather well. Moreover, introducing natural selection as below and taking account of the subsidiary hypothesis (see Sect. 8.1.4 above) would refine the model and make it closer to the real living world, although of course still oversimplified.

A drawback of the choice of the very short (7,4) Hamming code is that the average number of erroneous bits in a received word is small, hence subject to large statistical fluctuations. The stability of species in the simulated toy living world is thus much less than that of a model more realistically involving much longer genomes. Apart this important exception, we hope that the properties of the simulated toy living world shed some light on those which may be expected from the main hypothesis that genomic error-correcting codes exist.

The Hamming (7,4) code is the first invented nontrivial code in the late fourties (Hamming 1950). It uses the binary alphabet ($\alpha = 2$). The length of its words is $n = 7$. It comprises only $2^4 = 16$ words which are listed in Table 8.1 below and numbered from 0 to 15:

**Table 8.1** The 16 words of the (7,4) Hamming code

| | | | |
|---|---|---|---|
| 0: | 0 0 0 0 0 0 0 | 8: | 0 0 0 1 1 0 1 |
| 1: | 1 0 0 0 1 1 0 | 9: | 1 0 0 1 0 1 1 |
| 2: | 0 1 0 0 0 1 1 | 10: | 0 1 0 1 1 1 0 |
| 3: | 1 1 0 0 1 0 1 | 11: | 1 1 0 1 0 0 0 |
| 4: | 0 0 1 0 1 1 1 | 12: | 0 0 1 1 0 1 0 |
| 5: | 1 0 1 0 0 0 1 | 13: | 1 0 1 1 1 0 0 |
| 6: | 0 1 1 0 1 0 0 | 14: | 0 1 1 1 0 0 1 |
| 7: | 1 1 1 0 0 1 0 | 15: | 1 1 1 1 1 1 1 |

Notice that the number made of the first four bits of each word is the binary representation of the word number (with the less significant bits on the left), so this code is in 'systematic form'.

The constraints which tie together its bits $c_1, c_2, \ldots, c_7$ are the parity-check equations:

$$\begin{aligned} c_1 \oplus c_3 \oplus c_4 \oplus c_5 &= 0, \\ c_1 \oplus c_2 \oplus c_3 \oplus c_6 &= 0, \\ c_2 \oplus c_3 \oplus c_4 \oplus c_7 &= 0, \end{aligned}$$

where $\oplus$ denotes addition modulo 2.

This code has been designed according to a precise algebraic structure, but to understand its error-correction ability, it suffices to notice that its minimum distance is $d = 3$, so it can definitely correct any substitution of a single binary symbol for its binary complement, or the erasure of any 2 binary symbols (see Chap. 5).

As an illustration of a hereditary process where a 'genome' is successively replicated several times starting from an ancestral one, and of the introduction of an error-correcting code in this process, we consider a 'toy living world' whose 'genome' is binary and of very short length, namely 7. We assume that the error rate is very high, namely $p = 0.1$, with the errors drawn at random independently of each other. Then the probability that an error pattern of weight $w$ occurs in a word of length 7 is $P(w) = \binom{7}{w} p^w (1-p)^{7-w}$. The numerical values of $P(w)$ are given in Table 8.2 below:

**Table 8.2** Probability $P$ of an error pattern of weight $w$

| $w$ | 0 | 1 | 2 | 3 | 4 | 5 | 6 | 7 |
|---|---|---|---|---|---|---|---|---|
| $P$ | 0.4782969 | 0.3720087 | 0.1240029 | 0.0229635 | 0.0025515 | 0.0001701 | 0.0000063 | 0.0000001 |

Assuming the ancestral 'genome' to be 1111111, we considered the evolution which results from successive replications of this 'genome' in the presence of mutations due to bit errors of probability 0.1 (simulated on a calculator), either without error-correction or when the (7,4) Hamming code is employed. This code corrects all error patterns with a single erroneous symbol. However, all error patterns involving 2 or more erroneous symbols result in a regeneration error pattern with at least 3 erroneous symbols. The 'genealogical trees' obtained in both cases are represented in the following two figures.

A sample of the simulated evolution of the population of 'genomes' when no error-correcting code is used is given in Fig. 8.4. The population of 'genomes' contains more and more different individuals as the number of successive replications increases. They are short-lived genomes and no distinct species can be observed. The initial 'genome' rapidly ceases to be a majority in this population which, after a few replications, looks like a set of random words. When the (7,4) Hamming error-correcting code is used, as depicted in Fig. 8.5, the initial 'genome' remains present during a larger number of regenerations-replications. Regeneration errors give rise to 'genomes' differing from the initial one by at least 3 symbols, which may be interpreted as other 'species'. They moreover exhibit the same permanence as the initial 'genome'.

Notice that Figs. 8.4 and 8.5 represent genealogical trees, the elements of which are individuals. We consider as a 'species' the set of identical 'genomes' (the model does not take into account the slight individual differences which exist in the real world). One can deduce *phyletic graphs* from the genealogical trees by merging into a single branch all branches which correspond to a same 'genome'. Doing so for the genealogical tree of the uncoded case, represented in Fig. 8.4, results in a rather messy graph. The phyletic graph in the encoded case, deduced from the genealogical tree of Fig. 8.5, is more interesting. It is drawn in Fig. 8.6. New species originate in

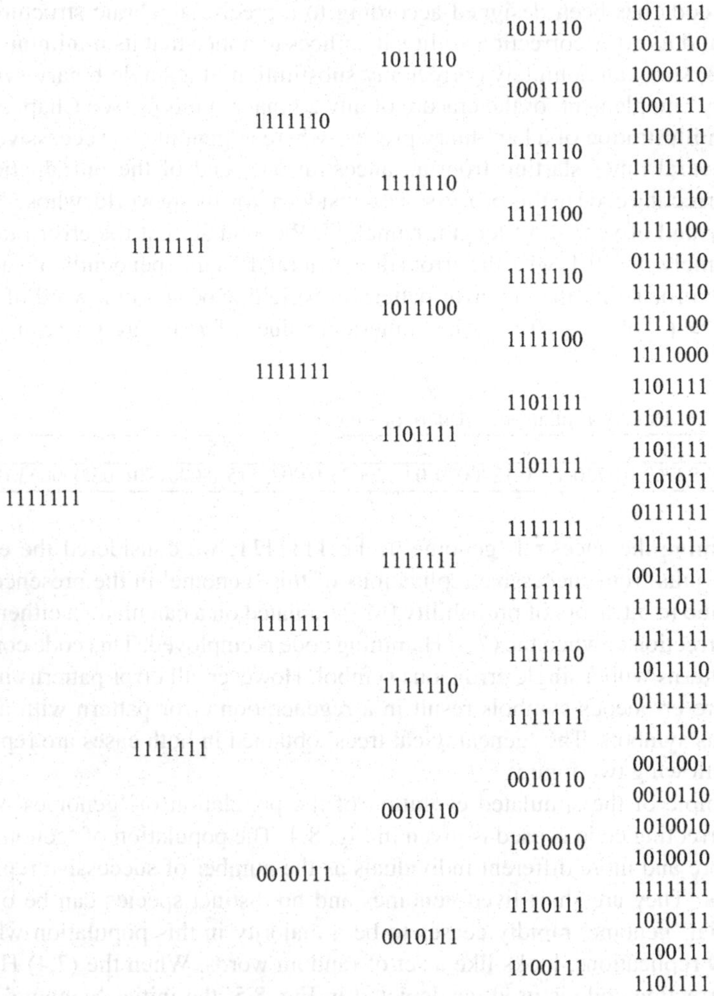

1011111
1011110
1011110
1011110
1000110
1001110
1001111
1111110
1110111
1111110
1111110
1111110
1111110
1111100
1111100
1111111
0111110
1111110
1111110
1011100
1111100
1111100
1111000
1111111
1101111
1101111
1101101
1101111
1101111
1101111
1101011
1111111
0111111
1111111
1111111
1111111
0011111
1111111
1111011
1111111
1111111
1111110
1011110
1111110
0111111
1111111
1111101
1111111
0011001
0010110
0010110
0010110
1010010
1010010
1010010
0010111
1111111
1110111
1010111
0010111
1100111
1100111
1101111

**Fig. 8.4** Genealogical tree of the toy living world, without error correction

regeneration errors, represented as dots in the figure. One notices that some branches converge, so this graph is not a tree. The code of the example contains however very few words so there is a high probability that distinct regeneration errors result in the same 'genome', making such convergences likely events. This probability is much lower for codes having more numerous words close to each other in terms of the Hamming metric, except maybe for the very redundant codes in the innermost layers of the nested system (see Sect. 8.1.4). This case excepted, it may be expected that branch convergences are highly unlikely so all phyletic graphs met in practice assume the shape of a tree: branch convergences are highly improbable, although not strictly impossible.

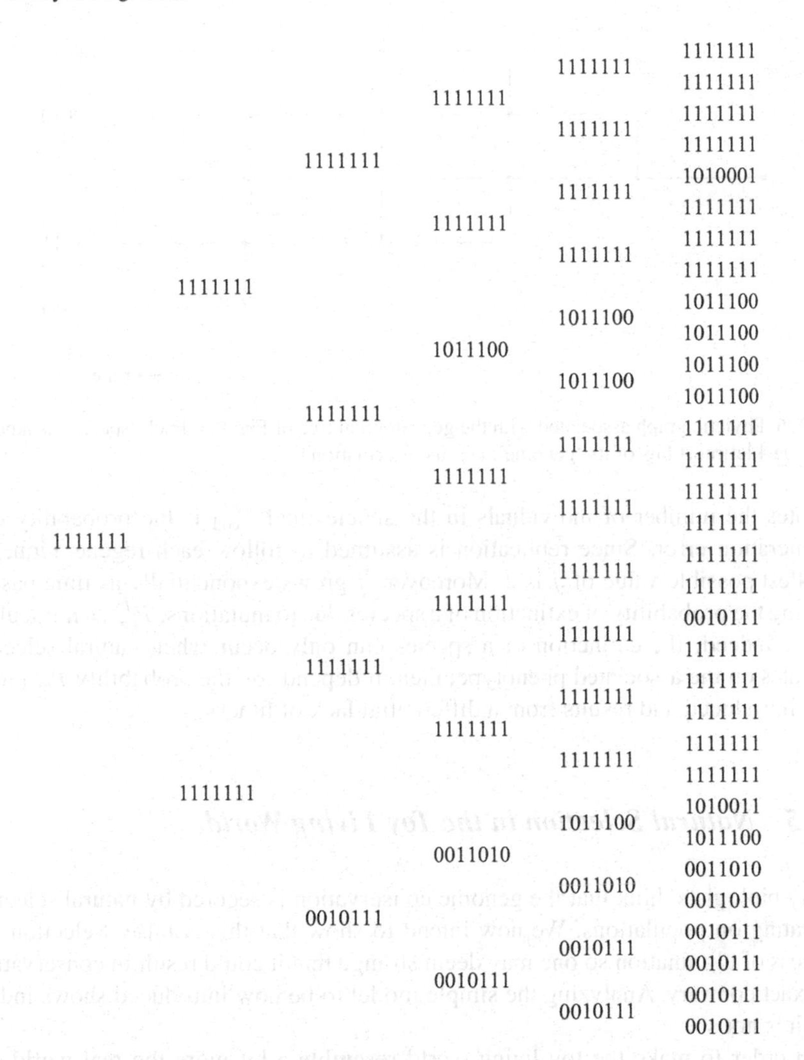


**Fig. 8.5** Genealogical tree of the toy living world using the (7,4) Hamming code

The graph in Fig. 8.6 illustrates the relationship between error-correction coding and evolution for the toy living world: new species originate in regeneration errors. If we ignore the branch convergences (represented by arrows) which occur due to the choice of a very short and simple code but would be very infrequent for a more realistic code choice, this graph depicts the evolution as a radiative process starting from the ancestral 'genome'. Due to a regeneration error, a chance event, the 'genome' of a new species is chosen among the codewords. Once a new species has been created, it potentially lasts indefinitely (just as the ancestral one) since its extinction due to mutations would imply the occurrence of simultaneous regeneration errors in all the individuals of the species. Its probability is thus $P_{\text{cod}}^{j}$, where $j$

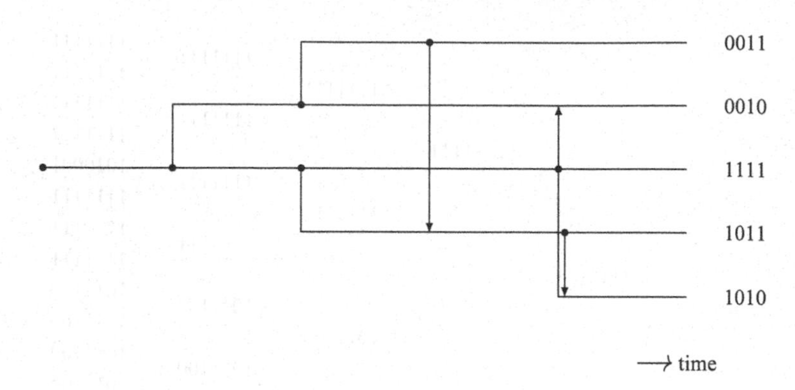


**Fig. 8.6** Phyletic graph associated with the genealogical tree of Fig. 8.5. Each 'species' is labelled with the 4 leftmost bits of its 'genome', i.e., its information bits

denotes the number of individuals in the species and $P_{\text{cod}}$ is the probability of a regeneration error. Since replication is assumed to follow each regeneration, the smallest possible value of $j$ is 2. Moreover, $j$ grows exponentially as time passes, making the probability of extinction of a species due to mutations, $P_{\text{cod}}^j$, a miraculous event. Indeed, the extinction of a species can only occur when natural selection operates on the associated phenotype; then, it depends on the probability $P_{\text{nsel}}$ to be now introduced and results from a differential lack of fitness.

### *8.3.5 Natural Selection in the Toy Living World*

Many biologists think that the genome conservation is secured by natural selection operating on populations. We now intend to show that this is false. Selection is a process of elimination so one may deem strange that it could result in conservation, its exact contrary. Analyzing the simple model to be now introduced shows indeed that it is not so.

In order to make the toy living world resemble a bit more the real world, we can try to take into account the effects of natural selection in its usual meaning. To this end we introduce the probability $P_{\text{nsel}}$ that a 'genome' is not selected, i.e., that another cause than incurring a mutation an event of probability $p_{\text{mut}}$, entails it is not conserved. This cause may be that its support or the phenotype which contains it is destroyed. Then the probability that a 'genome' remains identical to itself equals $(1 - P_{\text{mut}})(1 - P_{\text{nsel}})$ as the joint probability of two events assumed to be independent. The probability $P_{\text{ncon}}$ that a 'genome' is not conserved is the complement to 1 of this probability, namely:

$$P_{\text{ncon}} = P_{\text{mut}} + P_{\text{nsel}} - P_{\text{mut}} P_{\text{nsel}}. \tag{8.6}$$

The average lifetime of an individual genome would result from substituting $P_{\text{ncon}}$ for $P$ in (8.3), which diminishes the permanence if $P_{\text{nsel}}$ is strictly positive. The average population size after $i$ duplications becomes $[2(1 - P_{\text{ncon}})]^i$. Indeed,

$P_{nsel}$ would in the real world depend on many extrinsic factors like the amount of available resources, the population size of the considered species but also of all species ecologically related to it (predators, preys, commensals, pathogens, ...) and on physical factors of its environment. This multiplicity of factors would make $P_{nsel}$ not only very difficult to evaluate, but also rapidly variable as all species populations vary. The lack of a reliable estimate of $P_{nsel}$ would make the expressions of the average lifetime $L(P_{ncon})$ and of the average population size deriving from it of little practical interest. However, the species having the smallest values of $P_{nsel}$ would clearly benefit from natural selection as thriving at the expense of the others, given a finite amount of available resources, as expected.

Of course, this conclusion holds for the toy living world, but it shares with the actual living world the main properties that its 'genomes' are subjected to errors and replicated, so true genomes should obey Ineq. (8.5), too (up to a multiplicative factor close to 1 accounting for the actual alphabet size). Using for $p_{su}$ the Haldane-Kondrashov estimate referred to in Sect. 8.1.2, this condition becomes $n < 3.45 \times 10^7$. Many longer genomes exist in the actual living world and, moreover, there is no upper limit to the genome length of stable species: some of the oldest species have very long genomes. For instance, the lungfish genome contains about $1.4 \times 10^{11}$ base pairs, hence is 4,000 times larger than the limit expressed by (8.5). It may be concluded that a genomic error-correcting code must exist in the actual living world, as assumed in Sect. 8.1.3.

We may notice, however, that the genome of prokaryotes is generally shorter than the limit set by (8.5). This is consistent with species being less sharply defined in prokaryotes (Margulis 1998). Their genome is presumably less efficiently protected against casual errors than that of the eukaryotes, especially because the number of components in their nested system of soft codes is smaller.

There is another reason why natural selection alone cannot account for the conservation of genomes. As being many-to-one (or surjective in the mathematical parlance), the genetic mapping (depicted in Fig. C.3 in Appendix C) involves synonymous codons. For instance the 6 codons **UUA, UUG, CUU, CUC, CUA** and **CUG** 'code' for a single amino-acid: leucine. Only two single codons are in one-to-one correspondence with an amino-acid, namely **AUG** for methionine and **UGG** for tryptophan. Would the conservation of genomes rely only on the Darwinian selection of phenotypes, then they would be available only up to synonymy, i.e., genomes differing in synonymous codons would be found in the members of a same species. That it is not so shows that the means of genome conservation have to be found elsewhere, besides the stronger arguments given above.

## 8.4 Identifying Genomic Error-Correcting Codes

Up to now, we have shown that:

1. Genomic error-correction codes *must* exist (main hypothesis).
2. These codes must unequally protect parts of genomes, the oldest ones being the better protected. This is achieved by nested codes (subsidiary hypothesis).

3. *Assuming* that nested genomic error-correcting codes exist has consequences which turn out to be actual basic features of life left unexplained by mainstream biology. Other consequences provide decisive arguments in debated issues.

Genomic error-correcting codes are not described in this book because they have not yet been identified. The above reasonings rely on information theory and they do not imply detailed biological mechanisms. Identifying the means which implement error correction, in sharp contrast, is basically a problem of molecular biology. Information-theoretic reasoning is no longer an appropriate tool so only molecular biologists can solve the problem. Before they start searching for genomic error-correction means, however, they should be convinced that doing so can be useful to their discipline. Unfortunately, the only possible arguments to convince them are borrowed from information theory, not from biology. We meet here a big problem of transdisciplinary researches: each discipline developed its jargon and cultural habits in isolation up to becoming closed on itself, unable to receive any idea from outside. How is it possible to overcome the gap between increasingly specialized scientific disciplines and make them communicate? As intended to connect information theory and biology, the present book is hopefully a step in this direction.

It is possible that some experiments have already brought to light some of the functions pertaining to the genome regeneration. If it is so, the biologists who did these experiments were unable to identify them as elements of a solution, for lack of being aware of the problem.

The above remarks do not mean that we can say nothing about genomic error-correction codes. Information theory cannot be substituted for molecular biology, of course, in order to identify genomic error-correcting means. However, it turns out that some reflections inspired by information theory may help guessing what the genomic error-correcting codes look like. We may first think of them as being soft codes as defined in Sect. 5.5.11. Many biological constraints can indeed be interpreted as defining soft codes and we may guess that they are actually useful for ensuring the genome conservation. Even if these guesses are accepted as true, however, they give no clue for solving the most difficult facet of the problem, namely, how *regeneration* is performed by molecular machines.

It has been stated in Sect. 5.5 that an $(n, k)$ error-correcting code, with $k < n$, is a subset of $\alpha^k$ words among all possible $\alpha^n$ $n$-symbol sequences, where $\alpha$ is the alphabet size. This code is *redundant* since $k$ is smaller than $n$. It can be defined by the list of its $\alpha^k$ words (as in Shannon's random coding) or by an encoding rule which specifies constraints that the codewords obey (as in any practical coding system). The only constraints used in engineering are *mathematical*, being convenient as exactly defined and easily implemented by electronic devices. Any set of constraints, however, results in defining some subset of the $\alpha^n$ $n$-symbol sequences, and this subset is endowed with potential error-correction ability. Not only mathematical but physical-chemical or linguistic constraints have this result. It is just for convenience that engineering error-correcting codes are mathematically defined, but the basic reason why these codes work is that they impose *constraints* on sequences. Error-correcting codes defined by constraints of any kind, not necessarily mathematical, were considered as possible genomic codes, under the name of *soft codes*, in our paper (Battail 2001) and in later works. We have already shown in the first part, Sect. 5.5.11, that soft codes may be expected to be efficient error-correcting codes, although (and maybe even because) they are not designed for this purpose.

Let us give some examples of possible genomic soft codes. In eukaryotic cells physical-chemical (more precisely, steric) constraints are induced by the wrapping of DNA molecules around histone octamers, as shown by their autocorrelation and spectral properties (Voss 1992; Arnéodo et al. 1998; Audit et al. 2002). Other constraints, besides the previous ones, are induced on DNA molecules when they 'code' for protein substructures like $\alpha$-helices and $\beta$-sheets. Some sequences of amino-acids then become forbidden as not compatible with these structures (Branden 1991), which entails that the sequences of codons which code for forbidden amino-acid sequences are themselves forbidden. The constraints due to wrapping around histone molecules and due to forbidden codon sequences are cumulative, which means that they generate *nested* soft codes. Every time a 'natural convention' establishes a correspondence between two sets of objects otherwise unrelated, another layer of soft coding results from the constraints obeyed by the newly connected set. In other words, 'organic codes' in Barbieri's meaning (Barbieri 2003) result in endowing DNA with nested error-correcting soft codes. Moreover, as instructing the assembly of phenotypes, genomes necessarily use some syntax which involves linguistic constraints. Besides physical-chemical constraints, linguistic constraints too contribute 'coding layers' to the genomic nested soft codes. In order to account for the better conservation of the oldest information (e.g., that borne by the *HOX* genes), it was suggested in (Battail 1997) and in later papers that the various constraints were successively introduced during the ages so that the older is an information borne by DNA, the more numerous coding layers protect it against errors. Then layers of nested soft codes that successively appeared during the geological ages strikingly resemble Barbieri's organic codes, although their existence is inferred here by information-theoretic arguments based on how genomes can actually be conserved, while it is a deep reflection about the main steps of biological evolution which led Barbieri to infer it (Barbieri 2003).

Having thus extended the concept of error-correcting codes so as to include soft codes, the many constraints which affect genomes clearly entail that many genomic soft codes exist, which moreover are nested within each other as expected according to Sect. 8.1.4. An important question remains unanswered, however: are the constraints which define these codes actually exploited in order to regenerate genomes affected by errors, and how? It turns out, as stated in Chap. 5, that regeneration is the most difficult and critical step of the whole error-correcting process. As yet the molecular machines which perform these tasks remain to be identified, a research which obviously needs a close collaboration of communication engineers and molecular biologists. Before such a collaboration can be settled, the latter should be convinced of its interest, which implies they get some information-theoretic education (Battail 2006).

Another hypothesis regarding genomic codes is worth considering although no direct experimental proof of it has yet been given. We already mentioned the pioneering work of Donald Forsdyke suggesting that introns are made of check symbols associated with the message borne by the exons (Forsdyke 1981). This hypothesis is quite plausible, since the exons bear the information which controls the synthesis of the polypeptidic chain eventually becoming a protein. Then the introns are likely to gather the corresponding check symbols. One can easily check whether eukaryotic genes possess distance properties which may be expected from an error-correcting code by comparing the variability of exons and introns in different evolutive situations, a very general approach since it involves no further assumption than the existence of a code having some rather large minimum distance. This does not give any clue about the code which is used, let alone its decoding, but provides a strong argument in favour of Forsdyke's hypothesis. The literature states that introns are generally more variable than exons. A counter-example was however provided in 1995 by Forsdyke, who experimentally found that the exons are more variable than the introns in genes which 'code' for snake venoms (Forsdyke 1995). It turns out that the generally observed greater variability of introns and Forsdyke's counter-example can *both* be explained by the assumption that the system of exons and introns actually acts as a systematic error-correcting code where exons constitute the information message and introns are made of the associated check symbols. Interpreted as a decoding error, a mutation occurs with large probability in favour of a codeword at a Hamming distance from the original word equal to the minimum distance of the code or slightly larger. If the exons 'code' for a protein of physiological importance, which is by far the most usual case, it may be expected that only mutations with a few errors within the exons, hence having no (which is possible since several distinct codons may specify a same amino-acid) or little incidence on the protein, will survive natural selection. Few errors being located in the exons, most of them will affect the introns since the total number of errors is at least equal to the minimum distance of the code.

In the case of genes which 'code' for snake venoms, the Darwinian selection does not tend to conserve exons but on the contrary favours their mutation. The typical preys of snakes are rodents. Snakes and rodents are involved in an 'arms race': some rodents incur mutations which provide an immunity to snake venom, the population of rodents with such mutations increases as they escape their main predators, and the snakes are threatened with starvation unless mutations in their own genes make their venom able to kill mutated rodents (Forsdyke 1995). The genes which 'code' for snake venoms are thus under 'high evolutive pressure': natural selection favours mutated genes producing proteins as different as possible from the original ones. In terms of the Hamming distance, much of the difference should thus be located in the exons. The total number of symbol errors in exons and introns in the case of decoding error being roughly constant for a given code (equal to the minimum distance or slightly larger), introns are then much less variable, and it is what is actually observed. These properties are precisely those which can be expected from genes acting as systematic error-correcting codes in systematic form, which however remain unknown as well as their decoding process. This may be understood as an indirect evidence of their existence and an incentive to their experimental research.

## References

Arnéodo, A., d'Aubenton-Carafa, Y., Audit, B., Bacry, E., Muzy, J. F., & Thermes, C. (1998). Nucleotide composition effects on the long-range correlations in human genes. *European Physical Journal B, 1,* 259–263.

Audit, B., Vaillant, C., Arnéodo, A., d'Aubenton-Carafa, Y., & Thermes, C. (2002). Long-range correlation between DNA bending sites: relation to the structure and dynamics of nucleosomes. *Journal of Molecular Biology, 316,* 903–918.

Barbieri, M. (2003). *Organic codes*. Cambridge: Cambridge University Press.

Battail, G. (1997). Does information theory explain biological evolution? *Europhysics Letters, 40*(3), 343–348(Nov. 1st).

Battail, G. (2001). Is biological evolution relevant to information theory and coding? *Proceedings ISCTA '01*, pp. 343–351, Ambleside, UK.

Battail, G. (2006). Should genetics get an information-theoretic education? *IEEE Engineering in Medicine and Biology Magazine, 25*(1), 34–45.

Battail, G. (2008), *An outline of informational genetics*. San Rafael: Morgan & Claypool. doi:10.2200/S00151ED1V01Y200809BME023

Battail, G. (2010). Heredity as an encoded communication process. *IEEE Transactions on Information Theory, 56*(2), 678–687. doi:10.1109/TIT.2009.2037044.

Branden, C., Tooze, J. (1991). *Introduction to protein structure*. New York: Garland.

Dawkins, R. (1976). *The selfish gene*. Oxford: Oxford University Press.

Fabre, J.-H. (1880). *Souvenirs entomologiques*. Paris: Delagrave.

Forsdyke, D. R. (1981). Are introns in-series error-detecting sequences? *Journal of Theoretical Biology, 93,* 861–866.

Forsdyke, D. R. (1995). Conservation of stem-loop potential in introns of snake venom phospholipase $A_2$ genes. An application of FORS-D analysis. *Molecular Biology and Evolution, 12,* 1157–1165.

Gould, S. J. (1993). *Eight little piggies*. New York: Norton.

Hamming, R. W. (1950). Error detecting and error correcting codes. *BSTJ, 29*(1), 147–160.

Klein, E. (2010). *Discours sur l'origine de l'univers*. Paris: Flammarion.

Kondrashov, A. S. (2003). Direct estimate of human per nucleotide mutation rate at 20 loci causing Mendelian diseases. *Human mutation, 21*(1), 12–27.

Liebovitch, L. S., Tao, Y., Todorov, A. T., & Levine, L. (1996). Is there an error correcting code in the base sequence in DNA? *Biophysical Journal, 71,* 1539–1544.

Lolle, S. J., Victor, J. L., Young, J. M., & Pruitt, R. E. (2005). Genome-wide non-Mendelian inheritance of extra-genomic information in *Arabidopsis*. *Nature, 434*(7032), 505–509.

Margulis, L. (1998). *Symbiotic planet*. New York: Basic Books.

Nachman, M. W. (2004). Haldane and the first estimates of the human mutation rate. *Journal of Genetics, 83*(3) 231–233.

Rzeszowska-Wolny, J. (1983). Is genetic code error-correcting? *Journal of Theoretical Biology, 104,* 701–702.

Voss, R. F. (1992). Evolution of long-range fractal correlation and $1/f$ noise in DNA base sequences. *Physics Review Letters, 68,* 3805–3808.

# Chapter 9
# Information is Specific to Life

**Abstract** It is argued in Chap. 9 that information is specific to life. The concept of 'semantic feedback loops' is first introduced in order to explain the onset of living structures and their conservation. The basic fact is that certain proteins act as enzymes which catalyze the operations which are needed for their own synthesis: transcription of DNA into messenger or premessenger RNA, splicing out the introns (in the case of eukaryotic cells) and translation as operated by the ribosome, i.e., synthesizing proteins under the control of messenger RNA. This can be interpreted as a set of interwoven feedback loops which all involve the genetic 'code' (or mapping). This set of loops acts as a trap since, once closed, it keeps its structure. If any of its parts fails, however, the whole system ceases operating. We refer to it as 'semantic' because the genetic mapping consists of a set of semantic rules. Hence, the set of feedback loops implements the semantics which enables the assembly of proteins (and more generally of phenotypes) in terms of the information that DNA bears. This shows how information, as an *abstract* fundamental entity, controls the assembly of *physical* structures. 'Abstract' should be understood here as opposed to physical, since assuming that information is physical leads to results which contradict its very definition. The set of semantic feedback loops, although each loop is closed, does not prevent the lengthening of genomes by horizontal genetic transfer, hence is compatible with evolution. The last section of this chapter examines the achievements of Nature as an engineer and pleads for a collaboration of engineers and biologists.

## 9.1 Information and Life are Indissolubly Linked

That communication, hence information, has a paramount importance in the living world is rather obvious. Would no communications exist at all within cells, organs and organisms, life processes would shortly stop. One may wonder why understanding that communication is specific to life did not lead biologists to get deeply interested in it. The only possible explanation is that they did not (and still do not) realize that communication has become a matter of science, and information a scientific entity. Biology actually borrowed since its beginning its methods from already established sciences having objects less complex than life, especially physics and chemistry which successfully account for the inanimate world. The science of communication, originating in the middle of the XX-th century, had not yet a significant influence on

G. Battail, *Information and Life*, DOI 10.1007/978-94-007-7040-9_9,

biology because of the linguistic and institutional barriers which hinder transdisciplinary communication. Now in the cultural meaning of the word, communication is at the heart of this epistemological problem. Biosemiotics is an attempt to integrate a science of communication within biology, but semiotics fails to recognize that literal communication is a necessary prerequisite to semantic communication (Battail 2009a). Information theory appears here as an inescapable but yet missing link.

In support to this statement, we may quote Dawkins who wrote in *The blind watchmaker*, (Dawkins 1991):

> If you want to understand life, don't think about vibrant, throbbing gels and oozes, think about information technology.

Notice however that the semi-conductor hardware is quite foreign to the enzyme-catalyzed reactions which occur in the cell. It is not at the level of implementation means that information technology resembles life, but as regards the algorithms which are implemented. Thus, we can fully agree with the above quotation only if 'theory' is substituted for 'technology'. Apart from this reservation, we may claim that the present book complies with Dawkins' recommendation.

There is even a stronger argument which makes information theory mandatory for studying life. We'll see in Sect. 10.2 that any living thing behaves as a Maxwell's demon. Such a demon can be interpreted as a means for locally converting physical entropy into information (although the second law of thermodynamics remains globally in force). Since no Maxwell's demon exists in the inanimate world, the living world appears as the only place where information is recorded, processed, or used in any way. We may thus claim that *information is specific to the living world.* This statement is true only provided we include in the living world the artefacts created by humans for recording, processing, or using information, just applying to the human species Dawkins' concept of extended phenotype according to which the dam that the beaver builds belongs to its extended phenotype (Dawkins 1982). We then meet Henri Bergson's statement 'tools are like our organs *(les outils sont comme nos organes*)'. In such artefacts, the relationship to information may be considered as *delegated* to machines by humans. With this proviso, we may state that using or not information defines the border between the objects of the living world and those of the inanimate one (Battail 2009b). Stated otherwise, information delineates the border between the living and the inanimate.

The concept of 'semantic feedback', to be introduced in the next section, will help explicating the relation of information and life.

## 9.2 Semantic Feedback Loops

### *9.2.1 Semantic Feedback Loops and Genetic Mapping*

Emphasis has been laid in Chap. 8 on the necessity of genome conservation. Besides faithfully conserving themselves, the genomes instruct the assembly of phenotypes. They need for doing so a molecular machinery which is itself a part of the phenotype,

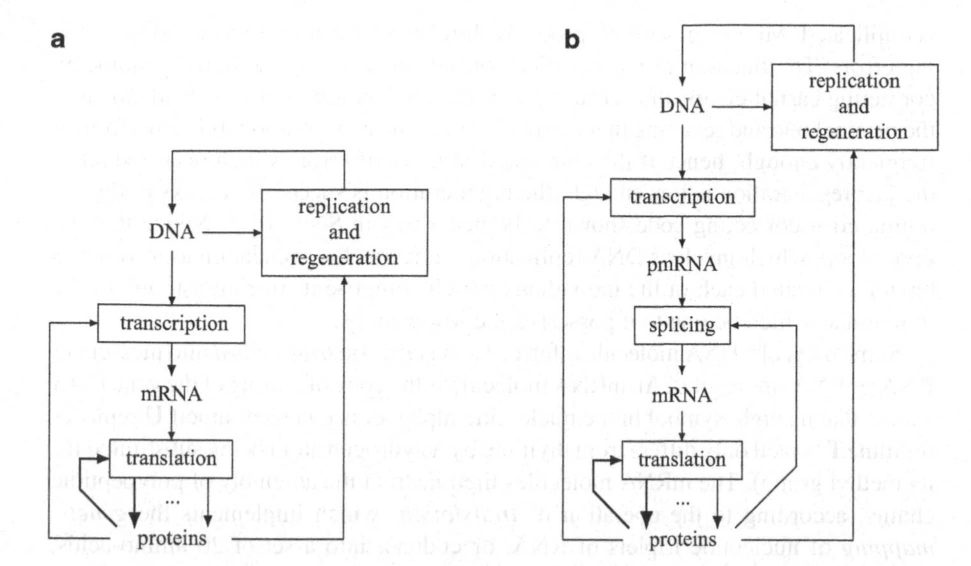


**Fig. 9.1** Feedback loops in the genetic process. The *arrows* originating in 'proteins' denote enzymatic actions. The semantic feedback loop which pertains to the genetic mapping is drawn in heavy lines. The scheme in **(a)** is the most basic one, pertaining to prokaryotic cells. For eukaryotic cells where the genes involve exons which actually specify amino-acids and introns which must be spliced out, the scheme in **(b)** derives from that in **(a)** by further including in it the function of splicing and the corresponding feedback loop

especially the *ribosome* which implements the translation of the messenger RNA into a polypeptidic chain through the agency of the 'genetic code' (which we preferably name 'genetic mapping'). The polypeptidic chains thus obtained eventually become proteins by an appropriate folding. The permanence of life demands the conservation of this molecular machinery as well as that of genomes. Moreover, the permanence and unicity[1] of the genetic mapping must be explained, as well as the fact that only very few types of ribosomes exist in the whole living world.

Figure 9.1 is intended to schematically describe the genetic machinery as a whole in two cases: the scheme at left (Fig. 9.1a) pertains to prokaryotic cells, while Fig. 9.1b at right is a more complicated variant which accounts for the transition towards eukaryotes. Both schemes pertain to a homogeneous population of living objects (say, a population of cells descending from a single ancestor), not to a single individual.

We first comment the most basic scheme of Fig. 9.1a. In the box labelled 'regeneration and replication', the function of *replication* is performed by the physical-chemical means which, given a DNA molecule, result in the synthesis of an identical molecule, thanks to its double-helix structure which enables using each of its strings as a template for assembling a string identical to the other one. Each newly assembled string is tied to the initial one, so the whole double-string molecule

[1] Except for a few very old variants like that found in mitochondria.

is duplicated. Moreover, several 'proof-reading' mechanisms correct possible copying errors. The function of *regeneration* consists of restoring the initial genome by correcting casual errors, thus ensuring that the regenerated molecule is identical to the original one and resulting in the expected genome conservation. If it is performed frequently enough, hence if the cumulated number of errors which occurred since the last regeneration is low enough, the regeneration is successful thanks to the genomic error-correcting code shown to be necessary in Sect. 8.1.3. Notice that the upper loop, which involves DNA replication, concerns the population as a whole as having generated each of the individuals which compose it. In contrast, any of the individuals which belong to it possesses the lower loops.

Some parts of a DNA molecule referred to as *genes* are *transcribed* into messenger RNA (mRNA) molecules. An mRNA molecule is the copy of a string of the gene DNA except that a single symbol in the nucleotide alphabet is changed: uracil **U** replaces thymine **T** (uracil only differs from thymine by a hydrogen atom being substituted for its methyl group). The mRNA molecules then instruct the assembly of polypeptidic chains, according to the operation of *translation*, which implements the *genetic mapping* of nucleotide triplets of RNA, or codons, into a set of 20 amino-acids. The successively selected amino-acids are linked to each other in the order of the codons in the mRNA. The generated polypeptidic chains eventually fold into proteins which lead to the construction of a phenotype, some of them acting as enzymes. The genetic mapping is represented in Fig. C.3 in Appendix C. The correspondences that this chart establishes between any codon and one among 20 amino-acids (or the stop instruction) are interpreted as semantic rules.

A fairly large number of proteins are produced by a living being if we except viruses (say, at least thousand or so for a prokaryote, a few tens of thousands for a eukaryote). Some of these proteins act as enzymes which perform all the enumerated functions: replication, regeneration[2], transcription, and translation. The very existence of these proteins thus depends not only on their specification by genes (through the agency of mRNA molecules) but on their own enzymatic action which makes possible the functions involved in the feedback loops of Fig. 9.1a. The genetic mapping is implemented in the box labelled 'translation' which is common to the loops of Fig. 9.1a involving 'replication and regeneration', 'transcription' and 'translation'. We now intend to discuss the significance of this fact.

Niels Bohr thought that the genetic mapping could be used as the starting point of any research about life. For Marcello Barbieri, it is the most basic *organic code*, an organic code being intended as a set of correspondence rules between sequences of elements otherwise unrelated (Barbieri 2003). Barbieri noticed that the genetic mapping implies 'coding by convention', where the correspondence rules do not result from any physical or chemical law, but appear as arbitrary as the relations a language establishes between words and outer objects. The genetic mapping, a very basic and universal biological fact, thus shares this property with the most

[2] The mechanisms which perform genome regeneration thanks to the genomic error-correcting codes have not yet been identified. However, all the functions of molecular biology need enzymes as catalysts; this one may safely be assumed not to be an exception.

emblematic 'code' of mankind, which is at the root of the human culture! This remark has obviously paramount scientific and philosophical consequences, but was rather ignored by mainstream biology. Barbieri moreover showed that the genetic mapping is only the first of a number of organic codes associated with the main events of life evolution, which all similarly imply 'coding by convention'.

We may think of the loops of Fig. 9.1a as implementing *'semantic feedbacks'* since the genetic mapping, which consists of semantic rules, is an integral part of them. Instead of stabilizing some parameter as in most engineering applications of feedback, the semantic feedbacks of Fig. 9.1a stabilize the genetic mapping itself, which explains its permanency and universality. The enzymes which enable the operation of the loops of Fig. 9.1a are as old as the genetic mapping itself, so the genes which 'code' for them are presumably very resilient to errors as encoded in the innermost layer of the genomic system of nested codes hypothesized in Sect. 8.1.4. Thus, their successful regeneration is a highly probable event. In the infrequent case where their regeneration fails, all the process is aborted, but the mechanism which implements the genetic mapping remains. Would a mutation change this mechanism, the production of the enzymes needed for the operation of the loops would cease, similarly leading to the process abortion.

The efficiency of the semantic feedbacks associated with the genetic mapping thus depends on two factors: the faithful conservation of the genome, and the high specificity of the enzymes. The first one is ensured by the genomic error-correcting code, and the second one by the semantics borne by the genome. Its improvement thus depends on increased redundancy on the one hand, on increased information quantity on the other hand, since according to the interpretation of information quantity given in Sect. 4.2.1 more information quantity actually implies more semantic specificity. Both imply a lengthening of the genome and suggest that their co-evolution resulted in the good conservation of the molecular machinery as well as that of the genomic message. This remark confirms our statement of Sect. 8.2.5 that the Darwinian evolution resulted in lengthening certain genomes, and suggests that it is beneficial not only to better conservation, but also to increased semantic content hence to more complexity. A more detailed account of semantic feedbacks will be found in the following section.

### 9.2.2 Semantic Feedbacks Implement Barbieri's Organic Codes

Engineered feedback loops designed in order to stabilize a parameter to some reference value typically use the difference between the reference and this parameter to control its variation in the sense which reduces the measured difference. The signal which controls the variation results from properly amplifying the difference, which thus can be made arbitrarily small by enough increasing the amplification gain. Some damping is often necessary for avoiding oscillations.

A semantic feedback loop has a similar stabilization effect, but it controls a set of correspondence rules, i.e., a *mapping*, not a parameter. At variance with the feedback loops used in order to stabilize a parameter, it exhibits an on/off behaviour since it implements, or not, the mapping which enables its own operation. Crick used the phrase 'frozen accident' to qualify the onset of the genetic mapping, and so did Barbieri to describe the onset of any 'organic code' (Barbieri 2003). We think that this phrase can acquire an operational significance with the help of the semantic feedback loop concept. Looking at Fig. 9.1, we see that the feedback loops work only if the message borne by the DNA string actually results, through the operations of transcription and translation, in the synthesis of proteins having the precise enzymatic properties which are needed to perform these very operations. Furthermore, repeating the process (hence sustaining life) also needs the enzymes involved in the replication-regeneration process. The loop which involves the operation of translation is crucial as implying the genetic mapping as a whole. If the enzymatic functions of the proteins are selective enough, any change in the mapping would modify the generated proteins, thereby aborting the process as ceasing to produce the enzymes necessary to its own implementation. A variant of it would be stable only if mutations of the DNA string would generate different proteins with enzymatic actions able to implement this very variant. The occurrence of a variant of the scheme of Fig. 9.1a such that *all* the enzymatic properties which are needed for its implementation are present again after the proteins it generates have been modified can be considered as miraculously improbable, which possibly suffices to account for the uniqueness of the genetic mapping.

The only alternative to faithfull reproduction of the semantic feedback loops is the abortion of the whole process. If the probability of abortion is small enough, the structure it specifies proliferates and makes up a population. Then once the loops are closed the genetic mapping is 'frozen' so that, in the present framework, this word acquires the meaning of 'stabilized by closing feedback loops'. In (Barbieri 2003, p. 235), Barbieri wrote: 'Even if the evolution of an organic code could take an extremely long time, the "origin" of a complete code is a sudden event, and this means that the great evolutionary novelties associated with that code appeared suddenly in the history of life.' Closing a feedback loop is indeed a 'sudden event' even if the building of its components and their assembly took an extremely long time. The feedback schemes of Fig. 9.1 are therefore in full accordance with Barbieri's concept of organic codes, and everything he wrote about these codes can be interpreted as involving semantic feedback loops.

Figure 9.1 will moreover be helpful to shed more light on the concept of 'nested codes'. Figure 9.1a has been drawn for prokaryotic cells, i.e., assuming that all the DNA of a gene controls the synthesis of a protein. If we wish to make it describe the case of a eukaryote where the genes involve exons (used for the synthesis of proteins) and introns (spliced out prior to the beginning of the synthesis process), what we need is just to replace Fig. 9.1a by Fig. 9.1b which involves means to perform splicing besides the functions performed in the scheme of Fig. 9.1a. Then, the transcription of a gene results in a 'pre-messenger RNA' (pmRNA) from which introns must first be removed. This task is performed in the box labelled 'splicing', which results in the actual messenger RNA (mRNA) eventually translated into a polypeptidic chain. Besides those already synthesized, splicing needs other enzymes than those needed in the operation of the prokaryotic loops. New constraints are thus added to those already existing, and we may think of them as defining a new soft code in the system of nested codes. Interestingly, the 'splicing code' is explicitly considered next to the genetic mapping in the hierarchy of organic codes established by Barbieri (Barbieri 2003, p. 233, Fig. 8.2). Besides the case of the 'splicing code', we can identify with Barbieri's organic codes the system of nested soft codes which we introduced in Sect. 8.1.4 . We may thus interpret Fig. 8.3 above as resulting from projecting along the time axis Barbieri's Fig. 8.3, *ibidem* p. 235, and moreover a third axis, that of time, is implicit in the scheme which illustrates the fortress metaphor (Fig. 8.3) above and explicit in the text which comments it.

The existence of semantic feedback loops, moreover, solves a rather puzzling problem left open in Sect. 8.4: how constraints associated with specific features of a protein (e.g., assuming structures like $\alpha$-helices or $\beta$-sheets) can induce constraints, hence 'soft codes' according to our interpretation, on the genome which specifies their synthesis? The closed structure of a feedback loop entails that each of its elements, including itself, is simultaneously located before and after any of them. Then, although the genome instructs the synthesis of proteins, the synthesized proteins control the genome by the agency of the feedback. The closed topology of a loop has the rather strange property of *distributing causality* within all its elements. A seeming teleology results. Many semantic feedback loops exist in the living world, as many as biological problems resemble the dilemma of the egg and the chick. In (Wills 1989), Christopher Wills quotes E. von Brücke: 'Teleology is a lady without whom no biologist can live. Yet he is ashamed to show himself with her'. Would he feel less shame to show himself with semantic feedbacks? It may be moreover that the yet unknown genome regeneration mechanisms involve such feedbacks.

To summarize, as a *redundantly encoded* sequence the genome contains the means of its own conservation. As bearing semantics, it *specifies* the means for implementing its own regeneration. The semantic feedbacks of Fig. 9.1 involve both the genome and the phenotype it specifies, hence appear as responsible for maintaining functional biological structures, which can be reproduced only with absolute faithfulness due to the mapping they include: any discrepancy entails the loss of their functional ability. The abstract concept of mapping is then embodied into functional structures, which illustrates in this case how 'information bridges the abstract and the concrete' according to our statement in Sect. 6.4.

In order to illustrate the properties of semantic feedback loops, we introduce now, in Fig. 9.2, a more detailed but still very simplified scheme of the lower part of Fig. 9.1b.

The DNA string is represented at the top of Fig. 9.2. Its genes $G_1, \ldots, G_4$ code for proteins $P_1, \ldots, P_4$. The three proteins $P_1$, $P_2$ and $P_3$ exert the enzymatic actions $E_1$, $E_2$ and $E_3$ by the agency of which the functions of translation, splicing and transcription, respectively, are performed (we thus simplify these functions by assuming that a single enzyme suffices to implement each of them). Protein $P_4$ has no influence on these functions. For convenience, all functions are represented as simultaneously performed although they are dealt with successively. This does not affect the conclusions which can be drawn.

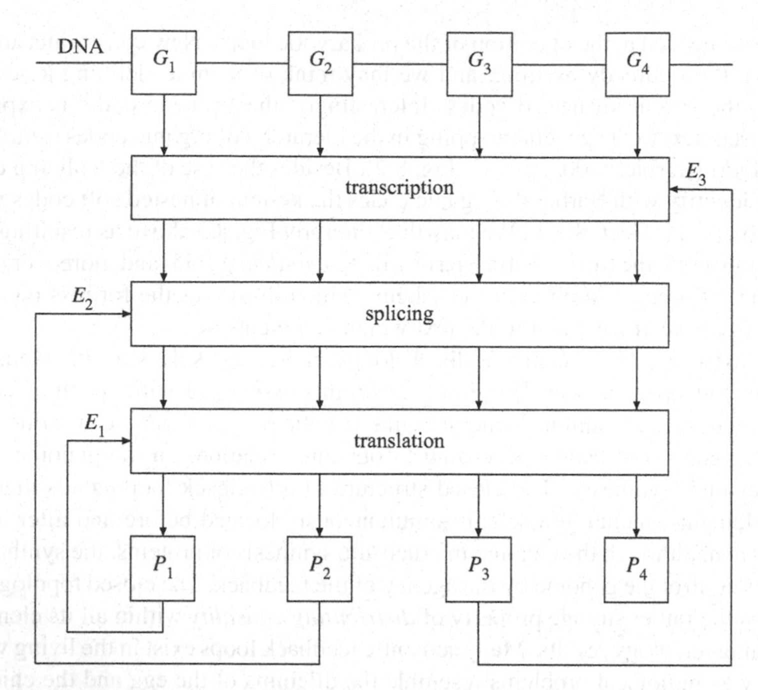


**Fig. 9.2** Simplified lower part of Fig. 9.1b. $G_1, \ldots, G_4$ denote genes which instruct the assembly of proteins $P_1, \ldots, P_4$; $E_1, \ldots, E_3$ denote the enzymatic actions of proteins $P_1$, $P_2$ and $P_3$ on the functions of translation, splicing and transcription, respectively

Specific structures of a protein, like $\alpha$-helices or $\beta$-sheets, are not compatible with some sequences of amino-acids, hence the corresponding sequences of codons *must* be forbidden in the genes which code for them. It is actually so because the enzymatic actions $E_1, E_2$ and $E_3$ of proteins $P_1$, $P_2$ and $P_3$ are necessary for performing the operations of transcription, splicing and translation which give origin to them. In the absence of these enzymatic actions, hence if the genes $G_1$, $G_2$ and $G_3$ would not code for these proteins, the whole scheme would not work. The constraints on the sequences of amino-acids in proteins $P_1$, $P_2$ and $P_3$ thus induce constraints on the genes $G_1$, $G_2$ and $G_3$ hence endow them with soft codes. In contrast, protein $P_4$ exerts no influence on gene $G_4$ because they are not linked by a semantic feedback loop. Although the genes $G_1$, $G_2$ and $G_3$ control the assembly of proteins $P_1$, $P_2$ and $P_3$, and not the other way round, constraints on these proteins do induce constraints on the corresponding genes. *Causality is not violated*, however, because the genes and the proteins are inserted in semantic feedback loops. Notice that a semantic feedback loop is necessarily oriented, which implies that at least one of its links is irreversible. Here, both the genetic code, implemented by translation, and the splicing are irreversible. A less simplistic scheme than that of Fig. 9.2 would lead to the same conclusion as regards the genes and proteins, provided they are involved in semantic feedback loops.

Figure 9.2 shows that the functions of transcription, splicing and translation are controlled by the enzymatic action of proteins $P_3, P_2$ and $P_1$, respectively. The synthesis of these proteins is controlled by genes $G_3$, $G_2$ and $G_1$, respectively, but the synthesis of any of the proteins demands that all the three functions are performed. Let $q_i$ denote the probability of regeneration error of the gene $G_i$, $1 \leq i \leq 3$. Then the whole structure is maintained with probability $(1-q_1)(1-q_2)(1-q_3)$. The probability that the semantic feedback loops cease to exist is its complement to 1. If the probabilities of regeneration errors are small enough, this probability approximately equals their sum $q_1 + q_2 + q_3$. The conservation of the whole structure thus depends on the good conservation of the genome. If $n$ enzymes are involved, each having the same small probability $q$ of regeneration error, the whole structure survives only with a probability of approximately $1 - nq$.

The above conclusions hold only because semantic feedback loops actually exist. The problem of how they came into existence is a hard one and, as often in such cases, only hypotheses can be made for lack of any fossil record of the objects from which they derived and of knowledge of their environment. However, a plausible scenario can rely on the fact that a semantic feedback loop acts as a trap, as we already stated, in the sense that if it is once assembled, maybe by chance, it ensures its own conservation (see Sect. 9.2.4).

### 9.2.3 Semantic Feedback Loops are Compatible with Evolution

Another capital feature of the system of semantic feedback loops depicted in Figs. 9.1 and 9.2 is that, although it locks its own structure, it remains compatible with evolution. Indeed, these feedback loops do not control the length of the DNA string and especially do not prevent its lengthening, which may result for instance from horizontal DNA transfer. Let us have a look at the transition which led to the splicing code in Barbieri's meaning, i.e., in the transition from Fig. 9.1a to Fig. 9.1b. The onset of the splicing function needs appending new genes, hence lengthening the genome. Then both the lengthened genome and the phenotype it specifies are subjected to natural selection. Let us assume that, as suggested by Forsdyke (1981), exons act as the information message of an error-correcting code while its introns are made of the corresponding redundancy symbols. Then, the resulting improvement in genome permanency gives an evolutive advantage to the splicing machinery as a whole *once it is inserted in the semantic feedback loops*, and the longer genome gives room to the specification of more phenotypic features. This may explain why the transition from prokaryotes to eukaryotes involves the simultaneous onset of several important novelties. Interestingly, one of the phenotypic features newly acquired is the formation of the nucleus within which the genetic material is separated from the remainder of the cell. We may think of this separation as strengthening the genome shielding against mechanical and chemical damages. It lowers the error rate at the nucleotide level, which in turn thanks to the hypothesized genomic code results in dividing by a much larger amount the regeneration error rate. Several genotypic and phenotypic features which result from closing the loops thus reinforce each other to the benefit of the genome permanency.

We already noticed that the efficiency of the semantic feedbacks associated with the genetic mapping depends on two factors: the faithful conservation of the genome, and the high specificity of the enzymes. The first one is ensured by the genomic error-correcting code, and the second one by the semantics borne by the genome. Its improvement thus depends on increased redundancy on the one hand, on increased information quantity on the other hand, since more information quantity actually implies more semantic specificity. Both imply a lengthening of the genome and suggest that the co-evolution of the genomic error-correcting code and of the enzyme specificity resulted in the high conservation of the molecular machinery as well as that of the genomic message.

Do semantic feedbacks answer Wittgenstein's question quoted at the beginning of Sect. 8.1: 'why would the fact that the world once began to be be a greater miracle than the fact that it continued to be?', at least as regards the persistence of the living world? A set of semantic feedbacks like those of Fig. 9.1 behaves as a *trap*. Once the loops are closed, they remain so. It is extremely unlikely that the conditions for their closing are fulfilled. However, if by chance they are once fulfilled, the structure maintains itself and, thanks to its reproduction ability, strives at the expense of its environment and tends to consume the available resources. But it has the remarkable feature that its closeness does not hamper the genome lengthening, so it can evolve while remaining closed. However, this is true provided the molecular mechanisms which implement the necessary functions of transcription, translation, regeneration and replication already exist. How these structures once began 'from the scratch' remains quite mysterious, so we may think that we plausibly answered only the second part of Wittgenstein's question.

### 9.2.4 Conjecture About the Origin of Semantic Feedback Loops

As regards the very origin of these structures, one may imagine an initial situation where many molecules of different types are mixed, some of them mainly acting as memories of symbolic sequences like modern DNA or RNA, others endowed with enzymatic properties like modern proteins, some possibly having properties of both kinds. Their random interaction will generally have no stable result, unless a semantic feedback loop is assembled by chance. Maybe this suffices to initiate the process which leads to life. The probability that a rudimentary semantic feedback loop is assembled may be extremely small but, if this probability is not exactly zero, hence if this event is not absolutely impossible, it may occur sooner or later. If favourable conditions are met the mixture of molecules of the required types may occur very often, and the experiment is repeated in many places during geological time intervals (while the interactions of molecules most often result in short-lived products). Then, once assembled, a feedback loop keeps its structure. If a set of semantic rules is by chance fitted to its operation, the semantic feedback survives. The closing of feedback loops having occurred once results in stabilizing their structure for a time much longer than the average lifetime of the products of the interaction of molecules. Not only these structures survive but they reproduce themselves. Other sets of semantic rules result in fleeting structures which vanish and leave room for other ones. Even if the probability of the set of rules which stabilizes the loop is very low, it has with respect to other assemblies the advantage of lasting and of perpetuating itself by cloning. Its proliferation will eventually capture all the resources locally available. Moreover, this scheme possesses the remarkable property of being able of evolution while remaining locked, thanks to the compatibility of its structure with the genome lengthening. Then, 'horizontal transfer' brings at the same time more information and more redundancy. This scheme enables trying variants of the initial set of feedback loops from which the Darwinian selection will choose better and better fitted ones, thus initiating the whole process of biological evolution. Extremely small probabilities are involved in the process assumed to lead to life, so the above scenario remains highly questionable. Maybe Yockey is right when he doubts that it will ever be possible to know the origin of life (Yockey 2005).

## 9.3 Information as a Fundamental Entity

### *9.3.1 Information is an Abstract Entity*

The great physicist Erwin Schrödinger, one of the founders of quantum physics, published in 1941 a short essay entitled *What is life?* (Schrödinger 1943). Introducing the concept of 'aperiodic crystal', he anticipated the structure and properties of DNA, which at that time was not identified as the physical medium of heredity. We devoted a paper to a proposed answer to Schrödinger's question based on concepts of information theory (Battail 2011) and much of the content of this paper has been incorporated in this book. Schrödinger hypothesized that yet unknown laws of physics should be necessary to explain life. Barbieri (2008) and myself (Battail 2011) argued that the discovery of such laws is rather unlikely, but that physics failed to recognize *fundamental entities* which are at the root of life. For Barbieri, these entities are 'nominable entities' as defined in Sect. 2.4 above, and I deem that it is information, as an *abstract* concept which includes any nominable entity, which should be recognized as *the* missing fundamental entity. The great difference of this statement with respect to Schrödinger's hypothesis lies in the abstractness of information, which we definitely deny to be a physical entity.

We have shown in Sect. 6.3.4 that defining the information quantity as the negative of the physical entropy results in measuring by $nH$ the quantity of information borne by $n$ clones bearing each an information quantity of $H$. Since mere copying does not produce any novelty, this conclusion is not acceptable, which led us to reject the derivation of information quantity as the negative of the physical entropy which was suggested by Schrödinger and Brillouin. Instead, we proposed that information, which exhibits the properties pointed out in Sect. 2.2, should be taken as the fundamental entity from which the physical entropy can be derived, understood as measuring the lack of information associated with changing the scale from microscopic to macroscopic (see Sect. 6.3). Then the ability to proliferate, which is specific to information and is not shared by physical entropy, is also not merely a property of life, but the one which the most obviously distinguishes the living from the inanimate. Notice that the ability to proliferate is a property of abstract information which is overlooked when the information and its support are confused, as did Landauer. Using his definition of information would thus fail to account for the most specific property of life.

Physicists like Schrödinger, Bohr and Brillouin were keenly interested in life. Rather strangely, many contemporary physicists do not share this interest and claim that *their* science is the *whole* science, which leaves no place for a science of life. They do not even attempt to rationally legitimate this exclusion. Although living beings are subjected to the laws of physics as other material objects, what makes them specifically living entirely escapes physics. Excluding biology from science is a rather unfair means to get rid of problems about which physics has nothing to say! We think that the reason why physics ignores life is its lack of realizing that information is an abstract fundamental scientific entity. Physics dealt with matter as a fundamental entity since its very beginning. It took centuries to elaborate the concept of energy but it is now recognized as a fundamental entity; the theory of special relativity furthermore states its equivalence with matter according to the famous Einstein equality $E = mc^2$. Information, a recent concept at the slow pace of the evolution of ideas, must be similarly recognized as fundamental for building a science of life with standards of rigour similar to those of physics, but how long will it take? The intrinsic abstractness of information strongly differentiates it from the usual physical entities, and this will not help recognizing it as a fundamental entity. We suggested above that information is specific to the living world and can even delineate the border between the living and the inanimate. That physics denies life can be considered as the negative facet of this statement: it is because physics has not an adequate information[3] concept in its toolbox that physicists have nothing to say about life.

The abstract character of information is underlined in (Avery 2012, p. 90) with arguments close to those presented here. The present book interprets indeed the relationship of information to life and evolution in a way similar to that of Avery, which makes our views fully compatible with his. However, the scope of the present book is more narrowly focused on information. It has been inspired by the engineering practice and more emphasis has been laid on the status of information and the means of its conservation.

### 9.3.2 On the Epistemological Status of Information

The compartmentalization of science into more and more specialized disciplines unfortunately results in the upholders of one of them having at best a scant and

[3] Similarly to biology, the word 'information' is far from unusual in physical texts. What lacks is an adequate scientific *concept* of information.

out-of-date vision of the many others, when they do not simply ignore them. Even when they are aware of their existence, they seem to be unable to imagine that the other disciplines progressed just like theirs. Moreover, although they are aware of the borders which separate their own discipline into subdisciplines, they have a monolithic vision of the other disciplines, as if they miraculously escaped the mechanisms which entail scissions in their own field. As a result, researchers in a discipline are extremely reluctant to accept solutions to their own problems when they come from outside. Symmetrically, if their results can provide solutions to problems met in another discipline, they will most often not be aware of this possibility for lack of knowing the very existence of these problems.

Of course, specialization appears as a necessity for teaching. However, the trend towards increasing specialization tends to make disciplines closed on themselves and eventually unable of any exchange with others. Even the unity of science is threatened since nothing warrants that the separate disciplines do not reach mutually incompatible conclusions. We thus believe (and try to illustrate it in this book) that transdisciplinary research has become a vital necessity for the future of science. Probably the best thing to do is to gather in a single institution many specialists of sciences ranging from pure mathematics to biology, engineering, social sciences and philosophy and let them talk together, each of them patiently trying to explain his/her own research activity to the others. Formal constraints would be counter-productive and should be avoided. For sure, valid results would be obtained although they would be largely unpredictable. Some high ranking institutions already operate this way.

The importance of conservation laws in physics is obvious. Many physical laws express that different equivalent forms of some fundamental entity (mass, energy, ...) are *exchanged* so as to keep constant an overall quantity. In sharp contrast, information is not a physical entity although it has no existence unless it is inscribed on some physical support. It does not obey any conservation law. It can be annihilated, especially if its material support is destroyed. However, being 'immaterial' for Bergson (see the quotation in Sect. 10.2), or 'abstract' as we qualify it, it can also be *shared*. As coextensive with life, symbolic information resides in that part of the physical world referred to as *living*. Being present in the physical world, it can interact with physical objects. We may thus think of life as resulting from the interplay of physical entities which can only be exchanged and abstract symbolic information which can be shared.

From *living to physical*, the symbolic information borne by the genome acts as instructing the assembly of the material structures of a living thing. Its proliferation results from repeating this action. Information has thus a constructive role and accounts for the life organization.

From *physical to living*, the entropy increase manifests itself by cumulative symbol errors in the genomic message which progressively reduce its informational capacity. The entropy law has thus here its usual 'deconstructive' role. As most often succeeding in maintaining the integrity of the genetic message, however, the genome is necessarily endowed with error-correction ability. It succeeds insofar as it is frequently enough regenerated. Then it turns out that regeneration errors have the constructive result of providing very infrequent variants which feed the natural selection process and play thus a capital role in evolution (see Sect. 8.2.4).

## 9.4 Nature as an Engineer

Reviewing in (Benner 2008) a book by Regis (2008), Steven Benner wrote:

> Because building something requires a deep understanding of its parts [and of their mutual relationship], synthesis also stops scientists from fooling themselves. Data are rarely collected neutrally during analyses by researchers, who may discard some, believing the data to be wrong if they do not meet their expectations. Synthesis helps manage this problem. Failures in understanding mean that the synthesis fails, forcing discovery and paradigm change in ways that analysis does not. (The phrase in brackets has been added by me.)

Synthesis being the engineers' job, this remark is an excellent plea for a close collaboration of biologists and engineers. It is originally intended to genetic engineering but actually applies to any instance where nature and engineers are faced with the same problems. It puts in the forefront the necessary *implementation* of biological functions. Indeed, assuming the existence of some biological function without caring about how it is implemented pertains to wishful thinking. The engineering approach advocated in the above quotation should avoid it. Besides a renewed *understanding* of biological facts, another benefit of an engineering approach regards methodology: it often makes possible *quantitative* assessments, especially by using easily tractable yet realistic models for computation and simulation. It makes the engineers' toolbox available for studying life. Then, hypotheses made for explaining biological facts can be tested, validated or invalidated.

Mainstream biology manifests little interest in engineering, however. In the course of evolution, Nature had to solve many engineering problems. Although she uses very different means, her achievements often outperform those of human engineers and may be deemed *outstanding*. Both biologists and engineers should be deeply interested in Nature's achievements, so there is no objective reason why engineers and biologists ignore each others. Their present divide appears as a mere legacy of the past.

Technology is probably the sole human activity where the concept of progress is unquestionably meaningful. The improvement of machines as time passes is an objective fact since the figures which measure their performance steadily increase. Moreover, newly invented machines perform tasks that earlier machines were unable to do. In the middle of the XIX-th century, the products of engineering had already revolutionized the human society. However, they did not compare with living things as regards their perfection and flexibility. The self-reproduction and self-repairing ability of living things had no engineering equivalent. They make an efficient and parcimonious use of energy, while maintaining the range of their internal parameters (especially pressure and temperature) close to that of their environment, unlike machines like the steam engine or other motors. Moreover, extremely complex but tiny functional living structures were observed, much smaller than any manufactured object, up to the molecular scale. These features of life outperformed so far the achievements of human engineers that the technology of the time could not provide models or analogies to help biologists. The gap between technological and natural products could then be thought of as unbridgeable. Nowadays, this gap still exists but technology made immense progresses which significantly reduced it, and researchers seek means, both material and conceptual, for further reducing it. For instance, nanotechnologies enable building devices of extremely small size, at the same scale as many basic biological structures. Moreover, life can no longer be thought of as eluding the laws of physics, hence as radically foreign to the physical world. Concepts like 'artificial life', although very ambitious, are no longer deemed impossible and have become research topics.

That machines steadily improve does not mean that an overall progress results, however, because the criteria which assess their performance are specialized and possibly conflicting: simultaneously improving guns and armors does not result in any overall advantage except for those who sell them. A global comparison shows that, as regards their flexibility and self-repairing ability, the products of Nature still widely outperform those of human engineering. Besides working softly and quietly, they have the decisive advantage of making up together a *lasting* world. Nature proceeds according to cycles which produce no wastes, and its processes are fuelled by no other energy than the inexhaustible solar radiation. In contrast, we now bitterly realize that the Earth obviously contains a finite quantity of raw materials, especially of fossil fuels. Consuming fossil energy and littering the planet with our wastes hinders sustaining the technological development which, in its present form, is doomed to last at most a few centuries, a negligibly small duration at the geological timescale. Life's engineering is fully sustainable while the way humans presently use technology is properly suicidal and even threatens life itself.

Nature's method is extremely different from that of human engineers. Rather than engineering, François Jacob refers to Nature's approach as 'tinkering' (Jacob 1981). It consists of letting proliferate a number of individuals some of which incur random genomic variations which entail somatic changes. Then, Darwinian selection chooses those of the variants which will similarly proliferate and incur random mutations and selection, according to a branching and pruning process. From the point of view of human engineers, this is a very slow and costly process with moreover very uncertain results, all the more it is not intended to any specific purpose. Nature, however, suffers no constraints of cost and time: living material appears as very cheap, and the process of evolution lasts for billions of years. Yet the products of evolution are beautifully engineered and often outperform human achievements. Nature's method has indeed, besides its obvious drawbacks of cost and slowness (according to human criteria), an advantage that no human engineer may claim: *exhaustivity*. This method randomly explores the field of what is possible[4]. Its blindness warrants its objectivity. This process can eventually find the best solution, maybe after many turns and twists. In contrast, human engineers are unavoidably subjected to prejudices which limit their horizon. Besides being exhaustive, Nature's method is a permanently continued process, hence as flexible as to fit environmental changes. The solutions found by Nature are always challenged by new ones. The 'best' solution referred to above is thus merely a local and provisional optimum.

---

[4] Shannon's random coding alluded to in Sect. 5.4.2 may to some extent have been inspired by Darwin.

Nature's method based on random search approaches the limits of what is possible with the means available to her. Human engineers may try to implement by their own means solutions already found by Nature, in order to solve their own problems. Such *biomimetic* designs often meet success in various fields and are very promising (Benyus 1997; Bar-Cohen 2005, 2011). But biologists too should be conscious that Nature found solutions to engineeering problems. It is likely that certain of these achievements have not yet been accomplished by human engineers. For lack of interest of biologists in engineering, and because of the increasing specialization of disciplines which hampers their mutual understanding, it may even occur that solutions found by Nature and already invented by engineers remained unknown to biologists. It is what occurred indeed as regards error-correcting codes, which provide a solution to the genome conservation, a problem of capital importance although it was not perceived by biologists. In any case, technology should be a source of inspiration to biology. Knowing the solutions that humans have invented is often (maybe always) the only means for understanding the solutions to engineering problems found by Nature aeons ago.

Communication engineering benefits from the theoretical framework of information theory. Literal communication of symbolic sequences ('literal' meaning that semantics is ignored) is actually a mathematical problem, and information theory is just that branch of mathematics which deals with it. Information theory can bring to biology its concepts and methods as well as its results. One of its most important concepts is that of *channel capacity*, proven to set an impassable limit to reliable communication. Information theory actually proves that 'errorless' communication (more precisely, with an arbitrarily small error rate) is possible over a channel despite the symbol errors which affect the transmitted message, provided the information rate is less than the channel capacity, a quantity which decreases as the channel error rate increases. However, the very means which enable errorless communication prevent communicating beyond the channel capacity. Both the possibility of errorless communication below the capacity and its impossibility above it, although rather counterintuitive, are fully confirmed by the engineers' experience, besides being theoretically proven.

The hypothesized 'invention' of genomic error-correcting codes by Nature is especially important since the conservation of genomes is at the heart of the evolutive process. The improvement of decoding error probability due to lengthening the codewords, a proven although paradoxical result of information theory, appears as the only means for explaining the trend of biological evolution towards increasing complexity (as I argued in (Battail 1997) and following papers). A low decoding error probability is necessary for ensuring the survival of a genome since its permanency is inversely proportional to this probability as shown above. The performance of genomic error-correcting codes has thus been at stake of evolution since its very beginning.

Genetics is the domain of biology in which information theory can the most obviously be fruitfully applied but it can be in many others, especially neurosciences and immunity. Biosemiotics contributed in realizing how omnipresent is communication in the living world and, everywhere it is, information theory should first have its say.

## References


Avery, J. S. (2012). *Information theory and evolution, 2nd edition*. Singapore: World Scientific.

Barbieri, M. (2003). *Organic codes*. Cambridge: Cambridge University Press.

Barbieri, M. (2008). Biosemiotics: A new understanding of life. *Naturwissenschaften, 95*, 577–599.

Bar-Cohen, Y. (2005). *Biomimetics: Biologically inspired technologies*. Boca Raton: CRC Press.

Bar-Cohen, Y. (2011). *Biomimetics: Nature based innovation*. Boca Raton: CRC Press.

Battail, G. (1997). Does information theory explain biological evolution? *Europhysics letters, 40*(3), 343–348.

Battail, G. (2009a). Applying semiotics and information theory to biology: A critical comparison. *Biosemiotics, 2*(3), 303–320. doi:10.1007/s12304-009-9062-4.

Battail, G. (2009b). Living versus inanimate: The information border. *Biosemiotics, 2*(3), 321–341. doi:10.1007/s12304-009-9059-z.

Battail, G. (2011). An answer to Schrödinger's *What is life? Biosemiotics, 4*(1), 55–67. doi:10.1007/s12304-010-9102-0.

Benner, S. (2008). Biology from the bottom up. *Nature, 452*(7188), 692–694.

Benyus, J. M. (1997). *Biomimicry: Innovation inspired by nature*. New York: Harper-Collins.

Dawkins, R. (1982). *The extended phenotype*. Oxford: Oxford University Press.

Dawkins, R. (1991). *The blind watchmaker, new edition*. London: Penguin Books.

Forsdyke, D. R. (1981). Are introns in-series error-detecting sequences? *Journal of Theoretical Biology, 93*, 861–866.

Jacob, F. (1981). *Le jeu des possibles*. Paris: Fayard.

Regis, E. (2008). *What is life? Investigating the nature of life in the age of synthetic biology*. New York: Farrar, Straus and Giroux.

Schrödinger, E. (1943). In *What is life?* and *mind and matter*. London: Cambridge University Press (1967).

Wills, C. (1989). *The wisdom of the genes. New pathways in evolution*. New York: Basic Books.

Yockey, H. P. (2005). *Information theory, evolution, and the origin of life*. Cambridge: Cambridge University Press.

# Chapter 10
# Life Within the Physical World

**Abstract** Chapter 10 considers the inclusion of the living world within the physical world. The border between the living and the inanimate is poorly understood despite its obvious importance. We suggest that the living world is the only place within the physical world where information is generated, copied, and used, which enables clearly identifying the divide. This statement needs to be valid that the artefacts of the human industry intended to process information are considered as belonging to the living world (including these tools in an 'extended phenotype'). We may then interpret any living being as a kind of Maxwell's demon which counteracts the increase of physical entropy. At variance with the genuine demon, however, it does not violate the second law of thermodynamics since its operation needs energy, as provided by metabolism which ultimately originates in solar radiations. Analyzing how the ribosome sorts amino-acid molecules shows that the assembly of a polypeptidic chain actually decreases the cell entropy, but repeating this operation decreases the physical entropy by the same amount while not creating any more information, confirming that information is abstract and not physical. A sketch of a physical measurement as necessarily crossing the border between the living world of the observer and an inanimate object is interpreted as a variant of Shannon's paradigm. Then, the information-theoretic capacity of the channel limits the information quantity which can be acquired by the observer.

## 10.1 A Poorly Understood Divide

We have already stated in Sect. 9.1 that information delineates the border between the living and the inanimate. This divide is one of the most important aspects of reality. However, both biology and physics leave it essentially unexplained. Léon Brillouin wrote about it (Brillouin 1959, p. 98):

> Living ourselves, we are so much used to this strange world that we fail to marvel at it. And though: reproduction, birth, growth, heredity, thought, all are enigmas for a physicist or a chemist. Nothing similar can be observed in the inanimate world. What we can understand is only death and the living system decomposition. (*Vivants nous-mêmes, nous avons tellement pris l'habitude de ce monde étrange que nous oublions de nous en émerveiller. Et pourtant: reproduction, naissance, croissance, hérédité, pensée, autant d'énigmes pour un physicien ou un chimiste. Rien de semblable ne s'observe dans le monde inanimé. La seule chose que nous puissions comprendre est la mort et la décomposition du système vivant.*)

G. Battail, *Information and Life*, DOI 10.1007/978-94-007-7040-9_10,

According to my interpretation of information at the heart of the living world, the properties of life at which Brillouin rightfully marvels can only be understood if, acting on matter, information transfers to it some of its specifically abstract properties, especially the possibilily of its sharing which then explains why life can proliferate. Information must be recognized not only as a fundamental scientific entity, but as basically abstract.

As regards how life interacts with the physical world, Howard Pattee expressed an opinion close to mine (Pattee 1972). He wrote: 'life is matter controlled by symbols'. I would prefer 'life is matter controlled by information' since symbols are unimportant by themselves, being only elements of a sequence, which itself represents an information, a much more general entity provided it is abstractly defined as an equivalence class, as I did above.

In (Pattee 2005, pp. 524–540 in Favareau 2010), Pattee asked the following question, pp. 524–525:

> All signs, symbols, and codes, all languages including formal mathematics are embodied as material physical structures and therefore must obey all the inexorable laws of physics. At the same time, the symbol vehicles like the bases in DNA, voltages representing bits in a computer, the text on this page, and the neurons firings in the brain do not appear to be limited by, or clearly related to, the very laws they must obey. Even the mathematical symbols that express these inexorable laws seem to be entirely free of these same laws.

and he further wrote, *ibidem* p. 537:

> One of the oldest, non-religious arguments against Darwinian evolution is the apparent improbability of chance mutations producing any successful protein, let alone a species. [...] This argument is based on the assumption of the sparseness of functional sequences and the immensity of the research space.

Answers to these questions are suggested in the present book: genomic or linguistic error-correcting codes are efficient means to fight 'the inexorable laws of physics', and 'functional sequences' necessarily involve specific constraints which result in 'soft codes'. Moreover, the sparseness of functional sequences explains the efficiency of genomic soft codes, but the symbols of a sequence are actually not independent: since they are tied by mutual constraints, the probability of a sequence is not the product of the individual probabilities of its symbols, but much larger than this product. Furthermore, the probability of finding one of them by random search is not that of a single sequence but of any sequence which belongs to its nearest neighbourhood, since the genome regeneration means operating on it would exactly recover the proper functional sequence: the optimum regeneration operating on any sequence in the nearest neighbourhood of a codeword (its Voronoi region) would result in this codeword.

Physicists deal with information as physical and, doing so, cannot perceive that the specificity of life actually results from that of information. Landauer claimed in 1996 that 'information is physical' (Landauer 1996). We saw in Sect. 6.3.4 that, much earlier, Schrödinger and Brillouin tried to define information as a physical entity. According to Boltzmann, Planck and Szilard, the physical entropy is interpreted as measuring an inaccessible information, i.e., an uncertainty which cannot be resolved. Shannon's entropy, on the contrary, measures an uncertainty which is eventually resolved, prior to its resolution. Schrödinger and Brillouin concluded that the informational entropy is actually the negative of a physical entropy, and they renamed it 'negentropy'. Physicists claim without further elaboration that information has become a physical quantity.

For instance, Gilles Cohen-Tannoudji considers that Boltzmann's constant defines a quantum of information (we already gave our own interpretation of this idea in Sect. 6.3.2 above). He wrote (Cohen-Tannoudji 1998, p. 128):

> ... all the progresses of the XX-th century physics lead to consider information as a physical quantity at least as fundamental and irreducible as mass, duration or length. This quantity can be dealt with as a pure number only because Boltzmann's constant, which expresses the limit of its divisibility, is a universal constant which can be set to 1. (*... tous les progrès de la physique du XX-ème siècle conduisent à considérer l'information comme une quantité physique au moins aussi fondamentable et irréductible que la masse, la durée ou la longueur, une quantité que l'on ne peut traiter comme un nombre pur que parce que la constante de Boltzmann, qui traduit la limite de sa divisibilité, est une constante universelle que l'on peut poser à 1.*)

He further wrote, *ibidem*, page 133:

> ... the discovery of information as a fundamental irreducible physical quantity [... makes] referring to a knowing subject unavoidable. (*... la découverte de l'information comme quantité physique fondamentale irréductible [... rend] inévitable la référence à un sujet de la connaissance.*)

We agree as regards the importance of information as a fundamental entity and the necessity of referring to a knowing subject, of course, but we deny that information can be likened to a physical quantity. On the contrary, we think of it as basically *abstract*: an information is a nominable entity, not a quantity. What can be quantitatively measured is merely one of its attributes, just like a man has a weight (among other attributes) but is not reducible to it. Indeed, the dwelling of information within the physical world makes it a *bridge* between the abstract and the concrete, as already discussed in Sect. 6.4.

Likening information to the negative of a physical entropy entails that copying $n$ times an information of entropy $H$ diminishes by $nH$ the physical entropy, as shown when analyzing the operation of the ribosome in Sect. 9.1. Concluding from this fact that the information quantity has been multiplied by $n$, however, is not acceptable since copying does not provide any novelty, hence not any more information. What has been multiplied is only the number of supports of a same information. The possibility of being copied has been stated in the axiomatics of information outlined in Sect. 2.2 and can be referred to as its *sharing property*. It is not compatible with the status of a physical quantity but it enables the proliferation of an information. The proliferation property also mainly differentiates the living world from the inanimate one, in accordance with the prominent role we attribute to information in life. Because information controls life, proliferation of life is a mere consequence of the sharing property of information.

## 10.2 Maxwell's Demon in Physics and in Life

This section complements the analysis of the relation of physical entropy and information which began in Chap. 6, Sect. 6.3. It is devoted to Maxwell's demon, whose talent is far more modest than Laplace's. It will be very useful for contrasting our approach with that of physics.

Maxwell's demon is assumed to violate the second law of thermodynamics. Many variants of it can be imagined, but most of the examples of this demon given as yet follow Maxwell himself, considering the entropy of a gas made of identical molecules within some enclosure. This model is obviously inspired by the steam engine. An enclosure which contains a gas initially at thermal equilibrium is divided into two compartments by a partition with a small hole (see Fig. 6.1). Maxwell's demon is endowed with a shutter it can use for opening or closing this hole. The demon can sense the direction and speed of the molecules, so as to let only the fastest ones enter one of the compartments and the slowest ones leave it. Doing so, the demon tends to create a temperature difference between the two compartments, thereby decreasing the entropy of the system with respect to its initial state and violating the second law of thermodynamics. A plentiful literature has been devoted to this fictitious object (Leff and Rex 2003). Many physicists tried to save the second law by showing that Maxwell's demon could not be implemented, and most physicists deem indeed that it cannot exist. We may think that Brillouin successfully exorcised the demon: in an enclosure at a constant temperature the demon cannot see the molecules, so it can work only if it has a lamp[1], hence a heat source at a temperature other than that of the enclosure (Brillouin 1951). From a more general point of view, we may think that the size of the demon, which is necessarily microscopic, is not compatible with the complex tasks of measurement and control it is assumed to perform, all the more it cannot escape thermal motion. Let us accept the conclusion that Maxwell's demon does not exist in physics.

However, looking at life phenomena leads to a very different conclusion. Living beings obviously develop and maintain differentiated structures against the physical entropy increase although they inhabit Boltzmann's world. Death is maybe the most striking experiment in this respect: once the process of life is interrupted, but only then, the relentless entropy law takes over. *Any* living being thus resists the second law, hence Maxwell's demon dwells in it! More precisely, although physicists imagined Maxwell's demon as foreign to the physical system on which it operates, a living thing is both this system and the demon, as noticed by Norbert Wiener (Wiener 1961, p. 58).

Thus, what is wrong? Life seems to violate the second law of thermodynamics but a necessary and very important condition for its validity is not fulfilled. The increase of entropy stated by the second law concerns an isolated system, hence which does not receive energy from outside. The operation of living beings implies metabolism,

[1] It is assumed that view is the only means the demon uses for sensing the molecules, which is rather anthropocentric.

hence needs energy. If the source of its energy is appended to the living being, an isolated system results and it obeys the second law. This law does not forbid that the entropy of some parts of a system decreases, provided it is more than compensated by an increase of the entropy in other parts. For instance, a refrigerator maintains its internal temperature lower than the outer one without violating the second law of thermodynamics, provided this law is applied to a system which includes its energy source. Life similarly does not contradict the second law, although the own entropy of a living thing does not actually increase (Avery 2012). We may think of it as going against the trend asserted by the second law, just like a fish can expend energy to swim against the stream of a river. Far from tending to uniformity as if it obeyed the second law, life results in creating and maintaining increasingly important differences between parts of a system. Moreover, this is true not only for the development of an individual but still more so for its entire species, and even more for all species, i.e., for life in its entirety. We witness both the increase of uniformity in the inanimate world as predicted by the second law, and the increase in diversity, its exact contrary, when we observe life. Moreover, although individual beings resist the entropy law during comparatively short time intervals (e.g., a few years or decades for large animals), their species does so until it becomes extinct and life lasts since at least 3.5 billion years. The Earth constantly receives a flow of energy radiated from the sun, which suffices to maintain life and to sustain the paradoxical decrease of its entropy.

Let us now look at how a living object resists the entropy increase. We may rightfully deem that the steam engine is not a good model of it. But physical entropy does not only measure how disorderly is a gas made of identical molecules within some enclosure. It also measures, for instance, how uniformly chemically different molecules are mixed. A decrease of the entropy of mixtures results from sorting the molecules in terms of their chemical species. This can be done by means of catalysts which can bind to specific molecules. It is indeed catalysts which enable the ribosomal machinery, under the control of messenger RNA, to sort amino-acids from an initial mixture and to assemble them into polypeptidic chains. Not only the most trivial observation of life at the macroscopic scale hints at living things violating the entropy law, but how the ribosomal machinery, a fundamental molecular mechanism of life, actually uses information at the molecular scale so as to diminish the physical entropy is well-known. The ribosomal machinery sorts the molecules in a mixture of amino-acids as Maxwell's demon does with respect to gas molecules in an enclosure. As regards the result, the difference only lies in the sorting criterion. As regards the implementation means, they are experimentally identified in the former case while no physical means can do so in the latter. Metabolism provides the needed energy. Maxwell's demon does not need any, but does not exist.

Chapter 4 of the book *Information theory and evolution* by John Scales Avery sheds light on this process (Avery 2012). The main physical quantity to be considered is *Gibbs' free energy* of a system, defined as

$$G \triangleq U + PV - TS$$

where $U$ denotes the 'internal energy' of the system, $P$ the pressure, $V$ the volume, $T$ the absolute temperature and $S$ the entropy. This quantity is usually interpreted as measuring the energy available for producing mechanical work. However, Avery interprets it more generally as 'a measure of a system's content of thermodynamic information' (Avery 2012, p. 92). Then mechanical work is only one of the possible products of the system and we may think of the free energy as measuring the energy available for any 'orderly' (as opposed to thermal) use.

Gibbs, who was mainly interested in chemistry, has shown that the free energy must decrease in any spontaneous reaction taking place at constant temperature and pressure. He introduced the free energy of formation of molecules, i.e., the decrease in free energy which occurs in their formation. For given conditions of temperature and pressure, this free energy of formation can be measured and is precisely known. Avery gives two examples of such free energies of formation: that of water by combining hydrogen and oxygen, and that of burning glucose. The latter is especially interesting to us.

The reaction of glucose oxidation reads

$$C_6H_{12}O_6 + 6O_2 \rightarrow 6H_2O + 6CO_2.$$

As a combustion, it produces heat. Although 'spontaneous' in the sense that the corresponding free energy of formation is negative, this reaction occurs only when it is triggered by some external factor because potential barriers which block it must be overcome, so 'a lump of glucose can sit for years on a laboratory table' without being oxidized. Now, says Avery, assume that the glucose is eaten by a girl working in the laboratory. The digestive enzymes quickly catalyze the oxidation and a large part of the free energy is used for the synthesis of ATP (adenosine triphosphate, the main source of metabolic energy) in the girl's mitochondria. Then,

> the high energy phosphate bonds of the ATP molecules will carry the available thermodynamic information further. In the end, a large part of the free energy made available by the glucose oxidation will be used to drive molecular machinery and to build up the statistically unlikely (information-containing) structures of the girl's body.

Interestingly, the inverse reaction, namely the synthesis of glucose

$$6H_2O + 6CO_2 \rightarrow C_6H_{12}O_6 + 6O_2,$$

which is no longer spontaneous but needs absorbing energy, is precisely what *photosynthesis* performs, where the needed energy is supplied by the solar radiations. This is the very source of all life at the Earth's surface, as being at the origin of any available food.

Since information-receiving objects exclusively belong to the living world, and since only an information-receiving agent can violate the second law of thermodynamics, one may assert that

*living things, and only living things, can decrease the physical entropy.*

In other words, the living world is populated with Maxwell's demons: any living thing is both the demon and the system on which it operates[2]. Contrary to the genuine demon, however, a living being needs energy so as to counteract the entropy increase. It successfully performs the most difficult task that the demon does, i.e., sorting the molecules, but not for free: metabolism is another necessity of life.

We now illustrate how the very operation of life results in decreasing the physical entropy. As mere examples, we consider three instances: the synthesis of a protein, the self-reproduction of living beings, and the evolution of a population.

As a first and paradigmatic example, a clear illustration of the above statement is provided by the gene-instructed synthesis of a polypeptidic chain (eventually becoming a protein) which occurs in the cell. We assume that the 20 amino-acids which make up proteins are present, each in sufficient quantity. The number of distinct *a priori* possible polypeptidic chains of length $n_p$ is $20^{n_p}$, a very large number since a realistic order of magnitude of $n_p$ is a few hundreds. The physical entropy associated with a mixture of $n_p$ amino-acids is at most $S_p = \log_2(20^{n_p}) = n_p \log_2(20)$ (approximately $n_p \times 4.322$) binary units. Now consider the translation process occurring within a cell. A molecule of messenger RNA (mRNA) uniquely determines a sequence of amino-acids which is assembled by the joint action of a transfer RNA (tRNA) molecule corresponding to each codon of the mRNA according to the genetic 'code' (see Fig. C.3 in Appendix C), which binds itself to the amino-acid specified by this codon, and of the ribosomal machinery which binds together the amino-acids in the order of the mRNA codons which specify them. Then a unique polypeptidic chain is synthesized which replaces an initial mixture of $n_p$ amino-acids, thus cancelling an amount of at most $S_p = n_p \log_2(20)$ binary units in the physical entropy of the system. The length $n_p$ of each synthesized polypeptidic chain is actually determined by the position of a 'stop' codon in the mRNA molecule. Every time a polypeptidic chain of length $n_p$ is synthetised, the entropy of the initial mixture of amino-acids decreases by at most $S_p = n_p \log_2(20)$. The cellular machinery results in decreasing the physical entropy because it controls individual amino-acid molecules. This may be thought of as a kind of Maxwell's demon. At variance with the system of Fig. 6.1 which contains identical molecules, however, the physical entropy results here from the mixing of different molecules and the ribosome decreases the entropy by assembling a definite polypeptidic chain made of $n_p$ of these molecules.

The information quantity brought by the choice of one among $M$ objects, for instance with the same probability $1/M$, is $H = \log_2(M)$. Letting $M = 20^{n_p}$ provides the informational entropy associated with the choice of a polypeptidic chain of length $n_p$, namely, $H = n_p \log_2(20) = S_p$, i.e., the translation of a gene by the cellular machinery provides a quantity of information which equals the physical entropy of the initial mixture of $n_p$ amino-acids. Then *the quantity of information has increased by the same amount as the physical entropy has decreased*, which substantiates the equivalence of informational entropy with physical negentropy in this particular case (Brillouin 1956).

---

[2] Of course, artificial devices may behave so, e.g., a device can mimic the ribosome operation. Remember that we include human-made artefacts within the living world.

The example just given illustrates the relation of physical and informational entropies. It is simple enough to be quantitatively dealt with. The following examples are much more complicated, but they can be understood as instances of the paradigm provided by the first example. As a second example, the self-reproduction of living beings diminishes the physical entropy for just the same reason as the synthesis of a protein does. When a living thing bearing an amount of structural information of $H_{\text{ind}}$ is duplicated, the physical entropy decreases again by the same amount (the subscript 'ind' in $H_{\text{ind}}$ stands for 'individual'). As a result, the setting up of a population of $N$ identical individuals from successive replications of an ancestor bearing an amount $H_{\text{ind}}$ of structural information results in a decrease in the physical entropy of $NH_{\text{ind}}$. Notice that, although there is no more information in a set of $N$ identical objects than in each of them, the decrease in physical entropy equals the product of the information quantity that each object bears by the number of these objects. This is the main reason why we deny that information can be likened to a physical entity.

A third example is again a population of $N$ individuals descending from a single ancestor, as already considered in Sect. 6.2. Each individual bears a quantity of structural information equal to $H_{\text{ind}}$, but we now assume that replication errors occurred. A set of $N$ identical objects bears no more information than each of its elements, but the replication errors now result in more and more differing objects as their number $N$ increases, either with small differences frequently occurring (in the absence of a genomic error-correcting code), or with much more infrequent but larger differences in the presence of such a code (Battail 2008a, b). Hence the existence of mutations in the population results in *increasing* the quantity of information contained in the population *as a set of individuals* when the size of this population itself increases. This increase of information quantity results from noise-generated errors, hence is again taken out from the physical entropy. The ability of individuals to convert symbolic into structural information results in an increase of the information quantity associated with the population diversity. Notice that the filtering operated by natural selection *diminishes* this quantity of information, in contradiction with Ronald Fisher's statement that 'Natural selection is a mechanism for generating an exceedingly high degree of improbability' (Fisher 1930).

The living world is organized as nested sets of objects. The molecular constituents of living tissues are mostly polymers made of small molecules, such polymers make up organelles, a cell is an assembly of organelles, an organ is a collection of cells, an individual is an assembly of organs, a population is a collection of individuals, etc, to name only a few. Besides the information borne by its individual constituents, each of these sets possesses its own information, borne by the differences between these constituents. Information thus exhibits the property of *emergence*. At any level, the information has been acquired at the expense of the physical entropy.

That life goes against the entropy increase has been noticed more than a century ago by the philosopher Henri Bergson, who wrote in (Bergson 1907), as quoted by Brillouin (Brillouin 1959, p. 132):

> This reality [the second law] strides towards a direction which suggests to us the idea of something which *becomes undone*; here is likely one of the essential features of materiality. What can be concluded from that, except that the process by which this thing *becomes done*

is directed towards the opposite of the physical processes and that it is then, by its very definition, immaterial? [... Life] cannot reverse the direction of the physical changes, as it is determined by Carnot's principle. At least it behaves absolutely as would a force working by itself in the opposite direction. Not being able to *stop* the course of material changes, it nevertheless succeeds in *delaying* it. (*Le sens où marche cette réalité nous suggère l'idée d'une chose* qui se défait*; là est sans doute un des traits essentiels de la matérialité. Que conclure de là, sinon que le processus par lequel cette chose* se fait *est dirigé en sens contraire des processus physiques et qu'il est dès lors, par définition même, immatériel ? [... La vie] n'a pas le pouvoir de renverser la direction des changements physiques, telle que le principe de Carnot la détermine. Du moins se comporte-t-elle absolument comme ferait une force qui, laissée à elle-même, travaillerait dans la direction inverse. Incapable d'*arrêter *la marche des changements matériels, elle arrive cependant à la* retarder.)

Noticing that life tends to counteract the second law of thermodynamics, Bergson qualified it as 'immaterial'. However, he interpreted this word as meaning 'spiritual', while what we oppose to material (or concrete) is 'abstract'. Doing so leads to very different conclusions since we remain in the field of science: we leave physics for information science, not for metaphysics.

## 10.3 A Measurement as a Means for Acquiring Information

Boltzmann's vision of a gas contained in some enclosure at the macroscopic scale is a kind of three-dimensional billiard containing a huge number $N_a$ of balls, say $N_a = 10^{23}$ as an order of magnitude. For ideal monatomic gases, the simplest case, collisions are perfectly elastic and the energy of atoms entirely lies in the kinetic energy of their translational movement. It turns out that billiard is a good example of deterministic chaos (Ruelle 1991). This means that the slightest uncertainty about the initial conditions results after some time interval in a sizeable uncertainty on the position and speed of the balls, which moreover increases with time without limit. The only possible description of the system is thus random, and the only measurable parameters are statistical means, like pressure and temperature for a given volume. Instead of, say, the $N_a$ positions and speed vectors, hence of the $6N_a$ coordinates which are needed for completely describing the system, only three quantities are then available to the physicist. Between the actual situation and what can actually be measured, there is an *immense loss of information*. Thermodynamics names *entropy* the quantity which measures this information loss (see Eq. (6.6) in Sect. 6.3.1). It is why Schrödinger and Brillouin, interpreting information as the negative of the physical entropy, renamed Shannon's measure of information *negentropy*, meaning entropy affected with a minus sign.

Strictly speaking, the second law of thermodynamics which states that the physical entropy can but increase is not relevant to the physical reality, but only concerns the extent to which an observer can grasp it. This reminds us that more generally, contrary to a belief of classical scientists, the statements of physics do not pertain to the reality of objects but to how objects are observed. They actually refer to the *interaction* between objects and their human observer by the agency of a measurement apparatus.

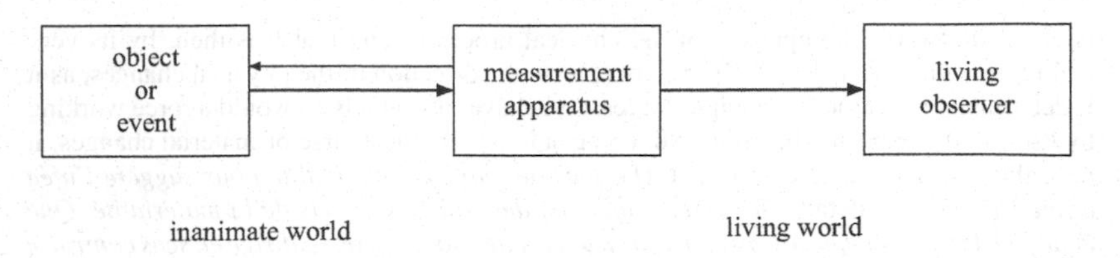


**Fig. 10.1** Scheme of a physical measurement. The *arrow* from the measurement apparatus to the observed object or event is intended to represent its possible stimulation from the observer's side

The information loss which is measured by the physical entropy is actually the necessary condition of any *observation or learning*. Not being Laplace's demons, humans cannot process too large amounts of information, so reducing the collected data to a few statistical means is a necessary condition for their perception by our minds. Indeed, learning in general does not mean merely acquiring information. It demands that a very large part of the acquired information be eliminated. We already quoted the statement of the neuroscientist Jean-Pierre Changeux: 'Learning is eliminating (*Apprendre, c'est éliminer*)'[3]. The immense potential information borne by any macroscopic physical system can only be dealt with after incurring a statistical processing which only keeps some average measured quantities available in the form of symbolic information. A similar remark made by James Collins as regards the human genome and its complexity is quoted in (Check Hayden 2010): 'We've made the mistake of equating the gathering of information with a corresponding increase in insight and understanding.'

Let us depict a physical measurement by means of Fig. 10.1 above. What is observed, at left, belongs to the inanimate world. The observer, at right, is a living being. The measurement apparatus is located between them, precisely at the border between the living and the inanimate worlds. The thick arrows indicate the flow of information between these entities. The arrow from the measurement apparatus towards the observed object or event represents its possible stimulation from the observer's side. As belonging to Boltzmann's inanimate world, the observed object or event bears a very large potential information. The measurement apparatus converts a fraction (a very small fraction indeed) of this *potential* information into *symbolic* information which the observer can use.

The scheme of Fig. 10.1 is clearly a variant of Shannon's paradigm (see Sect. 4.1), where the observed object or event is the source, the living observer is the destination and the measurement apparatus is the channel. This apparatus is subjected to thermal noise just as the other blocks of this scheme but the observer, as a living being, is assumed to be unaffected by thermal noise. (We assumed that the information received by the observer is symbolic, which seems to imply its digital display, but the notion of $\varepsilon$-entropy which pertains to any quantity known with an uncertainty of $\varepsilon$ (Kolmogorov 1956) enables dealing as well with measurements in analog form; see Sect. 5.2.4.)

[3] Contradicting Laplace, Brillouin and many others who implicitly liken learning to information gathering.

If the observed object or event belongs to the microscopic world, it is drowned in a sea of thermal noise, so we may reasonably think of the signal-to-noise ratio $\rho$ associated with the measurement as very small. Then we may use to express its capacity its small-SNR approximation $C \approx \rho/2\ln(2)$ Sh. If on the contrary it belongs to the macroscopic world, e.g., if the measurement consists of determining the position of a star by means of a telescope, the object can be considered as unaffected by thermal noise, but the measuring apparatus can assume, as belonging to Boltzmann's world, one among a huge number $W$ of possible 'complexions' which are indistinguishable at the macroscopic scale (see Sect. 6.3.1). In this case, the channel which models the measurement apparatus is again noisy and thus has a finite capacity, although the signal-to-noise ratio can be much larger. Then it should be expressed by Eq. (5.4) and no longer by its low-signal approximation. If it is very large, 1 may be ignored in this equality and the capacity is then expressed as $(1/2)\log_2(\rho)$ Sh. In any case, the noise may with an excellent approximation be dealt with as Gaussian because of the central limit theorem.

Let us illustrate a physical measurement with an example borrowed from Lewis Carroll's *Alice in Wonderland*. The Red Queen plays croquet using pink flamingoes as mallets and hedgehogs as balls. This game illustrates the experimental physics in Boltzmann's world if we liken the mallets to the measuring apparatuses and the balls to the observed objects. As living things, both a flamingo and a hedgehog can assume many possible states and the one they actually assume is unpredictable. In other words they can be dealt with but as random. Regardless of the nature of the balls, that of the mallets alone justifies dealing with any measurement as a random event. For this reason, any physical measurement can be modelled as a noisy channel. It is meaningless to ask whether the observed objects are deterministic or random, since the means for their observation are irremediably random: the Queen's croquet remains a random game if the hedgehogs are replaced with inert balls, as far as the mallets remain living flamingoes.

Interpreted as a channel, a measurement apparatus has an informational capacity which sets a limit to the information quantity which can be extracted from the inanimate world. The physical reality, thus, is not only seen in the special perspective of a given measurement apparatus, according to Carlo Rovelli's relational concepts (which we fully endorse) (Rovelli 2004), but it is perceived only up to an impassable horizon set by its capacity, which meets the concept of horizon-limited measurement of Cohen-Tannoudji (Cohen-Tannoudji 1998). The information-theoretic capacity enables quantitatively dealing with such a horizon.

## References

Avery, J. S. (2012). *Information theory and evolution, 2nd edition*. Singapore: World Scientific.

Battail, G. (2008a). Genomic Error-Correcting Codes in the Living World. *Biosemiotics, 1*, 221–138. doi:10.1007/s12304-008-9019-z.

Battail, G. (2008b),*An outline of informational genetics*, San Rafael: Morgan & Claypool. doi:10.2200/S00151ED1V01Y200809BME023

Bergson, H. (1907). *L'évolution créatrice*. Paris: Presses Universitaires de France.
Brillouin, L. (1951), Maxwell demon cannot operate: Information and entropy, I. *Journal of Applied Physics*, *22,* 334–337. (Reprinted in Leff and Rex 2003, pp. 120–123).
Brillouin, L. (1956). *Science and information theory*. New York: Academic Press.
Brillouin, L. (1959). *Vie, matière et observation*. Paris: Albin Michel.
Check Hayden, E. (2010). Life is complicated. *Nature, 464,* 664–667.
Cohen-Tannoudji, G. (1998). *Les constantes universelles, New Edition*. Paris: Hachette.
Favareau, D. (2010). *Essential Readings in Biosemiotics*. Dordrecht: Springer.
Fisher, R. (1930). *Genetical theory of natural selection*. Oxford: Clarendon Press.
Kolmogorov, A. N. (1956). On the Shannon theory of information transmission in the case of continuous signals, in (Slepian 1974, pp. 238–244).
Landauer, R. (1996). The physical nature of information. *Physical Nature A, 217,* 188–193. Reprinted in (Leff and Rex 2003, pp. 335–340).
Leff, H. S., & Rex, A. F. (2003). *Maxwell's demon 2, 2nd edition*. Bristol: IoP.
Pattee, H. (1972). *Laws and constraints, symbols and languages*. In C. H. Waddington (Ed.), *Towards a theoretical biology* (Vol. 4, pp. 248–258). Edinburg: Edinburg University Press
Pattee, H. (2005). The physics and metaphysics of biosemiotics. *Journal of biosemiotics, 1*(1) 281–301. (Reprinted in Favareau 2010, pp. 524–540).
Rovelli, C., (2004). Relational quantum theory. In N. Kolenda & A. Elitzur (Eds.), *Quo Vadis Quantum Mechanics?* Dordrecht: Springer.
Ruelle, D. (1991). *Hasard et Chaos*. Paris: Odile Jacob.
Slepian, D. (ed.). (1974). *Key papers in the development of information theory*. Piscataway: IEEE Press.
Wiener, N. (1961). *Cybernetics, 2-nd edition*. New York: Wiley.

# Chapter 11
# Conclusion

**Abstract** The book deals with information as a fundamental abstract scientific entity and attempts to popularize Information Theory so as to enable its application to biology. It is shown that the conservation of genomes must be ensured by error-correcting codes. If they are assumed to exist, Information Theory explains many basic biological facts yet unexplained. Life results from information controlling matter through biological mechanisms, and the use of information delineates the border between the living and the inanimate.

Since more than six decades, information theory has been extremely successful in communication engineering, its originating field, but its applications to natural sciences remained marginal. It turns out that the main properties of information are definitely foreign to those of the usual entities of physics, which entails that applying information outside its originating field demands a kind of mental revolution.

Information theory is a mathematical discipline which unfortunately is ill-known by the general public and by scientists of disciplines foreign to communication engineering. Its abstractness makes its popularization difficult and, in my opinion, it has never been successful, if even attempted. Its main object—information—is moreover rather elusive.

My initial motivation for applying information theory to life came from recognizing, as a communication engineer, that how biology accounts for heredity is absolutely inadequate. Information theory is the sole possible framework for dealing with such problems, and it shows that the conservation of genomes at the geological timescale demands that they are endowed with error correction means. Assuming that such means exist suffices to account for a lot of basic properties of life that biology leaves unexplained. Having become conscious of the shortcomings of biology as regards heredity, maybe the most important feature of life, I attempted in the present book to make information theory accessible to biologists.

Although applying information theory to engineering problems needs only a definition of the quantitative measure of information, not of information itself, I looked for such a definition in order to facilitate its application outside engineering. As my work progressed, it was more and more evident to me that information is central to life, up to delineate the border between the living and the inanimate: living things contain, use, receive, record, process and communicate information. Inanimate things do not, except for artefacts of human origin which are intended to this

G. Battail, *Information and Life*, DOI 10.1007/978-94-007-7040-9_11,

purpose. Moreover, I realized that defining as I did information as an *abstract* entity necessarily residing in the *physical* world made it appear as providing a *bridge* between the *abstract* and the *concrete*, and life as resulting from the interplay of both. Provided its abstract character is fully recognized, information then appears as an entity intrinsic to life, hence fundamental to biology. This conclusion is far more general than recognizing that genomes must be endowed with error correction means. It brings a new insight on the relationship of biology and physics, and paves the way to a theoretical biology with high standards of rigour.

Suggesting that a yet overlooked entity should be taken as fundamental in order to revolutionize a science established for centuries may look foolish. It is nevertheless what this book dares, proposing to refound biology on information. The scientific use of information is limited as yet to the field of communication engineering, which may seem narrow although it resulted in one of the most important transformations of daily life ever witnessed by humanity. But who knows that it heavily relies on information theory, a mathematical discipline which solved the basic problems of communication? The semi-conductor technology, which is more visible, simply implements its solutions. Can biology still ignore information, the fundamental entity which pertains to any communication, when it becomes increasingly clear that communicating is an essential function in the living world?

Dealing with information as an *abstract* scientific entity is mandatory in order to account for properties of life entirely foreign to those of physical entities, especially its ability to proliferate. Understanding life exclusively by means of physical concepts appears *a contrario* as hopeless. It is why I claim, contrary to the opinion of most physicists, that information is not physical.

Using as a fundamental scientific entity the abstract concept of information defined in this book results indeed in a coherent vision of life and of its place within the physical world. Many features of life left unexplained then become mere consequences of information theory.

Basing on information the divide between the living and the inanimate illuminates the relationship between biology and physics, which may be beneficial to both disciplines. Researchers in the sciences of Nature are invited to revisit their foundations so as to account for an overlooked though familiar entity: information.

# Appendix A: Tribute to Shannon

## A.1 Introduction

Claude Shannon died on 24 February 2001 of the after-effects of Alzheimer disease. With him, one of the greatest scientific minds of the century, and even of all times, disappears. His work exercised a deep influence, although often ill-known, in the communication techniques hence in the world where we are living, as well as in the thoughts of the XX-th century. I shall try, after having evoked the carrier and work of Shannon, to show in what his approach was extraordinarily innovative and also, which is more risky, to bring out promises for the future that this work contains. The use I make of the first person should be understood as intended to claim a deliberate subjectivity. I do not indeed pretend to evoke all facets of Shannon's genius but only those which my experience and my reflection enabled me, I hope, to grasp. Beyond the anecdotes and picturesque details I chose mainly to evoke the creator of *information theory*.

Shannon's papers were collected by N.J.A. Sloane and A.D. Wyner (Sloane and Wyner 1993). For convenience, I shall cite Shannon's works by reference to this collection. My main source of historic information, except for the few biographic data contained in (Sloane and Wyner 1993), is the excellent and monumental doctoral thesis of Jérôme Ségal (Ségal 1998, 2005).

## A.2 His Life

Claude Elwood Shannon was born on 30 April 1916 in Petoskey, Michigan, the United States. His father, businessman and for a time Judge of Probate, was a descendant of early New Jersey settlers; his mother, a daughter of German immigrants, was a language teacher and Principal of Gaylord High School in Gaylord, Michigan, where he spent all his childhood. He then admires Edison, a distant cousin of the family, and exhibits ingenuity in tinkering and inventions in mechanics, electricity and radioelectricity. He leaves Gaylord High School in 1932 and enters Michigan University at Ann Arbor. He obtains in 1936 the degree of Bachelor of Science in

G. Battail, *Information and Life*, DOI 10.1007/978-94-007-7040-9,

Electrical Engineering and Bachelor of Science in Mathematics. He then becomes a research assistant in the Department of Electrical Engineering at the Massachusetts Institute of Technology (MIT) near Boston, a part-time position which enables him to continue studying. His Master's thesis is devoted to the application of Boolean algebra to relay and switching circuits. It is published in 1938, meets a very great success and Shannon is awarded in 1940 the Alfred D. Nobel prize, an award given each year in the United States to an engineer less than 30 (do not confuse . . .).

In 1938, he leaves the Department of Electrical Engineering for the Department of Mathematics, at the instigation of the vice-chairman of MIT Vannevar Bush (who will become a consultant to President Franklin Roosevelt). Bush was an engineer of visionary imagination who invented machines predating the computer but failed by the technology of the time. He was just named as chairman of the Carnegie Institution in Washington, a branch of which was studying genetics (and eugenics, which will be discredited only after the war has revealed the monstrous usage made of it). With Shannon's memoir, the design of switching circuits passed from the status of an art to that of a science, thanks to a mathematical formulation of the problem, and Bush hoped that a similar approach by the same Shannon would be fecund to genetics. Back to MIT after a stay at the genetics laboratory of the Carnegie Institution at Cold Spring Harbor, Shannon wrote, under the supervision of the algebraist Frank L. Hitchcock, his thesis entitled 'An algebra for theoretical genetics'[1]. Incidentally, Shannon's work was examined by Barbara Burks, a psychologist expert in the 'genetics of geniuses', member of the American Eugenics Society. Her diagnosis was devoid of ambiguity: the young Shannon is a genius she compares, in a letter to Bush in 1939, with Blaise Pascal re-inventing Euclid's geometry at the age of twelve (Ségal 1998).

Shannon obtains his Ph.D. degree in the spring of 1940. He spends the summer of the same year at the Bell Telephone Laboratories (Bell Labs) where he successfully applies the method of his 1938 memoir to simplify switching circuits (an important stake in the design of telephonic exchanges). After he worked during the academic year 1940–1941 at the Institute for Advanced Studies in Princeton, under the supervision of Hermann Weyl, he comes back to the Bell Labs in 1941, called to integrate a research team (the main members of which were H.S. Black and H.W. Bode) working on anti-aircraft defence systems: a pressing problem in this war time. The works of this team eventually resulted in perfecting and manufacturing the fire control system M6 which enabled England to limit the damage due to the German missiles V1 and V2, and helped the Allies to get the mastering of the airs, a decisive step towards their victory. The war context is the reason why Shannon worked also as a consultant in cryptography to the National Defense Research Committee (NDRC), created even before the United States entered the war and chaired by Vannevar Bush. For this reason, he had the opportunity to meet several times Alan Turing. It seems that cryptography has been for Shannon a source of inspiration but also mainly a mask, honourable in war time, for the studies he already undertook on communication theory and information: they did not contribute to the war effort and their possible usefulness could not be justified but *a posteriori* (Ségal 1998).

[1] Notice that Shannon applied here mathematics to genetics, and that the constraints of the war reoriented his activity towards communication and cryptography.

Bell Labs were a very fecund assembly of researchers and engineers in all domains of physics and mathematics. Information theory and many other works by Shannon, the main ones to be found in (Sloane and Wyner 1993) are not the least production of Bell Labs. The invention of the transistor is another one, miraculously complementary to information theory, to which it provided, as well as to computer technique, implementation means which badly lacked in 1948. Shannon remained 15 years at Bell Labs which he left only to get teaching at MIT.

Reliable witnesses met Shannon in the corridors of Bell Labs riding a unicycle and juggling. Beyond the anecdote, this attests the immoderate taste for playing which was characteristic of Shannon's personality, his interest to precarious equilibria and, of course, a nonconformism he dared display. Maybe it was a paradoxical means to protect himself from inquisitive people: Shannon did not open up easily and lived retired. For instance he used to get rid of the journalists who tried to interview him by letting them visit his collection of 'toys'.

Shannon indeed loved play, all plays. Gambling, chess, music (he played clarinet and collected instruments of all kind) and, maybe still more, the sophisticated toys he constructed himself. His deep interest in roulette made him undesirable in casinos. Should we consider financial investments as gambling? Shannon was successful here to the point of making a fortune, which enabled him no longer to financially depend on the Bell Labs. He was an excellent chess player (during a trip in USSR, in 1965, he brilliantly resisted the world champion Mikhail Botvinnik, just missing the draw), which naturally led him to get interested in chess playing machines. His 1950 paper "Programming a computer for playing chess" (Sloane and Wyner 1993, pp. 637–656) made him a pioneer in this field.

This theoretician of genius was also a great handyman, constructing himself play machines, very diverse gadgets having only in common their almost surrealistic gratuitousness. Here are a few: computing machine entirely operating in Roman numerals, mind-reading machine, cybernetic turtles and mice learning to direct themselves in a maze, cycles with eccentric wheels, cycling and juggling robots, ... The 'ultimate machine' alone deserves to be described: it is a coffin-shaped box with a switch on one face. If turned on, an angry buzz rings out, the lid slowly rises, a hand emerges from beneath and turns off the switch, thus ending what may hardly be called the machine activity!

Let us go back to Shannon's career. Invited professor at MIT in 1956, he became there a permanent teacher in 1959, supervising doctoral dissertations of researchers (many of them made a brilliant carrier in information theory and coding). He formally remained at MIT until 1978, but with a progressively reduced activity. He then retired in a large house near a lake at Winchester (Massachusetts), where he could devote himself to his favourite pastimes. His last papers in the field of information theory and coding were published in 1967, cosigned by R.G. Gallager et E.R. Berlekamp (Sloane and Wyner 1993, pp. 385–423 and 424–454). One of his last papers published under his sole name, in 1959, "Probability of error for optimal codes in a Gaussian channel" (Sloane and Wyner 1993, pp. 279–324), besides being outstanding, maybe contains a key as regards Shannon's behaviour with respect to information theory. He uses several times a word which is unusual in scientific literature: 'tedious'. Shannon actually had to make lengthy and non-fascinating calculations so as to obtain bounds on error probabilities (with the help of his wife Betty he explicitly acknowledges), in contrast with the exaltation which he obviously felt with the discoveries of the beginnings. There is no doubt that the fear of boredom was a major motivation of this passionate lover of plays.

I feel nevertheless an impression of mystery as regards the behaviour of this man in front of the continuation of the researches stemming from his own work, which reminds Moses gazing at the Promised Land without entering it. His famous theorem of channel coding stated the existence of codes making the error probability arbitrarily small provided the source entropy is less than the channel capacity, a result he proved using an extraordinary process which left no hope of an actual implementation: random coding, probably inspired by cryptography and maybe also by Darwin. Why did not Shannon contribute to the search for explicitly defined coding means, as opposed to random, although efficient according to the criteria of information theory? His co-authors of the aforementioned papers, Gallager and Berlekamp, were both eminent actors of this search which was to look like the Holy Grail quest[2]. Was it the awareness that many efforts had still to be made, that this work could only be a collective, slow and rather boring task? Maybe, too, the reluctance of Shannon to any utilitarian finality, as illustrated by the gratuitousness of the 'toys' he constructed?

## A.3 His Work: Information Theory

I shall restrict myself to *information theory*, generally considered as Shannon's main contribution. His contributions to other fields are however by no means ignorable, but I do not feel competent to deal with them. Moreover, I do not wish to depart from what I believe the essentials.

It happens that a science originates in a founding text so obscure that it needs the work of many exegetes before it is eventually understood, and then years and efforts in order to exploit the ideas it contains only in germ. Far from being so, Shannon's seminal text ("A mathematical theory of communication", (Sloane and Wyner 1993, pp. 5–83)), issued in July and October 1948 in the prestigious journal of Bell Labs, the *Bell System Technical Journal*) looks like a popularization work. Not only did he discuss with great clarity the premises of this new science, but he developed it so coherently and so completely that he left little to find to his successors. The great ease of his discussion was not devoid of casualness in the mathematical treatment. This way of introducing a new science was disliked by some keepers of the mathematical orthodoxy, especially in the United States. Greater mathematicians, but in the Soviet Union, became enthousiastic of this emerging science despite the

[2] It still continues nowadays, although it enjoyed a decisive progress with the invention of turbocodes; truly, the way which led to this invention was sinuous and far from the initial directions of this research.

cold war. Thus, Khintchin undertook more rigourously proving Shannon's theorems (Khintchin 1957). Kolmogorov, who made no mystery of his admiration for Shannon, later introduced, with the concept of 'algorithmic complexity', a variant of information theory which was more complementary than competing with Shannon's (Kolmogorov 1965, 1968): instead of letting the measure of information depend on probability distributions assumed to be known, Kolmogorov introduces information as a basic concept[3], freeing it from that of probability the philosophical bases of which are rather weak. Its practical usefulness, however, is restricted since the quantities of Kolmogorov's theory can not be computed, at variance with Shannon's information.

The best account I can give of information theory consists of the analysis of its founding text. In the very *introduction* of "A mathematical theory of communication", Shannon discards semantics from the field of his discussion, only considering as relevant the fact that the message to be communicated in just an element chosen in a certain set and that the communication system must work regardless of the chosen message[4]. The problem of communication is thus basically of statistical nature, the information brought by a particular message being in fact measured by the number of messages among which it is chosen. Explicitly referring to Hartley, he shows the interest of using a logarithmic measure, according to the practical, intuitive and mathematical viewpoints. The choice of the logarithmic base determines the unit used for measuring information. If the base is 2, he names this unit the *bit*, an acronym for *binary digit*. Among other possible bases, he considers also 10, directly consistent with decimal numbers, and Euler's constant e which is more convenient when integrations or derivations have to be performed (logarithms are then referred to as 'natural'). As a model of the communication process, he introduces the famous scheme (or paradigm)

*source—channel—destination*

the 'channel' being actually split into the transmitter which generates a signal, a medium subject to noise (which he names 'channel' in a restricted sense) and a receiver. He then distinguishes three kinds of communication systems: the *discrete* ones where both the message and signal are strings of discrete symbols or signals (a typical example of which is Morse telegraphy); the *continuous* ones where both the message and the signal are dealt with as continuous functions (as in radio and television); and *mixed* ones where both discrete and continuous variables appear (as in pulse code modulation (PCM), used for transmitting speech).

The *first part* is naturally devoted to discrete noiseless systems, the simplest case. He first considers the noiseless discrete channel the capacity of which he defines in information unit per time unit. It depends on the duration of the symbols and the constraints which determine their succession. He then considers discrete sources of which he gives many examples: 'natural' languages; discrete sources deriving from continuous ones by quantization; discrete sources mathematically defined by

---

[3] Which has been briefly expounded in Sect. 6.1 above.

[4] See the quotation in Sect. 2.1.

stochastic processes which specify the symbol choices and their possible mutual dependence, especially in the form of Markov chains where the probability of a choice only depends on the previous choices through the present state of the system. He shows how the models of discrete source he just introduced can provide a series of statistical approximations to English with an increased fidelity according to whether one takes into account the frequency of letters, digrams, trigrams, ..., or of the frequency of words, couples, word triplets .... He gives examples which look like what will become later the exercises of the Oulipo[5]. He graphically represents the Markov chains by transition diagrams which he then assumes to be ergodic (he briefly explains this term). In order to evaluate the average information rate of such a source, he states the axioms which should be satisfied by an information measure in terms of the probabilities which describe it and thus obtains the *entropy* function $H = -\sum p_i \log(p_i)$. He discusses its main properties and defines redundancy as the difference between the maximum possible entropy and its actual value. He notices that the possibility of cross-word puzzles depends on redundancy of the language. It is the more difficult to construct a grid, the stronger the constraints hence the higher the language redundancy. He also considers coding operations aimed at minimizing the average message length. He states that the lower bound on this minimum length is proportional to the source entropy, which is the fundamental theorem of source coding, and gives some examples of optimum source coding.

The *second part* deals with noisy discrete systems, i.e., where the channel input/output transition probabilities differ from 0 and 1. He then introduces the quantity compatible with the entropy definition which measures the average information quantity that the output variable provides as regards the input variable (it is called mutual information due to its symmetry), and he defines the channel *capacity* as the maximum of this quantity with respect to all information sources which can be connected to its input. He sets out the fundamental theorem of channel coding, which paradoxically states that errorless communication of a message is possible if a proper code is employed, provided only that the source entropy is less that the channel capacity. He sketches the proof of this theorem based on the extraordinary idea of random coding. Since he cannot exhibit a particular code with good enough decoding error probability, he considers a *probabilistic ensemble of codes*, computes the average error probability for this ensemble and shows that it can be made arbitrarily small by increasing the word length provided the above condition is satisfied. In the considered ensemble of codes, thus, there exists at least a code which is at least as good as the average. Still better, the error probability thus obtained with a peculiar code is almost surely (asymptotically as the word length approaches infinity) close to the average so it vanishes. From this point of view, one can thus say that 'all codes are good'. Shannon comments these results, recognizes that random coding cannot be actually implemented, insists on the role of redundancy in protecting against noise,

[5] Oulipo (*OUvroir de LIttérature POtentielle*) is a group of writers interested in mathematical games and combinatorics. It was founded by the mathematician François le Lionnais, and its most famous members were Raymond Queneau and Georges Perec.

gives examples of noisy channels and computes their capacity. Finally, he gives an example of coding which turns out to be the (7,4) Hamming code, still unpublished when Shannon's paper brings out.

The *third part* considers the information measure for sets of functions depending on random parameters, especially the set of band-limited functions. Shannon defines entropies simply derived from those of the discrete case by replacing sums by integrals and probabilities by probability density functions (what is now called differential entropies, a term that Shannon does not use). He states the properties of these entropies homologous of the discrete ones, but notices that these quantities now depend on the coordinate system, at variance with the discrete case entropies.

In the *fourth part*, he computes the capacity of a continuous channel. The mutual information is then expressed as a difference of entropies as defined in the third part. At variance which each of the terms in the difference, it remains unchanged with respect to a change of coordinates, so the definition of the channel capacity does not change. He considers the case of additive noise, and especially that of Gaussian and white noise which is the usual model of thermal noise, when the average received power is given. Applying his definitions, he gets the capacity per time unit of this channel, $C$, probably the most famous formula of information theory (if not the best understood), namely:

$$C = B \log\left(\frac{S+N}{N}\right),$$

where[6] $B$ is the bandwidth, $S$ the average received signal power and $N$ that of the additive noise. He mentions that this formula was independently found by other researchers, especially Norbert Wiener and W.G. Tuller. He also considers other cases where he could only give lower and upper bounds of the capacity.

The *fifth and last part* considers the extension of what precedes to a continuous source. The information rate cannot then be defined without introducing a fidelity criterion, two messages close enough for this criterion being considered as equivalent.

I gave a long summary of these papers, especially as regards the introduction and the first two ones, which contain the main innovations and the meaning of which is not made obscure by mathematical difficulties, in order to show the extreme richness of their content. Another paper by Shannon, maybe less famous, remarkably complements the preceding ones: "Communication in the presence of noise" published in 1949 (Sloane and Wyner 1993, pp. 160–172). Starting from the sampling theorem (often wrongly ascribed to Shannon, although he refers to previous works[7]) which states that signals of spectrum limited to a frequency band of width $B$ can be exactly recovered from the values (or samples) they assume at instants separated by a time intervals of $1/2B$, he introduces a geometric representation of signals and of additive white Gaussian noise as vectors in a high-dimensional Euclidean space. Some unclear properties of analog modulation systems found an obvious explanation in this

[6] I changed the original notation so as to comply with the one I used in Eq. (5.5) of this book.

[7] It seems that it was already known by Augustin Cauchy (1789–1857).

representation, which also was much later used for the design of systems combining modulation and coding. Shannon uses it to sketch a direct proof of the capacity of the additive white Gaussian noise channel, i.e., to prove that the information rate must be less than the above expression of this channel capacity so that no errors occur. Shannon also discusses his very smart solution, referred to as water-filling, to the problem of communication in the presence of non-white noise, i.e., where the noise spectral density is not a constant in the signal band.

## A.4 Shannon's Influence

Engineers and scientists interested in information theory formed under the banner of the Institute of Radio Engineers (IRE, later become after a merger with another society, IEEE) a working group which started in 1953 publishing a journal, the *IRE Transactions on Information Theory*. From a few hundreds of pages for each of the first years, its volume did not cease to increase up to more than 2,000 pages yearly now[8]. This quantitative increase has nevertheless coincided with a narrowing of the field which was covered. The issues of the first years were indeed much more eclectic than they are today. Problems of signal theory, automatics or psycho-plysics found a place in it, while these topics are now relevant to other journals. A reason of this trend is a reaction against a fashion effect, ephemeral by definition, I shall more lengthily deal with below. The problem was to avoid that works really useful to information theory be diluted in the flood of papers about more or less relevant applications, and the policy of restricting the scope prevailed, after debates between the members of the working group to which Shannon participated, as we shall see it.

I already mentioned the reluctance initially expressed by certain mathematicians with respect to Shannon's work. On the contrary, it was enthusiastically welcome by many researchers of other fields: genetics, neurology, psycho-physics, psychology, economy, linguistics, sociology ... It was unfortunately an irrational fad, the intensity of which was often matched by the lack of understanding. The vocabulary of information theory then had often a decorative role and papers like "Information theory, photosynthesis and religion" proliferated (this emblematic title, from an editorial of the *IRE Trans. on Information Theory* in which Peter Elias[9] mocked at this fad (Elias 1958), is of course fictitious but it is hardly caricatured). Even the undeniable influence that Shannon's work had on first rate linguists and philosophers, like Roman Jakobson or Claude Lévi-Strauss, remained limited to its more superficial aspects. The deepest and most innovative ideas of information theory, especially the possibility of errorless communication despite the channel perturbations and the extraordinary method of random coding to prove it seem to have escaped any comment by established philosophers.

[8] In 2000; this number is more than twice than that in 2011.

[9] One of the most important researchers in the field. He invented arithmetic source coding and convolutional error-correcting codes.

Shannon himself reacted against the fashion he unwillingly initiated. He thus wrote an editorial[10] in the same journal in March 1956 (Sloane and Wyner 1993, p. 462), which I shall more lengthily comment when dealing with the future of information theory. Despite its brevity, it seems indeed to me that it opens up a program much of which remains to be performed.

One said once that the fecundity of a work is measured by the number of misunderstandings it gave rise. For this criterion, Shannon's work is immense! The obituaries just published in the French newspapers eloquently witness it, if I may write so. They are few, short and, far from helping to know his work, show the misunderstandings it suffers[11]. This also shows how the importance of Shannon's work was underestimated: rarely the futility of the media was so obvious.

Certain of these misunderstandings have as sole origin an erroneous reading. Thus, a myth that almost nothing justifies sees in Shannon an exalter of the binary: the core of his theory would be the possibility of transmitting a message, regardless of its nature, by the means of binary symbols or signals. I wrote 'almost'. Shannon could not imagine how journalists would dress up his work and he imprudently proposed to name 'bit', an acronym for 'binary digit', the unit of information quantity which results of choosing 2 as logarithmic base (he also contemplated other bases, as we have seen it). A digit and a unit are objects of different nature and defining 'bit' as the acronym for *binary unit* could maybe have avoided the misunderstanding. It turns out that 'bit' is usually employed for binary digit in technical jargon, even when it bears no information or an information quantity less than the binary unit. Everybody aware of information theory knows that and distinguishes with no risk of error the two meanings of the word 'bit' (personally, I use the word 'shannon' for the binary unit, which avoids any ambiguity). Hasty and inexpert readers unfortunately fell in the trap which was unwillingly set. I do not know if Shannon has been angry or, more probably, amused at that.

Other misunderstandings have a much deeper origin which it is important to analyse. The role of semantics is a point of major divergence between certain of the authors who tried to apply information theory outside its original domain, and the now unanimous opinion of engineers. The wide development of information theory in the mathematical and technical fields amply justified the exclusion of semantics which is a premise in Shannon's theory. This exclusion actually appears as a *methodological necessity* which enables distinguishing the information from both the message which bears it and the meaning ascribed to it. On the contrary, many people coming from different horizons, especially biologists, have felt when reading Shannon the exclusion of semantics as a congenital defect to be repaired. This misunderstanding is not recent. Under the title "The mathematical theory of communication" (the definite article substituted for the indefinite one cancelled the modesty of the original title), the two 1948 papers were reprinted as a book as early as 1949 (Shannon and Weaver 1949). A lengthy postface by the biologist Warren

[10] A significant quotation of which may be found in Sect. 2.5 above.

[11] In contrast with the excellent obituary of Calderbank and Sloane published by *Nature* (Calderbank and Sloane 2001).

Weaver, then administrator of the Rockefeller foundation, has been appended to them. Shannon claims that he discards semantics in his very introduction, in a few sentences, arguing that the semantic content of a message has no incidence on how the messenger works. On the contrary, most of the comments by Weaver deplore the exclusion of semantics and suggest remedies for it. Rather strangely, the two authors of the book thus express irreconcilable points of view. Time did not attenuate this misunderstanding. I shall try later to analyse the reasons for it at the same time I shall outline perspectives for the future.

## A.5 Shannon's Legacy

I believe that Shannon theory has a great future outside the technical domain, as applied to the sciences of nature. I shall in the following restrict myself to uphold this opinion as regards biology. Many engineers do not share this opinion, which moreover contradicts the one which currently prevails among biologists.

Answering his (few) interviewers, Shannon claimed his atheism. He saw no fundamental difference between the machines and the living things, including humans. His tinkering was maybe intended to imitate nature (widely anticipating on François Jacob), in a pathetic way which made intuitively perceptible the distance between the technical means available in the middle of the XX-th century and that of nature, generally endowed by evolution of an extreme refinement. He thus could not be hostile in principle to the idea of applying information theory to biology[12].

The short editorial of the *IRE Transactions on Information Theory* I mentioned above is entitled 'The bandwagon'. Shannon has mocked there at speculations which referred to his work, calling for patience and modesty. Asserting as a personal opinion the rightfulness of applying his theory to sciences of nature, he suggests however that this approach will be fruitful only after information theory will be firmly enough established in its domain of origin. One cannot but admire how lucid is this editorial. Most of the speculations Shannon denounced fell indeed into a well deserved oblivion, whereas information theory has confirmed its validity and its fecundity in the mathematical and technical domains; at the same time, the attempts to apply information theory to other sciences became more and more infrequent. One may regret this withdrawal into an ivory tower, but hope that the reflection acquired in the technical domain will eventually enable applying it to the sciences of nature freed from the naivety and vague approximations of the first attempts.

Shannon's mentors Barbara Burks and Vannevar Bush, as well as Norbert Wiener who had a less direct but inescapable influence on him, were fervent advocates of interdisciplinarity. If one defines information theory as the science of symbol sequences (with Shannon and Kolmogorov) then it can obviously be applied to biology: Crick and Watson identified in 1953 the DNA molecule as the bearer of the

[12] All the more his doctoral dissertation dealt with genetics.

hereditary information, made of a string of quaternary symbols. That attempts aimed at applying information theory to biology failed until now is not a reason to give up (I would like to say: on the contrary). The initial fad having passed and the misunderstanding I mentioned as regards the role of semantics being stronger and stronger, the biologists turned away from information theory. After its too discreet triumph in the technical domain, I think it is now mature for eventually fecund applications to biology and perhaps physics. A mandatory condition for the success of this ambitious plan is a clear awareness of the origin of the misunderstandings which as yet hindered it.

The comments by Weaver in (Shannon and Weaver 1949) gave the first example of a misunderstanding between information theorists and biologists which got worse with time. Weaver worries about the congenital inability of information theory to take semantics into account. Shannon accepts it, on the contrary. All the subsequent development of information theory has shown he was right, since the exclusion of semantics never appeared as a drawback or a brake. Information plays with respect to semantics the role of a *container*, and should not be confused with its symbolic supports, i.e., messages, and still not with the physical supports of the messages. Much more than mastering the mathematical difficulties of some of its chapters (but the discrete finite case, the most important one, does not suffer such difficulties), it is the understanding of the status of information as an intermediate which is the key of its fruitful application. Information is indeed abstracted from the set of supports and messages which can bear it, but it is also the bearer of a meaning which is completely independent of it and not amenable to a quantitative measure. The difficulty of information theory is thus not so much intrinsic than conceptual, insofar as it is the epistemological status of the main quantity it deals with which is far from obvious. At the turning point between the abstract and the concrete, information revealed itself as an unexpected intermediate. This status is now well perceived by the engineers who have learned by experience that 'it works', but not at all by the upholders of other disciplines, especially physicists and biologists.

This reflection shows how deeply innovative is Shannon theory. With the discovery of a measurable quantity as fundamental as hidden, it is a new world that it opened to science.

## References

Calderbank, R., & Sloane, N. J. A. (2001). Claude Shannon (1916, 2001). *Nature, 410*(6830), 768.

Elias, P. (1958). Two famous papers. *IRE Transactions on Information Theory, 4*(3), 99.

Khinchin, A. I. (1957). *Mathematical foundations of information theory*. Dover.

Kolmogorov, A. N. (1965). Three approaches to the quantitative definition of information. *Problems of Information Transmission, 1*, 4–7.

Kolmogorov, A. N. (1968). Logical basis for information theory and probability theory. *IEEE Transactions on Information Theory, IT-14*(5), 662–664.

Shannon, C. E., & Weaver, W. (1949). *The mathematical theory of communication*. Urbana: University of Illinois Press.

Sloane, N. J. A., & Wyner, A. D. (Eds.). (1993). *Claude Elwood Shannon, collected papers*. Piscataway: IEEE Press.

# Appendix B: Some Comments about Mathematics

## B.1 Physical World and Mathematics

The relationship of mathematics with physics is rather ambiguous. On the one hand, modern physics (say, since Galileo) heavily depends on mathematics. The 'laws of physics' are invariably expressed by mathematical equalities which involve physical quantities and mathematical operations. For instance, Newton's law of gravitation reads

$$f = (G \times m \times m')/(d \times d) = Gmm'/d^2$$

where $m$ and $m'$ denote numbers which measure the masses of two material bodies which are at a distance $d$ apart, $f$ denotes the attractive force they exert on each other, and $G$ is a fundamental constant the numerical expression of which depends on the units of force, mass and length which are employed. The mathematical operations of multiplication and division are represented in the expression in the middle by $\times$ and $/$, respectively, while the usual formulation at right can be understood only if it is known that the sign $\times$ is left implicit and that $d^2 = d \times d$.

On the other hand, mathematics is a science of its own which deals with entities *abstractly* defined by sets of *axioms*, without any reference to objects of the physical world, hence foreign to the sensible intuition. Mathematical concepts often originated in practical problems (e.g., counting objects or measuring field surfaces), but mathematics evolved towards increasing abstraction. Among the mathematical abstractions which have no physical counterpart, we may cite *infinity* and the related concept of *limit*. Henri Poincaré noticed in (Poincaré 1902) that 'The essential character of recursive reasoning is that it contains, so to speak as compacted in a single formula, infinitely many syllogisms (*Le caractère essentiel du raisonnement par récurrence c'est qu'il contient, condensés pour ainsi dire en une formule unique, une infinité de syllogismes.*)'. It turns out that recursive reasoning is absolutely needed for introducing as basic a concept as that of natural numbers (see Sect. 2.4.2). Given a natural number, adding 1 to it results in a larger one, regardless of how large is the given number, which clearly shows that natural numbers are infinitely many. The non-physical concept of infinity is thus met at a very elementary step of the mathematical construction.

Explicating the relationship between mathematics and physics is indeed a very difficult philosophical problem which has no universally accepted solution. Most physicists consider mathematics as a *tool* which they daily use. It works extremely well, often beyond what can be reasonably expected. It occurs even that a purely mathematical result predicts the existence of some physical object (for instance, the existence of the neutrino was predicted from mathematical considerations decades before its experimental evidence could be established). Physicists thus generally do not question the use of mathematical concepts. However, they use them in a non-mathematical way, often ignoring distinctions made by mathematicians, especially as regards the condition of validity of their theorems, which may be quite subtle but are mandatory from a mathematical point of view.

On the other hand, mathematicians are often, explicitly or not, upholders of Platonic realism: mathematical objects have for them an intrinsic existence entirely foreign to the physical world. But then why can these objects be very efficiently used in physics? Eugene Wigner entitled a famous paper 'The unreasonable effectiveness of mathematics in natural sciences' (Wigner 1960), and most physicists deal with the effectiveness of mathematics as a mystery which is relevant to metaphysics, hence foreign to their domain of competence. This alleged mystery is probably made less obscure if we think of physics as a means for a human observer to acquire *information* on the natural world, information being as stated in Sect. 6.4 an entity which bridges the concrete and the abstract.

## B.2 On Numbers

In order to illustrate how far from the sensible intuition mathematics has become, we briefly describe the process of successive abstractions which resulted in introducing the several families of numbers starting from natural integers (signed integers, rational, irrational, transcendental, 'real', ..., numbers), each family extending the previous one. We can thus realize how mathematical objects are human-made abstract artefacts, far from any physical reality. We briefly expounded in Sect. 2.4.2 how natural integers are introduced by recursion. Despite their interest, these numbers have to be extended in many ways so as to be useful in many mathematical instances.

The first main motivation for extending the concept of number is to introduce new elements such that the inversion of the basic operations of addition and multiplication becomes possible. Let us first consider the addition. Subtraction, denoted by the minus sign $-$, results by definition in the difference $c = a - b$, such that $c + b = a$, where $a$ and $b$ are elements of the set $\mathbb{N}$ of natural integers (defined here as including 0; $\mathbb{N}^*$ will denote the same set deprived of 0). However, no element $c$ of $\mathbb{N}$ such that $c + b = a$ exists if $b$ is larger than $a$. Appending to $\mathbb{N}$ a set of elements such that this equality always results in specifying an element $c$ defines the set of signed integers $\mathbb{Z}$. This set consists of the union of $\mathbb{N}$ and the set of negative integers, each of which is associated with an element $a$ of $\mathbb{N}^*$ which is denoted by $-a$ ($a$ is then said positive). Then each element $a$ of $\mathbb{N}^*$ has an additive inverse $-a$ (referred to as its opposite) such that $a + (-a) = 0$.

Similarly, $\mathbb{Z}$ is extended into a set $\mathbb{Q}$ such that any non-zero element $a$ of $\mathbb{Z}$ has a multiplicative inverse, denoted by $1/a$ or $a^{-1}$, such that $(1/a) \times a = 1$. The numbers which belong to the set $\mathbb{Q}$ are referred to as *rational*. Then the quotient $a/b \triangleq a \times (1/b)$ of any two elements $a$ and $b$ of $\mathbb{Z}$ has a meaning provided $b \neq 0$. Each non-zero element of $\mathbb{Q}$ having an additive and a multiplicative inverse, it is said to possess the *field structure*.

Although the set $\mathbb{Q}$ now contains the additive and multiplicative inverses of the basic operations effected on natural numbers, it turns out that other operations cannot always be inverted within the set of rational numbers $\mathbb{Q}$. The squaring operation, defined as $a^2 \triangleq a \times a$, where $a$ is a natural number (i.e., belongs to $\mathbb{N}^*$) cannot be inverted within $\mathbb{Q}$ when $a = 2$, a case of historical significance since it ruined the Pythagorean system which was based on ratios, i.e., on elements of $\mathbb{Q}$. The inverse of the squaring operation effected on $a$, denoted by $\sqrt{a}$ or $a^{1/2}$, does not belong to $\mathbb{Q}$ for $a = 2$ and is therefore referred to as *irrational*. In other words, the equation $x^2 - 2 = 0$ defines a number $x$ which does not belong to $\mathbb{Q}$ and, more generally, the roots[1] of an algebraic equation $x^n + a_{n-1}x^{n-1} + \ldots + a_0 = 0$, where $a_{n-1}, \ldots, a_0$ belong to $\mathbb{Z}$, when they exist[2], do not necessarily belong to $\mathbb{Q}$ and are then irrational. (The left hand side of this equation is referred to as a *polynomial* and $x$ as an indeterminate.) Further researches showed moreover that certain numbers that mathematics can define, like $\pi$ (the ratio of the circumference of a circle to its radius), or e (the base of the natural logarithms; see Sect. B.3.1 below), are not roots of algebraic equations. They are referred to as *transcendental*. Mathematicians were thus led at the end of the XIX-th century to define *real* numbers so as to include transcendental numbers besides irrational ones. The mathematical 'measure theory' has later shown that the rational numbers are a set of 'measure 0' within that of real numbers, although they are dense everywhere (i.e., any interval between two real numbers, however small, contains infinitely many rational numbers). Real numbers have thus a very strange 'fine structure', absolutely foreign to the sensible intuition. Especially, the mathematical concept of real number completely disagrees with the intuitive feeling of continuity of, say, space or time intervals. Using algorithmic complexity arguments, Chaitin has shown moreover that the so-called real numbers are generally *uncomputable* (Chaitin 2005). They have nothing in common with any physical reality and, from this point of view, the word 'real' is a misnomer.

## B.3 Definitions and Notations in the Book

Far from these difficult issues, what follows is mainly intended to simply remind some definitions and to explicate some notations used in the book.

[1] The roots of an equation are defined as the values of $x$ such that the equality holds.

[2] Defining the set of 'complex numbers', such that any algebraic equation has always complex roots, resulted in another extention of the number concept. A complex number, earlier referred to as 'imaginary', can be interpreted as a couple of ordinary numbers endowed with specific addition and multiplication rules. We do not use complex numbers in this book. Beware that the adjective 'complex' in this meaning has no relationship with the algorithmic complexity of Sect. 6.1.

### B.3.1 Exponentials and Logarithms

Let $\ell$ be a positive integer, and $b$ be any positive number. The product $b \times b \times \ldots b$, in which the number of factors is $\ell$, is referred to as $b$ to the power $\ell$ (or as the $\ell$-th power of $b$) and denoted by $b^\ell$, where $\ell$ is referred to as the *exponent*. From the very definition, $b^{\ell+\ell'} = b^\ell \times b^{\ell'}$, where $\ell'$ also is a positive integer. This notation can be directly extended to non-positive exponents: $b^0 = 1$, and $b^{\ell-\ell'} = b^\ell/b^{\ell'}$. The methods of mathematical analysis enable extending these equalities to any real numbers $\ell, \ell', \ldots$ Moreover, given any positive real number $b > 1$ (referred to as the 'base') and another positive real number $a$, there always exists a number $\ell$ such that $b^\ell = a$. There is a one-to-one correspondence between $\ell$ and $a$ (for a given value of $b$), so $\ell$ is a function of $a$ which is referred to as its *logarithm* and denoted by $\log_b(a)$. The subscript $b$ is intended to recall that, besides $a$, it depends on the chosen base $b$. The main properties of the logarithmic function are $\log_b(1) = 0$, $\log_b(a \times c) = \log_b(a) + \log_b(c)$ where $a$ and $c$ are positive real numbers. As a consequence, $\log_b(1/a) = -\log_b(a)$. The main usefulness of the logarithmic function is to convert a product of numbers into the sum of their logarithms. The choice of the base $b$ is arbitrary. Changing it for another base $b' > 1$ results in dividing the logarithm to the base $b$ by the positive constant $\log_b(b')$, namely:

$$\log_{b'}(a) = \frac{\log_b(a)}{\log_b(b')},$$

a direct consequence of the definitions.

The function $\log_b(x)$ tends to minus infinity when $x$ approaches 0, so the function $y = x\log_b(x)$ assumes for $x = 0$ the indeterminate form $-0 \times \infty$. It can be shown that the limit of $x\log_b(x)$ when $x$ approaches 0 is 0, and it is why the entropy function $H(p_1, p_2, \ldots, p_n)$ assumes the value 0 when any one of its arguments is 1, which entails that all others are 0. In particular, the binary entropy function defined in Sect. 4.2.3 by Eq. (4.11), $\mathcal{H}_2(p)$, assumes the value 0 when its argument $p$ is 0 or 1.

In this book, we mainly use logarithms to the base 2, in accordance with the choice of the information unit as binary (see Sect. 4.2.2). A convenient logarithmic base generally used in mathematical analysis is the constant e, a transcendental number which approximately equals 2.718. Raising this constant to the power $x$ defines the exponential function $y = \mathrm{e}^x$ of the variable $x$, usually denoted by $y = \exp(x)$. This function has the remarkable property that it equals its derivative, i.e., $\mathrm{d}y/\mathrm{d}x = y(x)$. Let us recall that the derivative $\mathrm{d}y/\mathrm{d}x$ of a function $y(x)$ is the limit of $\delta y/\delta x$, for $\delta x$ tending to 0, where $\delta y \triangleq y(x + \delta x) - y(x)$ (this limit exists only if $y(x)$ is regular enough). The derivative of a function $y(x)$ measures how fast $y$ increases in terms of $x$. Being equal to its derivative, the exponential function $y = \exp(x)$ can thus be successively derived arbitrarily many times.

Logarithms to the base e are referred to as 'natural' and denoted by $\ln(\cdot)$. The logarithmic function $y = \ln(x)$ is the inverse of the exponential one, meaning that

$\ln[\exp(x)] = x$. As a consequence, the derivative of $\ln(x)$ is $1/x$, showing that the increase of $\ln(x)$ is the smaller, the larger $x$. Natural logarithms and exponentials are used in Sect. 5.5.8 above.

## *B.3.2 Representing Symbols and Sequences*

**Endowing the alphabet with the finite field structure** We now very briefly deal with how symbols and sequences can be represented. For dealing with its symbols as mathematical objects, an alphabet of finite size $\alpha$ must be endowed with some mathematical structure. The most convenient one is that of *finite field*. Remember that the two operations of addition and multiplication can be effected on rational or real numbers and that each element $a$ of it has an inverse $-a$ for the addition and, if $a \neq 0$, an inverse $1/a$ for the multiplication. A set of elements having these properties is said to possess the field structure. Similarly, it is especially useful to endow the $\alpha$ elements of an alphabet with the field structure, endowing its element with the same operations. This is not possible for any value of the integer $\alpha$. The simplest case is $\alpha = 2$; the binary alphabet is then $\{0, 1\}$. There is a single non-zero element, 1, which is its own multiplicative inverse. As regards addition, the rule valid for the integers can be applied when 0 is added since it results in an element of the alphabet. However, the ordinary sum $1 + 1 = 2$ is meaningless since 2 does not belong to the alphabet. Using addition modulo 2, such that $1 + 1 = 0$, results in an element of the alphabet hence endows the binary alphabet with the desired field structure. Notice that if the elements 0 and 1 are interpreted as the classes of equivalence of even and odd integers, respectively, the ordinary addition of integers is transformed into addition modulo 2.

Defining finite fields for alphabets with more than 2 elements is possible only if $\alpha$ is a prime, or a prime raised to an integer power. No finite fields exist for other values of $\alpha$. Thus, there exist finite fields with 2, 3, 4, 5, 7, 8, 9, 11, ... elements, but not with 6, 10, 12, ... elements. The addition rule is modulo $\alpha$ only when $\alpha$ is a prime, but not if it is a prime raised to an integer power larger than 1. Defining the addition and multiplication rules of a finite field when $\alpha$ is not a prime relies on algebraic properties which would need lengthy preliminary definitions. We omit them here since the only examples given in this book involve symbols of the binary alphabet.

**Representing sequences** As regards the representation of sequences, we use polynomials where the coefficients are the alphabet symbols (as elements of a finite field), instead of ordinary integers as in usual polynomials as introduced in Sect. B.2, or formal series which extend the definition of polynomials to an infinite number of terms. We denote by $D$ the indeterminate, as usual in the coding literature. Then a term $aD^i$ in the polynomial or the formal series indicates that the $i$-th symbol of the sequence is the coefficient $a$ (beware that if $a = 0$, the term is omitted, as in ordinary algebraic notation). We used this representation in Sects. 3.4.2 and 5.5.4.

### B.3.3 Probabilities

Games of chance are typical human activities which rely on the observation that repeating some gesture can in certain circumstances result in several possible outcomes, the actually realized one being unpredictable. Such an event, referred to as *random*, seems to defy (or deny) causality. As an example, think of playing dice. Perfectly controlling all the mechanical parameters when throwing a dice would actually result in a deliberately chosen outcome, but no one can avoid slight differences when repeating several times the same hand movement, and these differences are so to speak amplified in the dice trajectory up to make the outcome unpredictable. Gambling like roulette could be similarly analyzed. As regards playing cards, the precise position of each of the cards in a pack, after they are shuffled, cannot be deduced from their initial position. (Of course, there are cheaters or conjurers, but we assume that the play is fair, and especially that the dice or the card pack are so.) Non-scientific thought attributes the result of such a game to fate, an obscure and relentless supernatural entity. Probability theory originated in attempts to scientifically analyse games of chance.

Probability theory associates with a random event a number $p$, $0 \leq p \leq 1$, referred to as its probability, which measures how likely is its occurrence. Precisely defining probability is difficult because it is almost impossible to avoid logical circularity. However, it is intuitively clear that, when a fair dice is thrown or a card is drawn from a fair 32-card pack, each possible outcome has a probability of 1/6 or 1/32, respectively. It is a matter of experience that, if the event is repeated many times, the average frequency of occurrences is close to these probabilities. Moreover, the laws of large numbers tell how close to probabilities are the measured frequencies by assessing a probability to the difference between an actual frequency and the corresponding probability. Even if the theory is based on a questionable definition of probabilities, a consistent system of axioms enables dealing with probabilities as true mathematical objects. The wide applicability of the probability theory and its importance lie in the fact that the 'deterministic chaos', which pertains to statistical mechanics, cannot be distinguished from 'true' randomness. Probability theory thus provides the adequate tools of statistical mechanics.

**Joint, conditional and marginal probabilities** Let us consider two non-independent events. As an example, assume that a card is drawn from a 32-card pack and consider the following two events: the card is an ace (denoted by $A$), and the card is a spade ($S$). The event $(A, S)$ designates the realization of both $A$ and $S$, so then the chosen card is the ace of spade, hence the probability of this event is 1/32. The probability $\Pr(A, S)$ is referred to as the joint probability of the events $A$ and $S$. The events $A$ and $S$ have as probabilities $\Pr(A) = 1/8$ and $\Pr(S) = 1/4$ (referred to in this context as marginal probabilities). We may think of the event $(A, S)$ as realized in two steps: first, the card is an ace; if it is an ace, then it is a spade. The probability of the event 'to be a spade if the chosen card is an ace' is referred to as the conditional probability of $S$ given $A$, and denoted by $\Pr(S|A)$. Its numerical value is 1/4. Clearly:

$$\Pr(A,S) = \Pr(S|A)\Pr(A) = \Pr(A|S)\Pr(S), \tag{B.1}$$

where the second equality results from the obvious symmetry of the problem. Any of these equalities is referred to as Bayes rule, and is valid for any couple of joint events. It has been written as Eq. (4.7) in Sect. 4.2.2 and is used above in many instances.

**Discrete random variables** A random variable $X$ can assume $n$ values $x_1, x_2, \ldots, x_n$ ($n$ is assumed to be finite; the extension to countably many values is possible but we do not need it). The set of possible values $\{x_1, x_2, \ldots, x_n\}$ is referred to as the *sample space*. The probability that $X$ assumes the value $x_i$, for $1 \leq i \leq n$, is denoted by $p_i$: $\Pr(X = x_i) = p_i$. The set of probabilities $\{p_i\}$ is referred to as the *probability distribution* of the random variable $X$. The event $X = x_i$ is referred to as the realization of $x_i$. For instance, the random variable $X_{\text{dice}}$ associated with a dice can assume the integer values from 1 to 6, each with a probability of 1/6. The realizations of $x_1$, $x_2$, and $x_n$ are mutually exclusive events. The probability of occurrence of any one among several such events is the sum of their probabilities. Since one of the realizations of $x_1$, $x_2$, and $x_n$ is assumed to necessarily occur,

$$\sum_{i=1}^{n} p_i = 1. \tag{B.2}$$

The average value of $X$, referred to as its *mean* or its (mathematical) *expectation*, is defined as

$$\mathrm{E}[X] \triangleq \sum_{i=1}^{n} p_i x_i. \tag{B.3}$$

It is also denoted by $\bar{X}$. The random variable $X_C = X - \bar{X}$ is referred to as 'centred' since its mean is 0.

According to this definition, $\mathrm{E}[X_{\text{dice}}] = 3.5$, and the corresponding centred random variable assumes the values $-2.5, -1.5, -0.5, 0.5, 1.5$, and 2.5.

The *standard deviation* of a random variable $X$ measures how broadly distributed it is. It is defined as the square root of its *variance*, which is itself defined as the expectation of the square of the centred random variable $X_C$, namely:

$$\mathrm{E}[(X - \bar{X})^2] = \sigma^2 \triangleq \sum_{i=1}^{n} p_i (x_i - \bar{X})^2. \tag{B.4}$$

The variance of $X_{\text{dice}}$, for instance, equals 8.75/3 and its standard deviation is 1.707825 ...

Let us now consider another random variable, $Y$, which assumes the $m$ values $y_1, y_2, \ldots, y_m$ with probabilities $q_1, q_2, \ldots, q_m$, and such that $\sum_{j=1}^{m} q_j = 1$. The random variable $(X, Y)$ is defined as the joint realization of $x_i, y_j$ for all pairs of indices $1 \leq i \leq n, 1 \leq j \leq m$. The probability $\Pr(X = x_i, Y = y_j)$ is referred to as the joint probability of $X$ and $Y$ and denoted by $p_{i,j}$. If the events $X = x_i$ and $Y = y_j$ are independent, the probability of one of them is not affected by the realisation of the other one. Then their joint probability is the product of their individual probabilities: $\Pr(X = x_i, Y = y_j) = \Pr(X = x_i) \times \Pr(Y = y_j)$. However, the probability of the realization of, say, $X$ may depend of the realization of $Y$. This event is referred to as the realization of $X$ conditioned on $Y$, and denoted by $X|Y$. The probability of $x_i$ conditioned on $y_j$ is denoted by $\Pr(X = x_i|Y = y_j)$. According to Bayes rule (B.1), $\Pr(X = x_i, Y = y_j) = \Pr(X = x_i|Y = y_j)\Pr(Y = y_j) = \Pr(Y = y_j|X = x_i)\Pr(X = x_i)$ for all possible values of $i$ and $j$, namely $1 \le i \le n$ and $1 \le j \le m$.

**Continuous random variables** The reception of a binary signal in the presence of Gaussian noise led us to meet continuous random variables in Sect. 3.3. Then, instead of assuming one of some finite number of values, the value that the random variable $X$ assumes is some real number $x$. This case involves mathematical difficulties, so we suppose that the conditions for which the statements below are true are satisfied.

Instead of a set of discrete probabilities $\{x_i\}$, we consider now the probability that $X$ belongs to an infinitesimal interval $(x, x + \mathrm{d}x)$. In the simplest cases, this probability is proportional to $\mathrm{d}x$, hence is infinitesimal. It is then expressed as $p_X(x)\mathrm{d}x$, where $p_X(x)$ is referred to as the *probability density function* of $X$. This formulation is meaningful only if the function $p_X(x)$ is regular enough, and it is fortunately so if this function is Gaussian. For instance, we expressed in Sect. 3.3 the probability density function of thermal noise of variance $\sigma^2$ by the centred Gaussian function of Eq. (3.7):

$$p_X(x) = g(x;\sigma^2) = \frac{1}{\sigma\sqrt{2\pi}} \exp\left(\frac{x^2}{2\sigma^2}\right).$$

Instead of the finite sum of probabilities being equated to 1 according to (B.2), we have for a continuous random variable the integral

$$\int p_X(x)\mathrm{d}x = 1, \tag{B.5}$$

where the integration is effected on the whole interval where the probability density function is defined; e.g, it is in the Gaussian case $(-\infty, +\infty)$. The expectation of $X$ is now defined by means of an integral, namely,

$$\mathrm{E}[X] = \bar{X} \triangleq \int x p_X(x)\mathrm{d}x, \tag{B.6}$$

instead of the discrete sum (B.3). Its variance is similarly defined by an integral which replaces the discrete sum (B.4):

$$\mathrm{E}[(X - \bar{X})^2] = \sigma^2 \triangleq \int p(x)(x - \bar{X})^2 \mathrm{d}x. \tag{B.7}$$

In the case of two continuous random variables the Bayes rule reads

$$p_{X,Y}(x, y) = p_{X|Y}(x|y)p_Y(y) = p_{Y|X}(y|x)p_x(x). \tag{B.8}$$

Similar relations exist if one of the variables, say $X$, is discrete and the other one continuous. We met such a situation in Sect. 3.3, with $X$ being the discrete input of a channel and $Y$ its output in the presence of additive continuous noise.

## References

Chaitin, G. J. (2005). *Meta math!* New York: Pantheon Books.

Poincaré, H. (1902). *La science et l'hypothèse*, Paris: Flammarion.

Wigner, E. P. (1960). The unreasonable effectiveness of mathematics in natural sciences. In *Communications in pure and applied mathematics* (Vol. 13, No. 1). New York: Wiley.

# Appendix C: A Short Glossary of Molecular Genetics

Some readers may be unfamiliar with the chemistry of molecular genetics. Its most basic keywords are gathered in the following glossary. The reader should be warned that its content is very simplified and that many statements incur exceptions or other restrictions to their validity. Italic words inside a definition refer to another entry of the glossary.

**adenosine (A)**

See *nucleic base*.

**amino-acids** Any amino-acid is made of a carbon atom (denoted by $C_\alpha$) to which are attached a hydrogen atom, an amino group ($NH_2$), a carboxy group (COOH) and a radical named 'side chain' R specific to it, which may assume a variety of forms (see Fig. C.1).

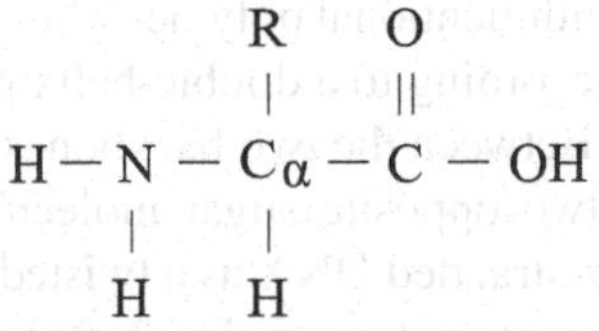

$$\begin{array}{ccccccc} & & R & & O & & \\ & & | & & \| & & \\ H - N & - & C\alpha & - & C & - & OH \\ | & & | & & & & \\ H & & H & & & & \end{array}$$

Fig. C.1 Chemical formula of an amino-acid. R denotes its side chain

Amino-acids are constituents of *proteins*, which contain only the 20 amino-acids listed in the following Table. The first column indicates the 3-letter acronym of the amino-acid, while the fourth column indicates the number of *codons* which correspond to it according to the *genetic 'code'*. The amino-acids have been put into 4 classes: hydrophobic (abbreviated as h-phobic), hydrophilic (h-philic), acid (bearing a negative electric charge), and base (bearing a positive charge). Except for glycine which has a single hydrogen atom as side chain, they all are chiral molecules (i.e., they can assume two forms which are symmetric with respect to a plane in the three-dimensional space, like an object and its image in a mirror), but only their L-form occurs in *proteins*. There is no known reason why these amino-acids were

G. Battail, *Information and Life*, DOI 10.1007/978-94-007-7040-9,

'chosen' rather than others nor why the L-form has been 'preferred'. Cystein, whose side chain is the group SH, can form disulphide bridges with another cystein molecule close to it in the 3-dimensional space due to the *protein* folding. A *protein* results from the proper folding of a *polypeptidic chain*, a unidimensional polymer made of a sequence of amino-acids listed in the Table.

| acr. | name | class | # | acr. | name | class | # |
|---|---|---|---|---|---|---|---|
| Met | methionine | h-phobic | 1 | Tyr | tyrosine | h-philic | 2 |
| Trp | triptophan | h-phobic | 1 | Ile | isoleucine | h-phobic | 3 |
| Asp | aspartic acid | acid | 2 | Ala | alanine | h-phobic | 4 |
| Glu | glutamic acid | acid | 2 | Gly | glycine | h-phobic | 4 |
| Asn | asparagine | h-philic | 2 | Pro | proline | h-phobic | 4 |
| Cys | cysteine | h-phobic | 2 | Thr | threonine | h-philic | 4 |
| Phe | phenylalanine | h-phobic | 2 | Val | valine | h-phobic | 4 |
| Gln | glutamine | h-philic | 2 | Arg | arginine | base | 6 |
| His | histidine | base | 2 | Leu | leucine | h-phobic | 6 |
| Lys | lysine | base | 2 | Ser | serine | h-philic | 6 |

**chromosomes** See *genome*.

**codon** See *genetic code*.

**complementary base pairs** See *nucleic base*.

**cytosine (C)** See *nucleic base*.

**deoxiribonucleic acid (DNA)** In its usual double-helix structure, the DNA polymer is made of two strands of a unidimentional polymer where phosphate groups alternate with deoxyribose, a sugar, according to a double-helix pattern. They are referred to as the 'backbones' of DNA. Between the two backbones, *complementary base pairs* **A–T** and **C–G** are tied to two opposite sugar molecules by covalent bonds. We thus may think of the double-stranded DNA as a twisted ladder, with the backbones as uprights and the *complementary base pairs* as steps. Its overall width is about 2 nanometers ($10^{-9}$ meter, abbreviated as nm), the distance between two base pairs is 0.34 nm and a complete turn of the double helix takes about 10 base pairs. Fig. C.2 is a plane representation of this structure as a ladder, i.e., ignoring its helix shape.

The two *complementary base pairs* **A–T** and **G–C** have a plane structure depicted in Fig. C.4 below. Their plane is orthogonal to the axis of the double helix.

**eukaryotes** As opposed to *prokaryotes*, they are living beings with cells where the *genome* is contained inside a nucleus physically separated by a membrane from the remainder of the cell, and having many other distinctive features like the presence of *mitochondria*. They may be unicellular like yeasts or amiboeae, but all multicellular beings like plants and animals are eukaryotes. Moreover, *genes* organized according to the *exons-introns* model generally belong to eukaryotes, and only a fraction of

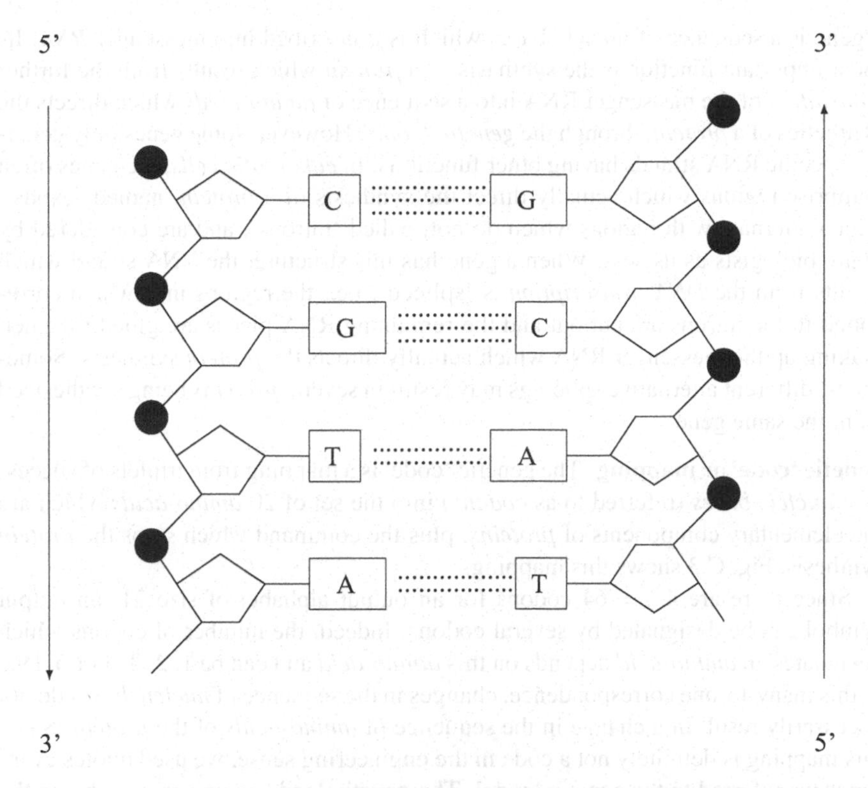


**Fig. C.2** Ladder representation of double-strand DNA. The black disks represent phosphate groups and the pentagons, deoxyribose molecules (sugars). The polymers made of alternating phosphate groups and sugars are referred to in the text as 'backbones'. Rectangular boxes represent the nucleotides **A**, **T**, **G** and **C**. The dotted horizontal lines between nucleotides of opposite strands represent hydrogen bonds. The vertical arrows indicate the direction of reading

the whole *genome* is made of *genes*, i.e., contributes to the synthesis of *proteins*. Again at variance with *prokaryotes*, the double stranded *DNA* is packed into nucleosomes, which consist of a *histone* octamer acting as a spool with almost two turns (165-nucleotide long) of double stranded *DNA* wrapped around it. (*Histones* are *proteins* associated with *DNA* in *chromosomes*.) Moreover, nucleosomes themselves are packed together in higher order structures generally referred to as 'fibers'. Another distinguishing feature of the eukaryotic cell is the presence of a 'cytoskeleton', i.e., a kind of molecular armature in a state of dynamic instability. The eukaryotic cell is much more complicated that the prokaryotic one and probably appeared almost 2 billion years later.

**exons** See *gene*.

**gene** Precisely defining a gene is not easy. At the very beginning of genetics, a gene was thought of as an 'atom' of Mendelian heredity but its nature was unknown. Since the discovery that *DNA* is the bearer of genetic information, we may say that a gene is a sequence of nucleic bases which is transcribed into messenger *RNA*. Its most important function is the synthesis of a *protein* which results from the further *translation* of the messenger RNA into a sequence of *amino-acids* which directs the synthetics of a *protein*, through the *genetic 'code'*. However, some genes only generate specific RNA strands having other functions. In *eukaryotic cells*, the genes often comprise regions which actually direct the synthesis of a *protein*, named 'exons', which alternate with regions which do not, called 'introns', and are considered by many biologists as useless. When a gene has this structure, the RNA strand which results from the *DNA transcription* is 'spliced', i.e., the regions in it which correspond to the introns are cut out and the remaining RNA pieces are glued together, making up the messenger RNA which actually directs the *protein* synthesis. Sometimes, different alternative splicings may result in several *proteins* being synthesized from the same gene.

**genetic 'code' or mapping** The genetic 'code' is a mapping from triplets of successive *nucleic bases* (referred to as *codons*) into the set of 20 *amino-acids* which are the elementary components of *proteins*, plus the command which stops the *protein* synthesis. Fig. C.3 shows this mapping.

Since there are $4^3 = 64$ codons for an output alphabet of size 21, an output symbol can be designated by several codons. Indeed, the number of codons which designates an *amino-acid* depends on this *amino-acid* and can be 1, 2, 3, 4 or 6. Due to this many-to-one correspondence, changes in the sequence of *nucleic bases* do not necessarily result in a change in the sequence of *amino-acids* of the *protein*. Since this mapping is definitely not a code in the engineering sense, we used quotes every times we referred to the genetic 'code'. The genetic 'code' is universal as being the same for all living beings (with very few exceptions, the most notable one being slight differences which concern *mitochondria*). It is the standard genetic 'code' which is shown in Fig. C.3. The genetic 'code' shows some regularities but seems to some extent arbitrary. Although it is not completely random, it may have resulted, at least in part, from contingent events.

**genome** The total *DNA* that a living being possesses in (almost) each of its cells, generally organized in *chromosomes*. The genome is replicated every time a cellular division occurs. The *chromosomes* are structures where DNA is packed together with *proteins*, especially *histones*. *Prokaryotes* generally possess a single circular *chromosome*, at variance with *eukaryotes* which have several rod-shaped *chromosomes*. In eukaryotic cells, the word 'genome' generally refers to the nucleic *DNA*, but the *mitochondria* also possess their own *DNA* which is independently replicated.

**guanine (G)** See *nucleic base*.

**histones** See *gene*.

**introns** See *gene*.

**mitochondria** A *eukaryotic* cell possesses a number of organelles named mitochondria which provide it with ATP (adenosine triphosphate) molecules, the source of energy of the cells. There are strong reasons why mitochondria are probably former

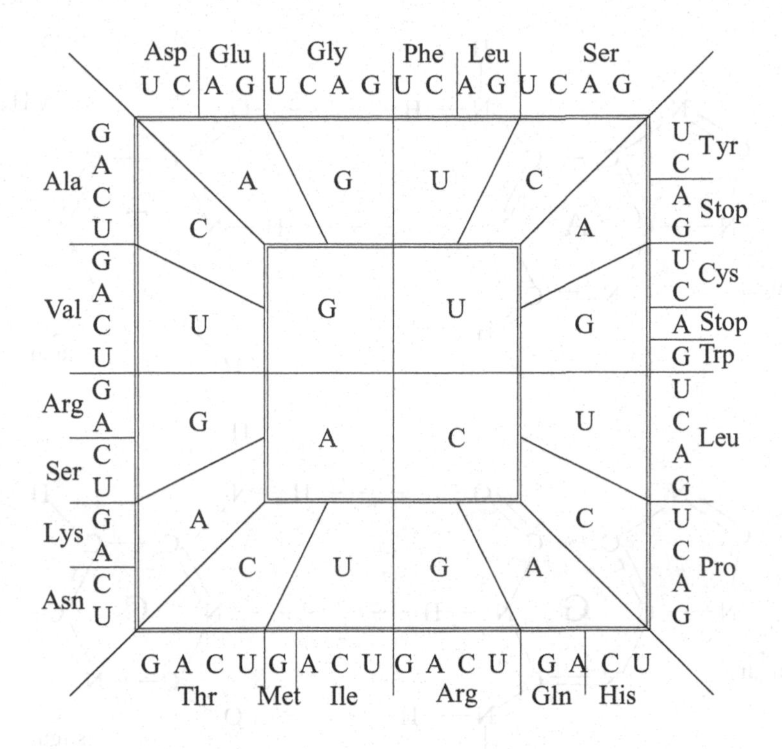


**Fig. C.3** The genetic 'code'. This chart indicates how codons of messenger RNA specify *amino-acids* and what codons stop the synthesis of a *protein*. The letters A, C, G, U denote the nucleic bases. Any codon is a 3-nucleic-base word read from center to periphery. The 3-letter acronyms at the periphery indicate *amino-acids*. For instance, UAC 'codes' for the *amino-acid* tyrosine (Tyr), while UAA stops the process. AUG both starts the synthesis process when it is preceded by a proper control message referred to as 'promoter' and 'codes' for the *amino-acid* methionine (Met)

*prokaryotes* which became symbiotic with other cells so as to make up the eukaryotic cell, about 1,500 million years ago. They still have their own *genome* which is replicated independently of the host cell, using a version of the *genetic 'code'* slightly different from the standard one.

**nucleotide or nucleic bases** The molecular elements which act as information bearing symbols in *DNA* and *RNA*. Three nucleic bases are common to *DNA* and RNA: adenosine, abbreviated as **A**, guanine (**G**) and cytosine (**C**). The fourth nucleic base is thymine (**T**) in *DNA* and uracil (**U**) in RNA. **T** and **U** have similar structure (a methyl group $CH_3$ of **T** is merely substituted for a hydrogen atom in **U**). As regards the chemical structure, **A** and **G** are purines, i.e., 2-cycle molecules denoted by **R**, while **C**, **T**, and **U** are pyrimidines, single-cycle molecules denoted by **Y**. A purine and a pyrimidine together can constitute a complementary base pair where the two molecules are tied together by two hydrogen bonds for the pairs **A–T** or **A–U**, by three hydrogen bonds for the pair **C–G**. The complementary base pairs **A–T** and

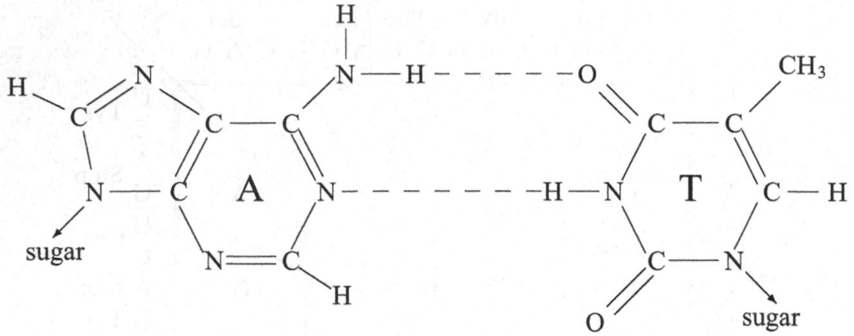

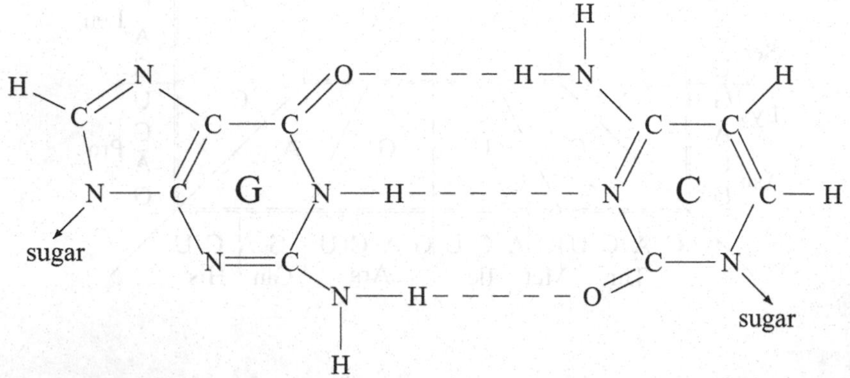

**Fig. C.4** Chemical structure of the complementary base pairs **A–T** (top) and **G–C** (bottom) of *DNA*. The purines (**A** and **G**) are at left and the pyrimidines (**T** and **C**) at right. In *RNA*, the *uracil* molecule (**U**) replaces *thymine* (**T**), which is identical to it except that the methyl group $CH_3$ at top right is replaced by a hydrogen atom H. Dashed lines represent hydrogen bonds. The mutual distances of the atoms are roughly respected, showing that the two complementary base pairs have almost equal lengths

**C–G** have almost the same geometrical dimensions, and the angles of their bonds to the *DNA* 'backbones' are close enough to enable their location anywhere in the double helix structure of *DNA*. Their chemical structure is shown in Fig. C.4 above.

The reason why precisely these molecules are used for bearing information in *DNA* and *RNA* is unknown. The words 'nucleic base' or 'nucleotide' may be thought of as misnomers since the *DNA* of a *prokaryotic* cell is made of nucleotides although it has no nucleus. These words have merely a historical origin since these bases were discovered in the nucleus.

**nucleosome** See *eukaryotes*.

**polypeptidic chain** A unidimensional polymer made of a sequence of *amino-acids*. How two amino acids are linked together after a water molecule is eliminated is represented in Fig. C.5.

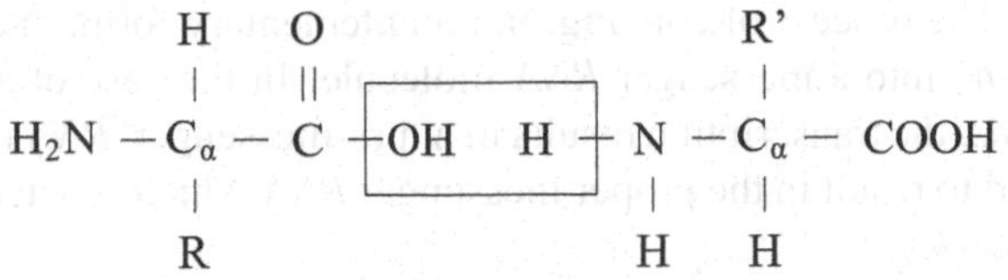

$$\begin{array}{cccccccccc} & H & & O & & & & R' & & \\ & | & & \| & & & & | & & \\ H_2N & - C_\alpha & - & C & \boxed{-OH \quad H-} & N & - & C_\alpha & - & COOH \\ & | & & & & | & & | & & \\ & R & & & & H & & H & & \end{array}$$

**Fig. C.5** Two *amino-acids* and their linking according to a peptide bond. R and R' denote their side chains. Their linking results from eliminating the water molecule shown inside the box. Other amino-acid molecules can be similarly linked at the left and the right, thus resulting in a polypeptidic chain

**prokaryotes** Living beings where the *genome* is not physically separated from the remainder of the cell. They all are unicellular beings: bacteria or archaea. Almost the full length of their *genome* contributes to the synthesis of *proteins*. They have most often a single circular *chromosome*. There is evidence that they predated the much more complicated *eukaryotes* by about 2 billion years.

**protein** A protein results from a *polypeptidic chain* generally made of a few hundreds of *amino-acids* by an appropriate folding which gives it a unique three-dimensional shape. It is made of several substructures like $\alpha$-helices and $\beta$-sheets, which themselves form higher structures named 'domains'. Disulphur bridges between cystein molecules help maintaining the protein spatial structures, as well as the attraction of electric charges borne by polar molecules.

Proteins are the most important constituents of the living matter, having both a structural and enzymatic (catalytic) role. Especially, *DNA* replication as well as *transcription* and *translation* are catalyzed by proteins. Therefore, proteins catalyze their own synthesis.

**purine (R)** See *nucleic base*.

**pyrimidine (Y)** See *nucleic base*.

**ribosome** It is the part of the cell, made of a large complex of *RNA* and *proteins*, which recognizes the transfer *RNA* associated with a *codon* and puts the corresponding *amino-acid* at its place in the *polypeptidic chain* during the process of *translation*.

**ribonucleic acid (RNA)** The ribonucleic acid is similar to *DNA* except that the sugar molecule is ribose instead of deoxiribose, and that the nucleic base *uracil* (U) replaces *thymine* (T) (see Fig. C.4). Its double-helix structure is less stable than that of *DNA* and it is most often found in the single-stranded form. RNA assumes many functions: the most important are 'messenger RNA' which copies in complementary form the genetic message borne by a *DNA* strand (see *transcription*), 'transfer RNA' which associates the proper *amino-acid* with each *codon* of the messenger RNA in the process of *protein* synthesis, and 'ribosomic RNA' as part of the *ribosome* machinery (see *translation*).

**thymine (T)** See *nucleic base*.

**transcription** The process of copying, in complementary form, one of the strands of *DNA* of a *gene* into a messenger *RNA* molecule. In the case of a *gene* made of *exons* and *introns*, the transcription results in a 'pre-messenger' *RNA* molecule which has to be spliced to result in the proper messenger *RNA* which is actually used in the *translation* process.

**translation** The process of synthesis of a *polypeptidic chain* which, properly folded, becomes a *protein*. The polypeptidic chain derives from the sequence of *nucleic bases* of a messenger *RNA* which itself resulted from the *transcription* of the genetic information borne by the *DNA* of a *gene* according to the *genetic 'code'*. It involves a transfer *RNA* associated with each *codon* and the operation of the *ribosome*.

**uracil (U)** See *nucleic base*.

# Index



G. Battail, *Information and Life,* DOI 10.1007/978-94-007-7040-9,

Printed in the United States
By Bookmasters

Printed in the United States
By Bookmasters